A Revolution in Music

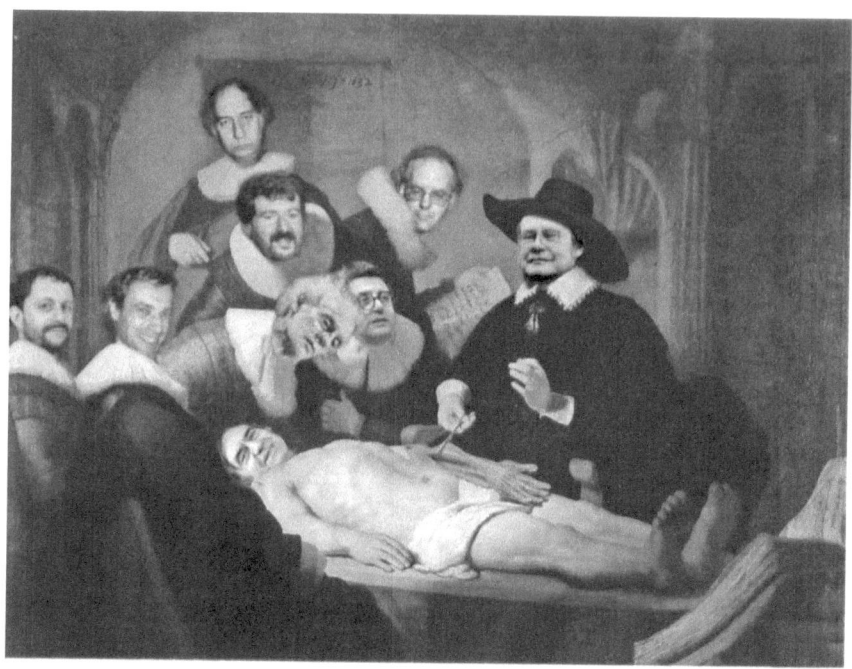

"Music Lesson at the GRM," after Rembrandt's *Anatomy Lesson of Dr. Nicolaes Tulp*. Pierre Schaeffer (lying down). From left to right, foreground: Yann Geslin, François Donato, François Bayle, Christian Zanési, François Delalande. Behind: Daniel Teruggi, Michel Chion. Above: Jean-Christophe Thomas. © Emmanuel Favreau (2005).

A Revolution in Music

The History of the Groupe de Recherches Musicales

Évelyne Gayou

Translated by David Vaughn

UNIVERSITY OF CALIFORNIA PRESS

The publisher and the University of California Press Foundation gratefully acknowledge the generous support of the Constance and William Withey Endowment Fund in History and Music.

University of California Press
Oakland, California

© 2025 by Évelyne Gayou

Originally published as *Le GRM: Groupe de Recherches Musicales (Cinquante ans d'histoire)*, by Évelyne Gayou. © 2007, Librairie Arthème Fayard.

All rights reserved.

Library of Congress Cataloging-in-Publication Data

Names: Gayou, Évelyne, author. | Vaughn, David (Translator), translator.
Title: A revolution in music : the history of the groupe de recherches musicales / Évelyne Gayou ; translated by David Vaughn.
Other titles: GRM, Groupe de recherches musicales. English
Description: Oakland, California : University of California Press, [2025] | Includes bibliographical references and index.
Identifiers: LCCN 2024020842 (print) | LCCN 2024020843 (ebook) | ISBN 9780520409767 (cloth) | ISBN 9780520409774 (paperback) | ISBN 9780520409781 (ebook)
Subjects: LCSH: Radiodiffusion-Télévision française. Groupe de recherches musicales—History. | Groupe de recherches musicales de l'O.R.T.F.—History. | Institut national de l'audiovisuel (France). Groupe de recherches musicales—History. | Music—France—20th century—History and criticism. | Electronic music—France—History and criticism. | Musique concrète—History and criticism.
Classification: LCC ML32.F82 G76413 (print) | LCC ML32.F82 (ebook) | DDC 786.70944—dc23/eng/20240605
LC record available at https://lccn.loc.gov/2024020842
LC ebook record available at https://lccn.loc.gov/2024020843

34 33 32 31 30 29 28 27 26 25
10 9 8 7 6 5 4 3 2 1

CONTENTS

List of Illustrations   vii
Preface to the English Edition   ix

    Introduction   1

PART ONE. ORGANIZING FORGETTING: A THEMATIC APPROACH   7

1. Before 1948: Prehistory   9
2. A Name—a School—a Style of Music   46
3. Concepts—Pedagogy—Tools   55
4. Space—Concert—Audience   103
5. In Search of Music Writing   129

PART TWO. MEMORIALIZING THE FACTS:
A CHRONOLOGICAL APPROACH   157

6. 1948–1958: The Avant-Garde of Musique Concrète   161
7. 1958–1968: Birth of the GRM   190
8. 1968–1978: End of the Schaeffer Era   219
9. 1978–1988: Real and Nonreal Time   260
10: 1988–1998: Innovation   288
11. 1998 and Beyond   311

    Postscript   347

Notes   349
Bibliography—Discography   379
Index   393

ILLUSTRATIONS

FIGURES

Frontispiece   "Music Lesson at the GRM," after Rembrandt's *Anatomy Lesson of Dr. Nicolaes Tulp*
1. The horseshoe-shaped studio of the Club d'Essai, rue de l'Université in Paris   29
2. François Bayle at the morphophone   80
3. The universal phonogène   81
4. The Coupigny "Cube" synthesizer still in use at the GRM   83
5. The famous Syter ball interpolator   90
6. Emmanuel Favreau and the trophy awarded for the new GRM Tools ST   91
7. Geneviève Mâche and the music archives at the Centre Bourdan   98
8. The "spatial projection" device devised by Jacques Poullin   110
9. The Acousmonium on the stage of the Grand Auditorium of Radio France   113
10. The Acousmographe version 3 software to assist with graphic transcription. Screenshots of the graphical interface   142
11. Pierre Schaeffer at the controls of the spatial device   167
12. Pierre Schaeffer at the keyboard phonogène; on his right, the slide phonogène   168
13. Francis Coupigny at the three-track tape recorder that was used for the "spatial relief" of Olivier Messiaen's *Timbres-Durées* during its premiere on May 21, 1952   169
14. Pierre Henry at the GRMC studio   188

viii ILLUSTRATIONS

15. First public appearance of the GRM logo on a poster   *191*
16. Letter of recommendation concerning Iannis Xenakis written by Olivier Messiaen and addressed to Pierre Schaeffer   *192*
17. Iannis Xenakis at the GRMC   *193*
18. File card of sound examples, one of thousands at the Research Department   *204*
19. Studio 54 at the Bourdan Center, with the synthesizer console, during the composition of François Bayle's *Grande Polyphonie*   *230*
20. Syter 2 wired microprocessor   *235*
21. Preparatory notes for the entrance exam for the GRM course at the Paris Conservatory   *251*
22. Computer training course at Studio 123 led by Bénédict Mailliard   *264*
23. Jean-François Allouis in Studio 123 in front of Syter 3   *266*
24. Évelyne Gayou in Studio 116 A, devoted to radio production   *267*
25. Trio GRM+: Laurent Cuniot, Denis Dufour, Yann Geslin   *286*
26. Syter workshop under the direction of Daniel Teruggi   *291*
27. Vinegar test by François Donato, Diego Losa, and Solange Barrachina using testing strips on the GRM archive tapes   *316*
28. The all-digital Studio 116 C; in production with François Bayle as "guest" composer   *321*

TABLES

1. Standards for indexing analog and digital audio documents held at the GRM on physical media   *100*
2. Charles S. Peirce and François Bayle's sound nomenclature   *139*

PREFACE TO THE ENGLISH EDITION

As I finished writing this book in 2007, I was already conscious of the necessity of safeguarding a heritage: the history of the origins of musique concrète and of the "concrete" approach applied to musical composition. I was well aware that for more than thirty years I had been witnessing an exceptional history that was in danger of being lost forever because all the trailblazers I encountered on a daily basis were one or even two generations older than myself—Pierre Schaeffer, Iannis Xenakis, Pierre Henry, Bernard Parmégiani, Luc Ferrari, François Bayle, and Beatriz Ferreyra, all of whom came from the GRM in Paris, but also leading figures of the other global music avant-gardes: Karlheinz Stockhausen from Germany, Luciano Berio from Italy, and the Americans of computer music, John Chowning and Max Mathews, as well as other important musical figures of the second half of the twentieth century: John Cage, Earle Brown, Christian Wolff, Robert Moog, Don Buchla, Pierre Boulez, Eliane Radigue, and Jean-Claude Risset, among others.

This book describes the emergence and the apogee of the GRM, presented in its ambient environment, which extends well beyond its Parisian studios, an effervescence already expressed from the 1950s in the framework of what would later be called globalization: airplanes, telecoms, and international exchanges, all contributing to encounters where artists are the direct beneficiaries and protagonists.

For "concrete" musicians, this effervescence means seizing all sounds, calling upon attentive listening to the world surrounding us. And even if the technologies at their disposal evolve at lightning speed, the art of music remains at the center of their concerns.

Today the fruit of all this research off the beaten track finds an even greater echo via the net and computers. Music has freed itself from the formatted universe of conservatories and other musical institutions to visit without filters all the interstices of the perceptible world, sometimes quite distant, at the limits of space and galaxies, or inversely, deep within ourselves, as close as possible to our sensations. From synthesizers to software, new instruments and new ways of presenting music appear without ever disturbing the continuous intertwining between past, present, and future.

My thanks to David Vaughn for his English translation of this work. As always, examining translation issues together brought us back to core issues, accompanied by beautiful and passionate discussions and a good deal of additional researching.

My special thanks to the initiative of John Chowning, without whom this translation would not exist. May I add that I consider John Chowning to be "an American Schaeffer"—inventor, composer, and founder of institutions, especially the Center for Computer Research in Music and Acoustics (CCRMA) at Stanford University in California. In this regard, our collaboration takes on an even more exalted dimension, for the most fruitful benefit of the world of music.

Évelyne Gayou
2025

Introduction

This book aims at filling a gap. Up until 2007, when I wrote the French edition of this volume, no book gave an account of the history of the Groupe de Recherches Musicales (GRM), which for close to seventy years has regularly produced new musical works, radio broadcasts, concerts, books, records, software, seminars, training workshops, referenced archives, and in recent years internet material. The birth of musique concrète and then electroacoustic music in France in the wake of the Second World War, which led up to the GRM, is always cited as a major artistic event. But it is rarely analyzed. Even today, musicologists feel uncomfortable about this new musical tendency, which they consider significant but difficult to describe and name, and consequently to assimilate.

Pierre Henry, Iannis Xenakis, Luc Ferrari, Bernard Parmegiani, and François Bayle are just a few of the major twentieth-century composers who have profoundly marked the history of music. All of these composers were members of the GRM at some point in their careers, or even throughout their careers, as was the case with François Bayle. In his brief book *Journal de mes sons* (Journal of my sounds), Pierre Henry wrote, "in 1949, musique concrète, invented in 1948 by Pierre Schaeffer, was for me the revelation of what in music could be done better and what could achieve more."[1]

Pierre Schaeffer, the inventor of musique concrète, related the event in his journal:

> 19 April [1948]. Striking one of the bells, I had recorded only the sound following the initial attack. Deprived of its percussion, the bell became the sound of an oboe. I listened carefully.
>
> Could there be a crack in the enemy's position? Had the upper hand changed sides? . . .

> Where did the invention reside? When did it happen? I answer without hesitation: when I tackled the sound of the bells. Separating the sound from its attack was the generating act. The seed of all musique concrète was contained in this strictly creative action on the sound matter. I have no special memory of the moment I made the recording. At first, the breakthrough went unnoticed.[2]

In this book, I intend to reconstruct the adventure of the GRM and the galaxy surrounding it in hopes of creating a synthesis. But to be able to do so, it is first necessary to analyze it, choosing the pertinent issues. So we will examine the history not only of technological developments but also of the musical concepts that underlie them. My position is particular since I am a member of the GRM, seeing things from the perspective of an insider. But this does not mean that my questioning will be any less rigorous. From the moment I arrived in 1975, I was surprised by the impression of a lack of coordination in the organization and functioning of the Groupe, while recognizing its great dynamism. This apparent paradox has never ceased to haunt me. It can only be explained by the drive engendered in each of the Groupe's members by their desire to create. This passionate commitment to creating music, inventing software, and seeking out new concepts is certainly the recipe for the productivity and richness of the GRM. After so many years, all these projects have come to constitute a gigantic corpus that retains its integrity thanks to the permanent support of the public institution Radio Television, and this musical undertaking is now supplemented by maintaining this cultural heritage. Consequently, I have drawn on the GRM archives—at the Institut National de l'Audiovisuel (INA)—in an attempt to add a human dimension based on my intimate knowledge of the subject and its actors. Laying out both the diverse aesthetic and the conceptual steps forward made by electroacoustic music will hopefully contribute to reflection about twentieth-century music, as well as about art in general.

INITIAL QUESTIONS

1948. We still struggle with this date for the birth of musique concrète. But there was a time before that! We must not assume that musique concrète emerged spontaneously in an artistic and conceptual vacuum. The first question that arises is therefore that of its origins. A number of other questions then follow, in no particular order:

—How to define the GRM, simply, in a few words?
—Can we associate the GRM's repertory with a specific musical genre?
—Is the GRM a school or a production studio?
—What has the GRM borrowed from the audiovisual sector, and what has it brought to it? Does it really belong in an audiovisual institution?

Why has it remained associated with it? Is this an explanation for its longevity?
— What is the difference between musique concrète, electroacoustic music, acousmatic music, and electro?
— The GRM was a nerve center for the musical avant-garde in the 1950s. Is this still true?
— Once the initial breakthroughs were made in 1948–49, why did Pierre Schaeffer persist in his desire to create a group?
— How has the GRM changed? Is it plausible that it has survived for more than sixty years by exploiting just one single discovery, musique concrète? What is the importance of its past in its present?
— What future for the aesthetics developed at the GRM?
— What are its key works?

INITIAL OBSERVATIONS

A first contradiction stands out when examining the GRM: the disproportion between its modest size (about fifteen people) and the volume of its productions, all linked to its considerable international reputation. In addition, it gives the impression of being a somewhat isolated entity, both within its parent institution, French National Radio, later the INA, as well as in the general milieu of music. The GRM is striking in its lack of institutional power compared to its immense creative force. I will try to define the parameters of this power, notably through the history of its development of compositional tools.

This power/force dichotomy leads to an evaluation of the musical aesthetics developed at the GRM. Is the concrète approach closer to classicism or to baroque eclecticism? In fact, in exploring sound matter with the intention of giving meaning to the insignificant, the concrète approach of "doing and hearing" places us in the present, immanency. Michel Maffesoli's words, as he described the process of the constitution of the baroque in our modern societies, apply perfectly: "Energy, instead of projecting itself, targets the depth of experience.... Focusing on the moment experienced leads to a condensation of time," which is transformed into another dimension, space.[3] My hypothesis is that in electroacoustics we find ourselves in a baroque aesthetic, which would explain why controlling space is so important in the work of the GRM. We shall see this through the analysis of the evolution of concepts concerning the concert and diffusion.

The spatiotemporal experience of contact with sound is not a new discovery, but it tends to be reappraised with the proliferation of media for storage and distribution. The emergence of techniques for *composition on a support* has, on the other hand, freed the composer from score writing.* Critical listening—associated with doing—is the only possible method for mastering the universe of musical sounds, in the purest oral tradition. We will examine the musical specificity of the electroacoustic genre, particularly on the basis of research into graphical transcription of musical listening.

The final test is the individual versus the collective. Since the origin of musique concrète, there has been a growing confusion between the status of composers, audiences, distributors, and producers of records and concerts. The classical schema of the composer as distinct from performers, audience, and publishers is becoming blurred. The electroacoustic community brings together almost indistinctly all enthusiasts, from the most to the least experienced. I will explore the entire repertoire of works created and/or performed at the GRM.

THE SOURCES

In addition to the exceptional quality of the GRM archives, one cannot fail to mention their great quantity: fifteen hundred inventoried works—accompanied by their entries, biographies of the composers, press reviews, and analytical comments written by researchers—as well as photographs, recordings, books, and more than two thousand hours of radio programs explaining ongoing research or presenting music for listening.

My challenge was to choose from this abundance, with all the arbitrariness implied. The works cited and described in this book nevertheless meet certain essential criteria: they are the most musically successful from the point of view of the public and/or their authors, or at the very least they reflect through their exceptional originality the innovative spirit of the period in which they were composed. I have also tried to mention each composer at least once, especially the younger ones. Finally, some works that do not belong to the GRM repertoire are included because of their aesthetic affinity.

GEOGRAPHY OF AN AVANT-GARDE

The avant-gardes can be analyzed within the time frame of their development, but also in the spatial frame of their diffusion. Proceeding from the hypothesis that "history is first and foremost geographical,"[4] I have explored the boundaries of avant-garde musique concrète and its descendants, using music as the focal point.

---

* Translator's note: In musique concrète, *support* refers to the physical media used to encode sound. In the early days this consisted of phonograph records, then tape recording, and more recently all the forms of digital media. Throughout the translation I have used the terms *support*, *media*, and *medium* interchangeably.

While the objective is indeed to explore the history of the GRM in search of clues to its longevity, the geography of its dissemination illuminates its relationship with the very subject of its raison d'être, music. Going "to the heart of sound" leads us within (and beyond) the limits of music: the spaces between music and radio, music and theater, music and dance, and music and visual arts.

OUTLINE OF THE BOOK

This book is divided into two parts: "Organizing Forgetting" and "Memorializing the Facts." The first part of the book tackles the history of the GRM transversally, following several thematic lines, with the avowed aim of organizing oblivion in the sense given by Paul Ricœur that "memory is organizing oblivion."[5] It is thus a work of mourning. For me, this mourning consists in the unavoidable loss of a large share of the numerous events, works, documents, and anecdotes that, taken together, constitute the substance of the combined actions of artists, researchers, technicians, and staff of Radio Television that gave life and meaning to the GRM.

The chapters of the first part describe the emergence of a new musical genre gaining an identity and recognition through a process of experimentation, interdisciplinarity, artistic creation, and constant innovation, largely technical. All of this centers on the GRM, which has been a trailblazer throughout the past sixty-plus years, but also touches on related research and experimentation across the globe. The dominant issues in music are brought together in a sort of arch where composers, technicians, and audiences are engaged in reimagining listening, hearing, composing, writing, playing, concert making, teaching, and communicating. Later the "musical fact" is explored even as far as the boundaries between musical and visual, by focusing on the question of notating and recording artistic composition: a major issue for all media arts today.

Under the title "Memorializing the Facts," the second part brings references and dates. It presents a chronology of the events marking the history of the GRM, from the most easily identifiable—works, publications, technological developments—to the less conspicuous: the elaboration of new concepts, research, and the evolution of the institution. This chronology is divided into chapters, one for each decade, starting from 1948. As much as possible, all the works are commented on and placed in the national and international artistic context of their time. The history of the École de Paris assumes its full significance when compared with that of the Cologne Studio in Germany, or the American school in New York with its figures such as John Cage, Morton Feldman, Earle Brown, and Christian Wolff. In a similar light, advances in computer music later in the 1970s cannot be fully understood without a look across the Atlantic at the American pioneers of sound synthesis, or at the evolution of the Institut de Recherche et Coordination Acoustique/Musique (IRCAM) in Paris. The most important musical works, presented in the order of their public premieres, are accompanied by commentary on the aesthetic and technical concerns of their authors. Except in certain specified instances, I have cited the dates of the first public performance, rather than the dates of composition. At the end of each chapter, the reader will find short biographies of a few of the most important figures in the vast galaxy of electroacoustic music, composers who for a few years (sometimes more) contributed their genius to the adventure of the GRM, before returning to their solitary paths: Pierre Henry, Iannis Xenakis, Luc Ferrari, Michel Philippot, Michel Chion, Guy Reibel, Denis Dufour.

The most astonishing thing about the entire GRM undertaking, focused on working with sound, is that it remains true to its past despite all the technical innovations that have emerged since the 1950s. The colossal effort of reflection about sound, imposed by Pierre Schaeffer on his collaborators in the Service de la recherche, which culminated in drafting the *Traité des objets musicaux* (1966) and the *Solfège de l'objet sonore* (1967), not only had repercussions on the GRM itself, but also gave it powerful impetus and the capacity to rebound aesthetically. The *TOM* and the *SOS* and the accompanying probing helped challenge the tradition of musical thought in our modern society. They have brought scientifically proven facts to the forefront, rather than the assumptions and impressions that so many musicians had long experienced and described regarding their perception of sound and, by extension, of music.[6] I cannot help drawing a parallel here with the research conducted in the same period into the atom and the understanding of matter in general.

Pierre Schaeffer as a fine physicist, in his quasi-mystical quest to understand musical reality through the systematic exploration of sound matter, profoundly disturbed the so-called learned musicians of his time, provoking rejection from most. In fact, he began a process of desacralization of the art of music by forcing it to "plunge its hands" into sound matter. This contradiction between the quest for what is commonly called musical pleasure, with all its spiritual dimension, and the trivial aspect of the manipulation of sound then became central in this debate. The transcendence of the art of music was directly confronted with the immanence of the world of sound. The monotheism of Western art music is increasingly submerged by the polytheism of "new music," in a flood of amateur musical productions using new technology, and later on the home studio and sound processing software. They all converge towards what Schaeffer and all the protagonists at the GRM had aimed for, creating *the most universal music possible*.

But first, let's back up a little earlier in time, as far as possible before 1948, tracing the roots of musique concrète from the futurist movements of the early twentieth century, as well as from Dadaism and surrealism.

I call this *prehistory*.

PART ONE

Organizing Forgetting

A Thematic Approach

1

Before 1948

Prehistory

Today most historical documents depict musique concrète as the fruit of spontaneous generation, appearing suddenly from nothing on the 5th of October, 1948.[1] On that day, Club d'Essai broadcast Pierre Schaeffer's *Concert de bruits*. Why did this *Concert of Noises* become the official founding act of musique concrète? As with any birth, there certainly must have been a gestation period! We are going to travel back in time, briefly retracing the roots of musique concrète and trying to understand why it is so often portrayed as the fruit of spontaneous generation.

It all began with the Studio d'Essai, created in 1942 as part of French National Radio. In this studio, rechristened Club d'Essai in 1946, Pierre Schaeffer would found the Groupe de Recherche de Musique Concrète (GRMC) in 1951 and then the Groupe de Recherches Musicales (GRM) in 1958. Schaeffer and his entire team were devoting themselves to extensive experimentation in radio at a time when the technical progress in sound recording and synthesis made since the end of the nineteenth century, notably in cinema, started becoming exploitable. Musique concrète also has its origins in the great whirlwind of avant-gardes that followed one another in rapid succession with the dawning of the twentieth century. From an aesthetic point of view, the lineage with the futurist movement, Dada, and surrealism seems to us to be indispensable to consider, even if Schaeffer himself never divulged the sources that influenced him.

We will therefore speak of *prehistory* in the literal sense of the term, that is to say, the period that precedes the appearance of *writing*. In this case, it is the period preceding 1948 that inaugurated, for Pierre Schaeffer, the first manipulations of *sound fixed on a support* to make music: in other words, the first occurrences of *writing sound*.*

PIERRE SCHAEFFER, A MAN OF EXCEPTION, A "REMARKABLE MAN"

Pierre Schaeffer's personality largely dominated the destiny of musique concrète, from his appointment as a young engineer at Radiodiffusion française (RDF, the French radio broadcasting corporation) in 1936 until his retirement from the Office de la Radio Télévision Française (ORTF) in 1974. Born on August 14, 1910, to musician parents, Pierre Schaeffer grew up in his hometown of Nancy until the age of eighteen. After obtaining his secondary school diploma, he entered prep school with the Jesuits at Sainte-Geneviève de Versailles, followed by the prestigious École Polytechnique. With his engineering diploma in hand, he furthered his studies from 1931 to 1933 at the École supérieure d'électricité et des télécommunications in Montrouge. At the same time, he studied music at the Conservatoire de Nancy (cello and solfeggio), and from 1932 onwards he attended Nadia Boulanger's class at the École normale de musique in Paris. During his formative years, he was also involved in the Scouts de France movement, then the Scouts Routiers, where in 1929 he discovered theater in joining the Comédiens Routiers troupe.

Describing his years of training might seem enough to grasp the essence of the man, but another element seems essential to complete the picture: the influence of Georges Ivanovitch Gurdjieff, which can never be underestimated in trying to understand Schaeffer, especially in cases where a simple rational analysis of events does not seem to be adequate. In this sense, we echo Abraham Moles when he asserted that Pierre Schaeffer saw in Gurdjieff a "condensed representation of the Sacred."[2] Let us also recall here that Schaeffer was a great mystic, a practicing Catholic, who went to mass every day,[3] at least until his meeting with Gurdjieff.

Pierre Schaeffer met Gurdjieff in Paris in 1941. At that time everything was going wrong for Schaeffer. The war and the beginning of the occupation forced

---

* Translator's note: In all arts, the term *écriture* (writing) has a broader meaning in French than is most common in English. For example, in music *écriture* may refer not only to notation, the act of writing down the notes, but also to the composer's art of composing, the artistic style, or aesthetic approach of the composer. In the present translation, *writing* may take on this larger definition. Especially in the early days of musique concrète, this terminology took on philosophical and practical importance, because detractors advanced the idea that since musique concrète most often did not possess a written score—unlike traditional serious music—it was in some way lacking, or even not legitimate as musical composition. Évelyne Gayou explores this idea in some depth throughout the present volume. Especially see chapter 5. Concerning the terminology "sound fixed on a support," *fixed* refers to the action of recording the sound, on a *support*, which may be a record, tape, or digital media. Also see *fixed sounds* at the end of chapter 3 (Michel Chion).

him, as a civil servant, to decide on a course of action.[4] He was also in a state of personal distress following the deaths of one of his daughters at an early age, and then of his wife, Elisabeth.[5] The encounter with Gurdjieff was a dazzling experience for Schaeffer and a driving force in his coming life.

Gurdjieff (1877–1949) had established himself in France in 1922, fleeing the Russian civil war, with a group of friends and disciples. After various stints in his native Russia, more exactly in the Caucasus, but also in a number of Eastern lands where he spent many years learning about Eastern philosophies, he had gained considerable renown in intellectual circles.[6] To summarize in a deliberately stark manner: Gurdjieff was a guru organizing the teaching of his spiritual ideas, based on artistic, gestural, and musical practices that were quite similar to the rituals of Turkish whirling dervishes as well as to those of yoga, and aimed at enabling his followers to liberate themselves from themselves and to awaken themselves to things and to the world of matter.

Since we cannot explain in detail the Eastern philosophies whose ideas Gurdjieff disseminated, we direct the reader to the numerous works on the subject, especially that of Ouspensky, his principal disciple, *Fragments of an Unknown Teaching*.[7] It is simply important here to underline that all these ideas were completely novel at the time for the many Western intellectuals who made the effort to discover them. Gurdjieff founded "groups" where candidates for initiation came together in "work" and "movement" sessions. He advocated a daily practice of "movements," a kind of gymnastics for the body and the mind during which the practitioner sought a form of self-surrender, a form of humility. In this period between the two world wars, still largely dominated by the churches, many inquisitive people were attracted by these new experiences of being, labeled *étriques* by Gurdjieff. The aim was to get closer to matter (this great whole, of which man is also a part) and to feel its energies. Schaeffer took part in these sessions, along with dozens of other intellectuals and well-known people of the time: René Daumal and the whole team of the Grand Jeu (of which François Bayle would later become a great admirer), Luc Dietrich, A. R. Orage (who prior to the war directed the English literary review *The New Age*), Lanza del Vasto (who after Gurdjieff's death in 1949 would carry on with Gurdjieff's teachings in France), Jean Paulhan, Henri Michaux, Louis Pauwels, Lord Mountbatten, among others.[8]

From this spiritual and human experience with Gurdjieff, with its synthesis of Western and Eastern cultures, Schaeffer retained many elements that he reinvested in his future research, all the more so as it was in tune with his finely structured mind, that of a Cartesian physicist. In no particular order, let us mention the experimental attitude that Schaeffer always advocated in terms of research as well as of artistic creation, condensed in his famous formula "doing and listening"; the assiduous pursuit of understanding sound matter, in particular its forms and energies, transposed in Schaeffer's vocabulary under the term of "dynamics"; and the notion of "working with sound" during the creative act. In addition, the gestural aspect of the "movements" advocated by Gurdjieff led Schaeffer to reconsider the relationship between sound and the body, and to broaden it on one hand towards analysis of auditory perception in humans and on the other towards exploring the concept of sound space.

Finally, Gurdjieff's influence on Schaeffer can be seen from a social point of view, notably in his use of the term *groupe* in Groupe de Recherches Musicales, as well as its mode of operation, which was directly derived from Gurdjieff with the foundation of the Groupe de Recherche de Musique Concrète as early as 1951. Gurdjieff's influence can also be seen in the mentor-like attitude that Schaeffer developed towards his collaborators; an attitude that went hand in hand with techniques akin to psychological suggestion, but also with the fairly systematic practice of a form of intimidation through sometimes violent or authoritarian judgment, or very calculated techniques of group dynamics such as the almost ritualistic spatial management of work meetings (arrangement of chairs in a room, placement of people according to a hierarchy of interventions), or putting certain members of the group in a tenuous situation—for example, if someone turned out to be a good musician, they were given administrative tasks. Ample testimony from those close to him confirm: Schaeffer also behaved like a guru or, at the very least, enjoyed being respected.

When he joined what would later be called the Direction de la Radiodiffusion Française as an engineer in 1936, Pierre Schaeffer was in particular tasked with installing a system for recording sound at the Paris Opera, since Radiodiffusion Nationale had decided to program live broadcasts of certain performances. But how to give listeners an idea of the visual dimension of the stage sets, costumes, lighting, and scenery? And Pierre Schaeffer had to solve the problem of placing microphones on the stage in such a way that they would be invisible to the audience but nevertheless effective for recording. These first difficulties were for him the opportunity to experience at first hand, and at length, the delicate relationship between art and technique. This first experience gave rise to a first publication, *Vingt leçons et travaux pratiques destinés aux musiciens mélangeurs* (Twenty lessons and practical work for mixing musicians), which in 1938 marked the beginning of his collaboration with *Revue Musicale*.

But the war came, then the defeat and the 1940 armistice. Schaeffer was assigned to the new Vichy administration. He was successful in introducing the project of Radio Jeunesse, but above all he participated with friends including Olivier Hussenot, Paul Flamand, Claude Roy, Albert Ollivier, Roger Leenhardt, and Emmanuel Mounier in the activities of Jeune France,[9] aimed at producing artistic performances in the Maisons Jeune France network, with a view to giving work to unemployed artists (actors, dancers, musicians, visual artists) and to offering access to culture to the general public. Schaeffer was the general director and was principally in charge of the administration, but his heart always leaned towards the theater. During 1941 alone, the association was responsible for 770 performances, half of all the performances given in the entire country. Schaeffer also established a network of theater groups in the Chantiers de Jeunesse created in the free zone. The young Pierre Arnaud, whose account appears below, emerged from this network.[10] After the war, he became a radio director and producer under the name Pierre Arnaud "de Chassy-Poulay."[11]

Despite its artistic success, Jeune France would last less than two years, from November 1940 to March 1942, the date of its dissolution. Schaeffer, considered an "ambiguous figure" by Action Française, was transferred to Radiodiffusion in Marseille.[12] He took advantage of this quarantine to write his first text on *L'esthétique et la technique des arts relais* (The aesthetic and technique of the bridge arts).[13] This short three-page text would later prove to be a cornerstone of his thinking.

For Schaeffer, cinema and radio are bridge arts (*arts relais*), situated between concrete and abstract.* He contrasts them with language, which in his view is more at ease in abstraction than in the concrete. Until now humans have had only language to "leave a trace of events, to escape the shortcomings of oral memory . . . and the betrayals of gesture." Schaeffer compares step by step the different stages in the evolution of language with the development of cinema and radio. In the beginning, language also evolved from the concrete to the abstract: images were first transcribed into ideograms, culminating in an alphabet. But today, while language "gets lost in endless and disappointing efforts, clever at defining, it has great difficulty in representing"; cinema and radio allow us to move from the concrete to the abstract, to go from the thing to the idea. "They express, through their forms, that which could not be spoken."[14] And according to Schaeffer, just as language in its time had progressively reached abstraction, to the point of adopting punctuation, style, and even layout, cinema and radio are in the process of developing their own writing.

THE BEAUNE WORKSHOP, STUDIO D'ESSAI, AND THEN CLUB D'ESSAI, 1942–1946

"Strange times, for a strange experiment," commented Pierre Schaeffer about the period 1942–46, in the preface to the 1990 edition of *La Coquille à planètes*, a radio work produced in 1942–43 and broadcast in 1946. In order to understand the importance and timeliness of this Beaune workshop, we have to go back in time to 1936, to discover another foundational experience for Schaeffer. That year, during

---

* Translator's note: A central concept in Schaeffer's writing is the concept of *arts relais*. I have rendered this term as *bridge arts*. The concept is that these arts form a link, a connection, a bridge between the concrete and the abstract.

a study trip to Germany in the context of his mission with the RDF, he discovered that Germany was clearly ahead of France in terms of radio expression: the Hörspiel (piece or play for listening), or creative broadcast for the radio, was already an established genre.[15] The term *Hörspiel* was used to describe theater broadcasts on the radio in Germany. According to Martin Kaltenecker, it first appeared in Nietzsche's *Zarathustra*, and was applied to radio from 1924 onwards.[16] When at the beginning of 1942, after writing his text on the aesthetics of the bridge arts, and after he was given the title of chief engineer in charge of staff training at the Radiodiffusion Nationale, Schaeffer already harbored the idea of a Studio d'Essai (experimental studio) where artists and technicians would strive together to develop a new dramaturgy tailored for radio, probably a bit like what he had seen in Germany before the war.

It was the spring of 1942. Taking advantage of the vagueness of the period of the occupation in terms of cultural projects, Schaeffer pushed forward his ideas. He contacted the renowned theater icon Jacques Copeau to propose the organization of a training workshop for the emerging radio arts in an ancient former monastery in Burgundy. The "famous" Beaune Workshop, which lasted one month from September 15 through October 15, today stands out as an essential experience in Schaeffer's career. For the first time, Schaeffer put his ideas into practice and obtained tangible results. Jacques Copeau himself was transformed. He who swore by the theater and devoted his entire life to it wrote in 1943, after the Beaune experience: "Radio is an instrument that speaks in your ear in a silent room . . . It is an instrument that allows the poet to seek out millions of listeners in their homes all across the world, and to confide in them, while the poet barely raises his voice."[17]

Schaeffer thus convinced Jacques Copeau to join him in his project. One thousand candidates responded to the press release he issued for the Beaune Workshop. Only six actors were to be accepted, along with a few actors "ready to sing." A voice quartet, one musician, and technicians were also included. They needed hybrid artists: actors who were also curious about the technical challenges of radio. The selection was based more on the candidates' capacity for imagination and innovation than on their curriculum vitae.[18]

At first Copeau was rather skeptical, and his first advice was to be modest: "Start by learning to read for radio. Don't start with Hörspiele or with complicated things."[19] Copeau was first and foremost a man of the theater, at the root of the French theater revival, an initiator of public readings of poetry and prose, taking to the stage solo from time to time at his Vieux-Colombier theater in Paris to read pages from Charles Péguy. From a professional point of view, he no longer had much to prove, and yet he embraced the challenge of working with microphones, exploring how to speak into one, how to deliver a written text, how to convey emotions, how to create a voice, how to express confidence. Copeau advocated a return to diction and to a "school of sincerity" in contrast with that of the fiction of theater.

For Schaeffer, the Beaune Workshop was a model of successful symbiosis between art and technique, between research and creation. Interdisciplinarity proved very productive. Throughout his subsequent career, he would never abandon perfecting this model.

The Studio d'Essai, 1942–1946
FROM THEATER TO RADIO, THE ADVENTURE OF *LA COQUILLE À PLANÈTES*

After the "dress rehearsal" that was the Beaune Workshop, the Studio d'Essai was inaugurated on November 12, 1942, and officially assigned to Pierre Schaeffer on January 19, 1943. Located at 37 rue de l'Université in Paris, it was to be both a radio arts laboratory and a professional training center.

Following a first small production by Jacques Copeau, *Lecture à une voix*, the members of the Studio d'Essai embarked on the colossal production of *La Coquille à planètes*, a "radio opera" by Pierre Schaeffer with music by the composer Claude Arrieu. This production kept them busy for more than a year (1943–44) but was only broadcast after the Liberation, in 1946. In the meantime, Schaeffer carried out other projects, in particular a long interview with Paul Claudel, in the company of Jacques Madaule.

La Coquille à planètes is a *suite fantastique pour une voix et douze monstres* (literally, a fantasy suite for one voice and twelve monsters), consisting of eight radio broadcasts. The original long version consisted of eight one-hour episodes. The short version lasts five hours.[20]

The twelve "monsters" are the twelve signs of the zodiac. The voice of the hero, Léonard, is that of Pierre Schaeffer, the author of the work, who played his own role. His name is Leonardo, as in Leonardo da Vinci, an artist and a man of science at the same time. Leonardo is a lonely man, trying to find his way in an imaginary—abstract—world, symbolized in the program by concrete sound material. When Pierre Schaeffer began writing *La Coquille à planètes* he had just lost his wife, Elisabeth. Was this his way of forgetting the harsh reality of his life in this troubled period of war and occupation? Throughout the work, autobiographical references are made, in particular in the extract from the seventh episode titled *Mon tout est l'amour* (My all is love), where Léonard confesses "his secret" (an astrologer pronounces Elisabeth's first name and Léonard replies, "You forced my secret on me. You opened my dead heart. You have said the magic word"). It should also be noted that in *La Coquille à planètes*, the astrologer resembles Gurdjieff; he advocates the same philosophy of existence—being both actor and spectator of one's own life.

Since the Beaune Workshop, Pierre Arnaud had been part of Schaeffer's team, as director, technician, and adviser. He explained, "*La Coquille* was the result of everything Pierre Schaeffer had accumulated while dreaming of 'total theatre.' It was the continuity of the ideas exchanged with Paul Claudel, and with Jacques Copeau. It was 'total art' with all these techniques—sound, music and text—pushed to the extreme."[21]

Schaeffer's approach is comparable to that of Erwin Piscator, who as early as the 1930s, together with Walter Gropius, had advocated for "total theatre" in the context of the Bauhaus in Germany. Was Pierre Schaeffer aware of that experiment? Probably so, but he never mentioned it in his writings. In *La Coquille à planètes*, the scenario is constructed like a play, with characters and dialogue. But to this the actors add almost all the possibilities of acting using a microphone, from close-ups to long shots, from whispered voices to bursts of laughter. Schaeffer believed that amateur voices stand the test of sincerity at the microphone better than the voices of actors who are used to reciting. He introduced this idea in a passage of *La Coquille à planètes*, through the anecdote of Léonard, the hero, who discovered that the voice of "The Backward Clock" is none other than that of a department store cashier whom he had met at the beginning of the adventure. Consequently, the dialogue was written in simple French, like the language spoken in everyday life. Finally, we note that the scenes are called "episodes," as they were on the radio. The piece is at the edge of theatrical form, on the verge of a mutation towards the shape of radio.

The instrumental music written by Claude Arrieu is interspersed between the scenes and also sometimes mixed in behind the dialogue. The rhythm of the sequence of scenes compensates for the lack of visual cues. The sound effects, or sound scenery, are always carefully selected. Schaeffer's recommendations for their execution are precisely described: "clicking noise," or "All this speech superimposed with musical verses in several languages and Leonardo's reflections," or "Music—Quartet of ondes,"[22] "Like distant plainchant." Unlike a theater set, which remains in place, the sound set "evolves over time, fades away, reappears; it addresses the ear, like the words, interacts with them, makes inferences; it is more than a background, it is a sound context."[23]

Certain lines of the text reveal Schaeffer's early explorations of the concept of the object. His investigation of morphology and metamorphosis is highlighted in the sixth episode of *La Coquille*, where he describes the change of state of a visual object (we know that for him, a visual object is comparable to a sound object).[24] The Astrologer says to Leonardo:

> Stop pouting. Listen.
> Before your eyes the form, color and state of a certain object will disappear, but you will never be able to see the permanence of its being behind its many different appearances. What do you say to that?

This is the sequence where The Astrologer and Leonardo watch a lump of sugar melt in a bowl of water, wondering when, precisely, it can be said that the lump of sugar has changed shape.

Still in this sixth episode of *La Coquille*, subtitled *Idylle aux machines*, Schaeffer tackled for the first time several other important themes that would resurface in his later research, in particular his fascination for machines. Schaeffer had Léonard say, "There is something touching and irresistible in the clicking of our machines, something that reflects my profound nature." Another important theme in Schaeffer's work is the voice, which for him has a special status in relationship to noise. It is directly linked to feelings and romanticism; it is the "messenger of the soul," as Jacques Copeau suggested. In *La Coquille*, Léonard declares, "And to be able to be in love with nothing but a voice, you understand, a voice without a face, without a body, without hands, without sight." Could it be the voice of Elisabeth, his deceased wife? In this first work by Schaeffer, as in all those that followed, voices almost always pronounce snippets of intelligible words. In his *Étude pathétique* from 1948, Schaeffer takes this process of repeated words in a loop to the extreme, so that they end up losing their original meaning and are perceived only as onomatopoeia, no longer belonging to the world of the intellect but rather to the world of emotions, words becoming only sounds, "fait dans la, faitdansla fédanla, fédanla," "y a le vieux moulin, yal vieux moulin, yalvieumoulin," "et qu'elle a, ékela, ékela." In this way, Schaeffer was following in the footsteps of the earliest proponents of radio art, the sound poets who had experimented with the explosion of language and words right from the birth of radio in the 1920s.[25] Today one can even argue that between 1910 and 1920 the Russian futurists (Maïakovski, Gontcharova, Burliuk, Khlebnikov, Livsic, etc.) were the first to take the decisive step towards the total emancipation of sound and of language, in favor of pseudo-language. Following these criteria, they invented imaginary languages such as Zaoum, where vowels symbolized space and time, consonants colors, and sounds, smells.

In *La Coquille à planètes*, Schaeffer also examined the question of diffusing music over the airwaves, music being *fixed on a support* and *acousmatic listening*. In the sixth episode, he has one of the monsters, Taurus, declare:

> We take the harmony of the spheres
> With a strand held in the paw.
> It is the wireless?[26] my dear
> That puts the music in a box.
> Sol, re, la, mi, sol, re, la

Throughout this adventure of *La Coquille à planètes*, one senses that Schaeffer is trying to develop a style specific to radio, neither journalism nor spectacle, but rather reminiscent of the novel—a form of expression that moves from description to action, from dialogue to commentary. Schaeffer links this analogy to the novel with the need to take into account the rhythm of radio, which structures time as music does; a sense of time that belongs to the listener, and not to the author or to the show. Schaeffer also twice plays with trying to speed up or slow down psychological time in *La Coquille*. In the fourth episode, he squeezes "an hour of time (measured by the four quarter-hour strokes of the clock) into a ten-minute sequence." In the last episode, in contrast, he suspends the twelve strokes of noon for as long as possible, succeeding in prolonging this midday moment for eight and a half minutes.

In a recent analysis of *La Coquille à planètes*, the researcher Giordano Ferrari underscored at last the presence of another reality, that of war. He tells us that "we feel this reality throughout the work, hidden among the surreal visions, as an underlying and unavoidable content.[27] For example, to go to the opera at midnight, Leonardo breaks the curfew imposed by the Germans." Giordano Ferrari also points out that there are references to "hunger, poverty, the violence of war, which becomes clear in Chapter V," especially in the scene of the "arrival of the Eagles" (the German air force). This final sound sequence, composed in the style of musique concrète, is described by Schaeffer in the script as "Music, sirens, this soundscape must be from a great distance."

Through all these experiments, Schaeffer raises the question of realism and the onset of dreaming because, as he remarked, "it is at the precise moment when the sound scenery disappears that the space within opens up."[28] As Andrea Cohen stated in her thesis titled *Les compositeurs et l'art radiophonique*, "radio becomes poetic when logic gives way to feeling."[29]

In this quest for the sensitive, Paul Valéry (whom Schaeffer read extensively) was the first to evoke the symbol of the shell in his 1937 work *L'Homme et la coquille*, the shell that, when held to the ear, recalls the sound of the sea or, in the case of *La Coquille à planètes*, opens the door to the chamber of wonders. The shell, like the radio, becomes a medium allowing us to cross from the real to the imaginary. In *Poétique de l'espace* (Poetics of space, 1957), Bachelard extended Paul Valéry's reflection about the shell:

> Inwardly man is an assemblage of shells. Each organ has its own formal causality, already tested during the millennia when nature was teaching itself to make man, by means of one shell or another. Function builds its abode just as the mollusk builds its shell.
>
> If we can succeed in reliving this partial life, in the precision of life giving itself form, the being that incarnates form will dominate the millennia. Every form retains life. The fossil is no longer simply a being that once lived, it is a being still alive, asleep in its form. The simple shell is the most manifest example of this universal shell-like life.[30]

Today we know, based on all his work and reflection on radio sound design, that Pierre Schaeffer conceived the idea of creating a "symphony of noises," which later led him to musique concrète. But another source of influence also intervened: the cinema.

THE CONTRIBUTIONS OF CINEMA TO SCHAEFFER'S THINKING

On numerous occasions in his commentaries, Pierre Schaeffer alludes to the cinema, which he considers to be a bridge art, in the same vein as radio. According to Schaeffer, cinema and radio have many points in common. In both cases works are produced collectively, in both cases the object is transformed into an image—visual in the case of cinema, auditory in the case of radio—and the techniques of production are similar: recording, editing, special effects, mixing, projection/broadcasting. It should be added that cinema and radio—in particular the radio practiced at the Studio d'Essai, or in Germany in the Hörspiel—have a common origin, the theater. Between theater and cinema, radio art finds its place.

In the late nineteenth and early twentieth centuries, magic lantern projection was most often performed in theaters or even in open air, interspersed with theatrical or pantomime performances.[31] Projection was accompanied by a pitchman or a commentator. This was also the case throughout the early period of the moving images of the cinematograph. In fact, as long as cinema was silent, or more precisely, unsynchronized, its operation was very similar to that of theater, including in terms of audience participation. The American researcher Rick Altman, in his 2004 book *Silent Film Sound*, even explains that so-called "talking" cinema shut the audience up, gradually forcing them to listen silently. Silent film stands at a point of transition between theater and cinema. But silent cinema was not silent; a speaker, and a musician or sometimes even a small instrumental group, had the task not only of drowning out the awful noise of the projector but also of attempting to enhance the power of the projected images. But while the presence of the announcer hindered audience identification with the subject being filmed, as the filmmaker René Clair showed in his 1946 film *Le silence est d'or*, music intervened on a completely different level, that of emotion.[32]

In 1942, from the very beginning of the Studio d'Essai, Schaeffer was well informed about the progress of sound in cinema. He was aware in particular that the composers Arthur Honegger and Arthur Hoérée had composed retrograde effects in the musical score of Dimitri Kirsanoff's 1931 film *Rapt*, and that as early as 1937 filmmaker Jean Vigo had conducted tests of backwards sound in his *Zéro de conduite*.[33] The technique of optical sound in cinema already offered all the advantages that would later be discovered with magnetic tape on a tape recorder, which did not appear in radio studios until 1950.[34]

In his thesis titled *l'Electroacoustique dans la musique de cinéma au XXe siècle* (Electroacoustics in twentieth-century film music), Philippe Langlois explored the argument that the experimental and exploratory cinema of the interwar period was at the origin of Schaeffer's inspiration for musique concrète.[35] According to Philippe Langlois, the issue arises in terms of technical progress: "It is in no way provocative to claim that the optical track allows for acoustic manipulations that are much more advanced on the technical level than those on disc."[36]

In my opinion, the inspiration for musique concrète comes not only from cinema, but also from theater, radio, and music. But indeed there is ample evidence for cinema's head start in the art of sound transformation during the interwar period. The articles gathered in a special issue of *La Revue musicale* in December 1934 (approximately twenty articles, under the title *Le film sonore, l'écran et la musique en 1935*) bear witness to this.

Excerpt from the article by Arthur Hoérée:

The work of sound film:
The possibility of connecting sounds in an almost inaudible way certainly does not allow work equivalent to that of editing images, but it can give rise to unexpected applications. When you bang out chords on the piano, the incisive attack and the diminishing resonance can be clearly heard. Remove all the attacks from the film and connect the remaining fragments and we end up with a series of very mellow chords that are quite similar to those of the organ, since they lack the characteristic attack. We could proceed in the opposite direction: remove the resonances, keep only the attacks and render a piano as dry as a banjo, which amounts to changing the nature of an instrument....

There's more. An engineer, by studying the design of several sounds, was able to establish complete scales, collections of timbres. To create a soundtrack for a cartoon, he composed a linear score on the actual film—or more precisely on a strip of paper ten times as large as the film, then reproducing it photographically on the film. Each series of strokes, curves and serrations gave rise to unusual intervals, new timbres, to new music, or more precisely, to literally unheard-of noise.[37]

Here Arthur Hoérée refers to the German researcher Rudolf Pfenninger (1899–1976), who in 1932 succeeded in recomposing (synthesizing) voices from waveforms hand-painted on strips of paper, sounds "from nowhere," as he called them.[38] These strips of paper were then photographed and placed in a reduced form on the optical track of the film and played back normally.

This discovery had a crucial virtue: it reconciled concrete sound recording (of a voice, for example) with classical predictive music composition. It gave composition total freedom, outside all the known scales, from whole step to half step, with all the possible microtones.

But in 1932, another German, Oskar Fischinger, was filming images, or rather drawings, to attempt to transform them into sound.[39] He wanted to explore the musicality of the graphic form. He called it "absolute film." His work was in the tradition of animated cinematic synthesis, represented in Germany by the directors Viking Eggling, Hans Richter, and Walther Ruttmann.[40] These two currents of research (Pfenninger and Fischinger)—more concurrent than competing, according to the American researcher Thomas Y. Levin—testify to the progress made working with sound in cinema during the 1930s.

Pierre Schaeffer, fully aware of the richness of the contribution of cinema in terms of sound production, asked the Institut des hautes études cinématographiques (IDHEC, under the presidency of Marcel L'Herbier at its creation in 1943) to accommodate sound engineering students from the Studio d'Essai in order to "broaden their horizons in related professions." He was even on the admission jury for the first entrance exam, together notably with Albert Camus.

But despite all this progress, Schaeffer put his reflection about the bridge arts on hold. He would only bring it back into the light of day after the war, around 1946, because other preoccupations mobilized him and all the members of the Studio d'Essai.

CLANDESTINE ACTIVITY

During the artistically prolific period of 1942–46, another venture was being played out in the background—resistance to the German occupation. We could say that Schaeffer managed to kill two birds with one stone. In the evenings, or on Sunday mornings, during the staff's time off, his colleagues made clandestine recordings of forbidden poets or musicians in the Studio d'Essai. Paul Eluard read the *L'Honneur des poètes* with Louis Aragon, Jean Tardieu, or Albert Camus, then all these texts were recorded on acetate discs. Copies were also made of the songs of the Red Army Choir and of Schoenberg's *Pierrot Lunaire*, also banned.

But the most heroic feat was the active part played by the Studio d'Essai at the time of the Liberation of Paris in August 1944. A clandestine 100-watt transmitter (weighing 200 kg!), appropriated by resistance fighters from the Thomson company, was installed in an attic room of the Studio d'Essai building at 37 rue de l'Université. It was from there that the *Appeal to Arms* by the leaders of the Parisian resistance was broadcast, repeated every half hour beginning on August 22, 1944, three days before the Liberation! In addition to the *Appeal*, one could hear an entire series of broadcasts, readings and information for Parisians, from the first reporting by Pierre Crénesse, from City Hall, the Hôtel de Ville, through the declaration of the Liberation of Paris by General de Gaulle.

And in these intense and tormented hours, Schaeffer had an idea of genius that was to mark history. The night of the 24th to the 25th of August, knowing that General Leclerc's troops had arrived, the (theatrical?) idea occurred to him to use the microphone to request that the bells of all the churches of Paris be rung together. And in the silence of the night, little by little, the concert of all these bells—set swinging—arose.

A few days after the Liberation, an American soldier appeared at Pierre Schaeffer's office presenting him two gifts. First was a transmitter that he had brought to launch a "free radio." But the French engineers of Thomson had already taken care of that. The second gift was a wire tape recorder, the first of its kind to be introduced into French radio. In civilian life, this soldier was none other than the president of the Radio Corporation of America (RCA), a private network founded in 1919 that had rapidly spread throughout the United States.

After the Liberation, this participation of the Studio d'Essai in the resistance was rewarded with the reconfirmation of its existence and even with a new impetus for its activities.[41] This renewal took form with a name change. The Studio d'Essai became the Club d'Essai, directed by the poet Jean Tardieu. Pierre Schaeffer was promoted to the post of general director of Radiodiffusion, but only under the lesser job title of "mission leader," because despite—or perhaps because of—his merits, people were afraid of him. He gave off the impression of wanting too much to change everything, in particular by advocating separating radio production from broadcasting, by entrusting production to the private sector, thus fully refocusing broadcasting towards its political dimension.

The Club d'Essai, 1946–1960

RESUMPTION OF THEORETICAL REFLECTION ON THE BRIDGE ARTS

The effervescent period immediately following the war gave Schaeffer the opportunity to write three articles for *La Revue du cinéma*, the first in the October 1946 issue, under the title "L'élément non visuel au cinéma, Analyse de la bande-son" (The nonvisual element in cinema, Analysis of the soundtrack). In the second issue of the journal, the continuation of the article was titled "Conception de la musique" (by implication, film music). Finally, in the third issue, Schaeffer examined the "Psychologie du rapport vision-audition" (Psychology of the vision-hearing relationship).

These three articles from 1946 build on his first reflections about the bridge arts from 1942, mentioned above. But they go much further than the first comparisons between radio and cinema, which Schaeffer had addressed in the form of an observation: language had power over the abstract, but cinema and radio were better placed to evoke it "magically." According to Schaeffer, there was "what embodies meaning" (language) and "what embodies the senses" (cinema and radio).[42]

In the 1942 article, several questions were left unanswered. For example, Schaeffer wondered how to explain the fact that in the cinema one can convey "Peter hits Paul," and yet one does not know how to express "Peter does not hit Paul." In the same way, with radio, "you want to make it clear that a quarter has fallen to the ground, you can clearly hear it fall. But how do you hear it being picked up?"

In the articles from 1946, inspired by the experiments carried out at the Studio d'Essai for the first four years, and then at the Club d'Essai, Schaeffer was able to begin to answer all these questions. First of all, Schaeffer drew attention to the superposition of the three elements of the film soundtrack—noise, speech, and music—elements that he considered to be heterogeneous. He quoted the director Jean Grémillon (1901–59), who established a hierarchy between these elements, from the most realistic (noise) to the most abstract (music). Schaeffer concluded that "noise is the only sound perfectly suited to the image, because the image can only show things and noise is the language of things."[43] But then, what place should be given to speech? Schaeffer proceeded from the idea that "the noise of people is words."[44] According to him, this speech should first be considered as the rumor of the characters and not as a text that they have to deliver. While "in the theatre we can understand the characters as being in their reality or in their symbolism through their behavior or through their text, in cinema speech is perceived only as an echo of reality." Speech emerges as a vehicle for narration, for linearity and, by extension, for order, for organization, leading to the impression in some films that the dialogue rings hollow. According to Schaeffer, this is one of the negative influences of theater on cinema, and he suggested giving more weight to the intonation of phrases and to the texture of the voice, rather than to the text. Here we find the influence of Jacques Copeau, who at the Beaune Workshop had advocated first learning to read for radio.

This brings us to the role of music, again in the context of the three 1946 articles in *La Revue du cinéma*. Schaeffer thought music was an "ideal plastic material" and that in the cinema it only needed to be correctly chosen for it to interconnect with the image. He regretted that for economic reasons and to simplify the work process, directors often made use of sounds catalogued in thematic music libraries (such as the well-known collection of the Italian Giuseppe Becce, or the Berlin Kinothek, where listings were labeled Ambiance, Action—chase, fight, disturbance, anxiety, terror, etc.). Schaeffer cited certain directors who went further in their use of music. Jean-Jacques Grünenwald, in *Les Anges du péché*, chose a few musical themes that he assigned to each of the characters or to particular moods in the film. These themes are consequently repeated at each appearance of their "subject," and during the film's unfolding their repetition ends up creating counterpoint that is abstract as well as emotional. Grémillon, for his part, went so far as to suggest that the composer of the music play emblematic themes, *with* the image, as well as sometimes *before*, in order to create emotion—and even to awaken emotions prior to the associated material. Schaeffer concluded that the question of music in cinema (and in radio, too, I would add) must be addressed.

The final important point raised by Pierre Schaeffer in these three foundational 1946 articles concerns the psychology of the relationship between vision and hearing. Aware that he was setting off into unexplored territory, he decided to work by analogy using established knowledge about acoustics. More precisely, Schaeffer tackled the visual image by evoking the phenomena of sound masking. This hypothesis of masking is fruitful, notably in explaining the feelings of enrichment or impoverishment of the image by way of sound, and reciprocally that of sound by way of the image (which is rarer). Schaeffer reminds us that in acoustics, "when two sounds are very close in frequency, they are not heard distinctly, but give the ear an overall impression known as a 'beat.' This beat has a frequency equal to the difference in frequency of the two sounds." And when these "two sounds are of similar intensity without having a very close frequency, they are no longer heard independently, but are accompanied by 'subjective sounds' known as 'differentials' and 'additives,' the pitches of which are defined by the difference or the sum of the frequencies of the two sounds in question."[45]

Schaeffer believed that these same concepts of beats, differentials, and additives between two sounds could also be applied to sound and image, concluding that "roughly speaking, the image is strong when it captures our attention" and that "music is strong when it imposes an essential theme on us and stirs our emotions."[46] The transition from the sensory to the emotional observed in the perception of sound is mirrored in the sound/image relationship: "The analogy of frequency, applied to sound and visual impressions, would therefore cover both tonality and tempo. It could refer to the 'rhythm' of the image in its relationship with the music as well as to their reciprocal 'tonality.'"[47] Here the word *tonality*, taken from the musical vocabulary, means, in terms of the image, a choice of a color palette, comparable to a choice of a key signature in music. Finally, Schaeffer echoed a 1936 article by Maurice Jaubert:

> "We don't come to the cinema to hear music." Jaubert advised music to "get rid of all its subjective elements, ... and recreate under the visual substance of the image, impersonal sound material, by means of a mysterious alchemy of associations which should be the very foundation of the film composer's profession."[48]

The management of Radiodiffusion Française (RDF) had assigned new responsibilities to Schaeffer. Since the end of 1946, he has been traveling the world on behalf of the International Telecommunication Union (ITU), notably to attribute radio and television broadcasting wavelengths to different countries. His numerous journeys allowed him to discover a wide range of music, both learned and traditional. But between two journeys Schaeffer returned to the Club d'Essai. As a tireless researcher, he was conscious of his role as a discoverer; like all the members of the Club, he had the feeling of being a pioneer.[49] And like an explorer who draws a map of the countries he visits, Schaeffer kept a diary, or at least took notes. This diary, for the period of 1948–49, would be published in 1952 by Seuil, under the title *À la recherche d'une musique concrète*. In addition, Pierre Schaeffer scrupulously preserved not only the radio and musical works, but also all the elements of montage and the production documents.

PRODUCTIONS FROM *10 ANS D'ESSAIS RADIOPHONIQUES 1942–1952*

10 ans d'essais radiophoniques 1942–1952 is the title of a collection of five vinyl discs published in 1955 by RTF, under the direction of Wladimir Porché, its general director, and Jean Tardieu, director of the Club d'Essai and the Centre d'études radiophoniques (CERT).⁵⁰ These discs were reissued twice, first in 1961 by RTF, then in 1990 by Marc Jacquin of the association Phonurgia Nova in collaboration with INA. It is a documented retrospective, with commentary by Schaeffer, of the work and theoretical advances carried out at the Studio and then at the Club d'Essai between 1942 and 1952.

In Schaeffer's view, radio art emerged from the moment the adaptation of literary works was granted the right to be heard on the radio. Output in the 1942–52 decade was intense: several hundred broadcasts, with relatively few original creations, that is, works written especially for radio, but many adaptations of works by often very prominent authors, such as Paul Claudel, Jean Cocteau, Gogol, Roger Martin du Gard, and Georges Duhamel. The collection of discs presents twenty twenty-minute montages, chosen from this large corpus, preserved and restored today by the INA's sound archives department. In addition to *La Coquille à planètes*, already mentioned, here are some significant examples of the discoveries and experiments of the time, on the road towards musique concrète.

The 1947 production titled *Christophe Colomb* was introduced by its author, Paul Claudel, with these words:

> An old French poet has just written a book which he called "L'œil écoute"—The Eye Listens.⁵¹ One could just as well say that "the ear sees". Behind the eye, as behind the ear, as behind each of our sensory organs, there is sensitivity itself in all its fivefold power. And behind sensitivity, there is imagination, and there is intelligence.

Text:	Paul Claudel
Adaptation:	Paul-Louis Mignon
Music:	André Jolivet
Mise en ondes:*	Maurice Cazeneuve
Actors:	Michel Bouquet, Habib Benglia
Production assistant:	Olga Lencement

In this drama, the composer André Jolivet, with the help of radio director Maurice Cazeneuve, attempted the experiment of playing previously scored music

---

* Translator's note: *Mise en ondes* is derived from the model of *mise en scène*, also used in English for stage direction, and which literally means "putting on stage." French National Radio adopted the neologism *mise en ondes*, "putting on the (air)waves," to refer to certain highly qualified experts working in sound who possess both technical and musical expertise. The term is still used today in 2023 at Radio France, but is not in general use elsewhere in France. Associated terms in English include *broadcast producer* and *director of sound production*, but these may carry less artistic weight. A variant on the term is *musicien metteur en ondes* (MMO), parallel in meaning with the German term *Tonmeister*, coined by Schoenberg and sometimes used in English. Also see https://en.wikipedia.org/wiki/Tonmeister.

backwards.⁵² This involved playing the recording backwards, after having had the musicians play the music beginning with the very last note.⁵³ Cazeneuve described the experiment as follows:

> The sound played backwards reveals an unprecedented dimension. It is born of silence, gradually enriched by all the harmonics, finally bursting out like "crystal petals struck by an ice cube." In order for this phenomenon to take on its full value, the music must, of course, be written with long held notes. The sound then produces the effect of lights seen through the window of a moving railway car at night. A point of light pierces the fog, grows rapidly and, just as we see it in all its brilliance, suddenly disappears from our field of vision. The sound takes on this character of forward thrust, seemingly charged with creative momentum.⁵⁴

By carrying out this experiment of retrograding recorded music, Jolivet demonstrated that by inverting the music and its support, the result is not at all the same as when one reverses the music by composition alone, as is commonly done in serial music. There is more information in the recording than in the score, or at least different information.

Adrienne Mesurat, 1948, another drama produced at the Club d'Essai, was the occasion to inaugurate a new use of musique concrète.

Text:	Julien Green
Music:	Jean-Jacques Grünenwald and Pierre Henry
Mise en ondes:	Georges Godebert
Actors:	Jeanne Moreau, Jean Topart, Louis Seigner
Production assistant:	Catherine Felix

To accentuate the dramatic character of the sound scenery during an action scene, the production team decided to simulate a fall. Schaeffer described the recording session:

> This fall—at the heart of the drama—was in fact performed in the staircase of the Club d'Essai, between the 2nd and 1st floors, with the help of Foley artist Robert Maufras, also an occasional stuntman. The director, microphone in hand, accompanied the fall, sliding down the staircase banister. After recording about fifteen takes the best were mixed together. The sound of an inflated tire was tacked on to create the supposed impact of the head striking the bottom step.

To stylize this tumble in the stairs and enhance its pathetic character, an excerpt from Pierre Schaeffer's early musique concrète studies was used in the mix to create a "special effect." The success of this experiment opened the door to broad and fruitful collaboration between radio directors and musique concrète composers.

Listening to the various excerpts from radio dramas presented in the recordings selected for *10 ans d'essais radiophoniques*, one notices that the principal themes of research, which will later span the entire history of *music fixed on a support* at the GRMC, and then at the GRM, have already been revealed, either in the form of specific examples in the sound settings and their staging, or in the form of interrogations in the commentaries accompanying the broadcasts. As an initial (nonexhaustive) summary, let us mention the passage from the concrete to the abstract, the concept of sound realism, intrusion into the listener's imagination, the quest to understand auditory perception, exploration of the concepts of volume and space, the function of text and language in relationship to sound, the use of the microphone and layers of presence, acoustic illusion, the interplay of sound colors, brilliance, filtering, masking effects, the different kinds of time, as well as the relationship with memory, the morphology and movement of sounds and their transformation. This brings us to the frontiers of learned music, theater, radio, cinema, and the visual arts. All these research themes will be discussed in detail in the following pages.

SOUND TECHNIQUES AT THE STUDIO AND CLUB D'ESSAI

From the beginning of the Studio d'Essai in 1942, Schaeffer and his collaborators focused on two contrasting approaches to sound, broadly speaking: the concrete method and sound synthesis. The concrete method is based on recording (concrete) sound and then transforming it by various manipulations. The method of sound synthesis begins with generating sounds from oscillators, and then also manipulating them.

The generally accepted notion is that Schaeffer was "opposed" to the idea of synthesis. It is true that he was very critical of the poor quality of the sounds obtained by sound synthesis, but in that he was no different from the majority of musicians of his time. The sinusoidal, triangular or square sounds of the early days were particularly impoverished from a musical point of view. Under those circumstances, Schaeffer preferred the concrete approach, working directly on recorded sound material. But this did not prevent him from keeping himself informed or even from taking a keen interest in research into sound synthesis. Remember that the office next to Schaeffer's at 37 rue de l'Université was occupied by Maurice Martenot, inventor of the Ondes of the same name. Arthur Honegger and Henri Tomasi, fervent users of the Ondes Martenot, were regulars at the Studio d'Essai. Schaeffer had used the Ondes in *La Coquille à planètes*, his first realization, in 1943. The search for new instruments and new sounds later materialized at the Club d'Essai, and then at the GRMC,[55] in the construction of prototypes of sound transformation devices (both acoustic and synthetic), the most famous of which were the phonogènes.

Sound Capture, Recording, and Playback

Background: In 1877 the French poet and scientist Charles Cros suggested the idea of sound reproduction by means of a long-lasting engraving, in a stamped envelope sent to the Academy of Sciences on April 30, 1877, and opened on December 3, 1877. But it merely evokes vague phonographic processes. In contrast, in July 1877, the American Thomas Edison actually demonstrated the use of his phonograph. The system was then refined, and finally, in 1885, Alexander Graham Bell and Charles Sumner Tainter inaugurated the use of wax engraving and thus enabled sound reproduction by electroplating. But in 2008, a group of American researchers, led by David Giovannoni, discovered that the Edison recording of "Mary Had a Little Lamb" was not the first. A Frenchman named Edouard-Léon Scott de Martinville had recorded "Au Clair de la Lune" on a phonautograph in 1860, without being able to listen back to it. So the discovery still remains to Bell. As for the first carbon microphones, they were introduced in 1920.[56]

At the Studio d'Essai in 1943, new equipment arrived for the brand-new Studio 38 (the largest) and Studio 39.[57] Pierre Arnaud describes

> two consoles enameled green, which were curved in the shape of a horseshoe. This was the first time that this system was introduced, and it was later adopted in all the studios. Two pick-ups on the left and two on the right with a "synchro catch-up" crank making it possible to continuously swap out two discs, the only method that existed before the use of the tape recorder. It worked very well.[58]

Until 1951 the recording/playback machine in use was the acetate disc engraver. This "flexible" disc was in fact a rigid aluminum wafer coated with a soft "wax" that could be cut with the engraving lathe. Calling it a "tender" disc—easy to cut—might be more appropriate.*

Each disc had a recording capacity of four minutes of sound. Early musique concrète was all recorded on this medium. In the very midst of the broadcast, studio technicians had to be virtuosos to synchronize the successive discs, recreating the sensation of sound continuity. The record players were fitted with a synchro crank that allowed the operator to launch a record so that it seamlessly segued from the previous one in the right rhythm, without any break in the perceived sound continuum.[59] The members of the Club d'Essai also explored the possibilities of the disc to produce their early music: reverse playback and speed variation to obtain transpositions, and most particularly, looping. All these techniques were accomplished by subverting the normal functioning of the devices.

---

* Translator's note: The *souple disc* is known as an *acetate*, *shellac*, or *lacquer disc* in English. In contrast to vinyl discs, which are pressed onto a lump of vinyl, acetate discs were *cut* with a recording lathe into a soft nitrocellulose lacquer coating. Used to make masters for mass-produced records, they were also used in radio broadcasting to prerecord programming or archive live broadcasts.

FIGURE 1. The horseshoe-shaped studio of the Club d'Essai, rue de l'Université in Paris. © INA (1943).

A question that often arises is whether there were not earlier experiments than those by the members of the Studio d'Essai. Often cited are experiments by Paul Hindemith,[60] Ernest Toch, and Ottorino Respighi, who in the 1930s had composed works for German radio.[61] In these Hörspiele they had notably superimposed several musical sequences on disc, as well as executing speed variations. These works, certainly among the earliest, would not be further developed by their authors. We know today that the most fruitful manipulations of sound on a support occurred with the optical sound process in cinema around 1930. Jean-Yves Bosseur, in his book *Musique et arts plastiques*, reminds us of this anteriority of German production, telling us, "Moholy-Nagy saw the disc as a precious tool for the music of the future, provided that it was not used solely for the purpose of preserving recorded works." From all these manipulations, Moholy-Nagy "believed that it was possible to obtain ... a variety of sounds that would make it possible to create music solely from a blank disc."[62]

Ultimately we notice that each time new tools appear, it is a given that artists always try to enlarge their capacities or to hijack them, taking them to extremes. This remark is important because it can help avoid the quarrels surrounding the search for authorship or anteriority in the use of new processes in art. From the moment that a technology is available, it is likely to undergo this same approach. But someone is needed to conceptualize the process and to establish its aesthetic scope.

The Closed Groove and Reduced Listening

The elaboration of the concept of reduced listening constitutes a fine example of this process of taking new tools beyond their original function. Reduced listening is a foundational and essential act in electroacoustic music, involving listening to a recorded sound without first looking for its cause. From the same single sound it is often possible to formulate several very different descriptions, for example, "it sounds like the noise of an engine," or "it is an iterative sound, with complex mass and regular intensity, in the low register." The first description borders on the anecdotal; the second, through a preliminary effort of concentration, opens to abstraction and then to invention.

At the origin of Pierre Schaeffer's invention of the concept of reduced listening, around 1948, we find an extreme use of a banal technical act linked to recording. At that time, recording was still done on an acetate disk (shellac), along a spiral engraved over the entire surface. At the end of the recording, a closed groove was created, like a "scratched disc," that is, the final spiral looped back on itself, preventing the playback head from falling off the disc when it reached the end.* On some acetate discs, recording was executed by going from the center to the periphery, unlike with vinyl discs. About one percent of the recordings in the INA archives were made on this type of disc. This method of cutting was quickly abandoned because it contributed to the confusion of the manipulators who no longer knew where they were with their *synchros*. Jean-François Pontefract, technician at the INA sound archives and virtuoso in the restoration of all types of sound recordings, explained the reasons for this experiment of recording from the center to the periphery of the acetate discs:

> Discs being played back were muffled by the slowing of the linear speed when approaching the center, and this made synchronization from one disc to the next perceptible during broadcasting. To render the sound transition between two discs more fluid, it was decided to engrave odd sides of discs from the edge to the center and even sides from the center to the edge (this disc was labeled the *escargot—snail*), or vice versa. But it was quickly discovered that when several different recordings were used during the same broadcast, there was a risk of error, and broadcasting would become very fastidious for the technician in charge of the "synchro." In addition, the bandwidth of medium and long wave transmitters was truncated in the high frequency range (cut-off at above 4–5 kHz), so the transition between the two sides

---

* Translator's note: The *closed groove* forces the needle to repeat the very last groove of the record, creating a sound loop repeating indefinitely either silence or a sound.

of the disc was practically imperceptible to listeners at the time. Eventually we returned to the simplicity of edge-to-center engraving, identical to commercial record engraving.

One day in 1948 at the Club d'Essai, a record was left spinning on its turntable for a long time, forgotten by the technician. Suddenly Schaeffer began listening "with another ear" to this final spiral repeating itself endlessly. It is important to note that unlike the vinyl record, which makes a "click" sound each time it encounters a damaged groove, the "closed groove" of the acetate record does not perturb the listening experience when it is looped. So when a closed groove is left to spin continuously, repeating itself many times, the listener's ear ends up detaching itself from the sound fragment being heard, to the point of perceiving it as a sound object in itself, with its own morphology. "Reduced listening" thus no longer involves being interested in identifying the source of the sound (its story), but rather escaping into the world of abstraction that is opening up. "Musical" composition of sounds then becomes possible. One passes from reality to the world of dreams, as the surrealists said. By describing this specific aspect, Schaeffer paved the way for musique concrète. In June 1948, he even filed a patent application for the invention of the closed groove.[63]

Sound Synthesis

Background: In 1906 the American Thaddeus Cahill (1867–1934) invented the Telharmonium, the first electroacoustic instrument of (additive) synthesis. The musical pieces thus produced were transmitted, via telephone lines (in New York), in restaurants, theaters, public gardens, and hotels.[64] Lee de Forest (1873–1961), also in 1906, finalized a version of the triode tube, capable of producing audio frequencies using heterodyning, which combines two high frequencies to create an audible frequency. Then came the Theremin, invented by the Russian Leon Theremin in 1919; the Melochord, by the German Harald Bode; Jörg Mager's Jörg Mager Sphärophon, an electrophone using beats; the Ondes Martenot, invented by Maurice Martenot in 1928; and Friedrich Trautwein's Trautonium in 1930.[65] Like the Theremin, the Ondes Martenot exploits the interference between two waves.[66] In the Ondioline, invented by Georges Jenny in 1941, smaller and therefore easier to transport than the Ondes Martenot, the fundamental sound was produced by the oscillation of a low-frequency tube.

Unfortunately these instruments, precursors of electronic music, never (or very rarely) made it to the orchestra, often because of their limited range of capabilities, or occasionally because of certain major technical defects. They remained associated with their respective inventors, who sometimes founded their own teaching centers. One of these was Martenot, who stood out as an exception with his course that opened at the Paris Conservatory in 1943. Nevertheless, the relatively modest price of these prototype instruments made their dissemination possible in specific environments: movie theaters, churches (for Hammond organs, the first of which dated from 1934), private parties, and research centers.

ROOTS AND RELATIONS TO MUSIQUE CONCRÈTE
Italian Futurists

Musique concrète offers many points of similarity with the bruitism of the Italian futurists. Luigi Russolo (1885–1947), in the wake of Filippo Tommaso Marinetti (1876–1944) and Francesco Balilla Pratella (1880–1955), had developed the idea as early as 1911 that noise was the manifestation of the dynamism of objects. Marinetti declared in his futurist manifesto of 1913: "Futurism has for principle the complete renewal of human sensibility through the action of the great scientific discoveries . . . Every single day we have to spit on the 'Altar of Art,' entering the unlimited domains of free intuition."[67]

It is, however, essential to insist on the fact that the Italian futurists had initially expressed certain extremist political ideals, through the voice of Marinetti, their principal founder. They called themselves anarchists, patriots, prowar, and even antifeminist. These statements tended to scare off a large number of would-be followers of the movement's new aesthetic advances.

But alongside this radical futurism, more moderate futurisms had developed, with their own manifestos and leaders. Francesco Balilla Pratella had published the *Manifesto of Futurist Musicians* in 1911, and Luigi Russolo *The Art of Noises* in 1913. This last "bruitist" manifesto, the best known today, focused on putting forward the values of the future—the city, industry, machines. The musicians Pratella and Russolo were bearers of this form of futurism.

Like Pierre Schaeffer in 1966 in his *Traité des objets musicaux*, Russolo had proposed a classification of sounds in his 1913 manifesto *L'Art des bruits*. Russolo, like Schaeffer, sought a form of abstraction of sounds, resembling sound objects. Let us also note that Russolo's classification offers analogies with the system proposed by the team of the laboratory of Musique Informatique de Marseille (MIM, Computer Music of Marseille) at the end of the 1980s, using *unités sémiotiques temporelles* (UST, temporal semiotic units). The descriptive words used by the MIM, "fall, momentum, stretching, floating, braking, by waves," resemble the following excerpt from the futurist manifesto of 1913, where Russolo also classifies his sounds from a typo-morphological point of view:[68]

> Here are the six families of the Futurist Orchestra that we soon intend to construct mechanically:
>
> 1—Growling, Bursts, Noise of falling water, Noise of plunging, Howling
> 2—Whistling, Snoring, Snorting

3—Whispers, Mumbling, Rustling, Grumbling, Growling, Gurgling
4—Screeching, Crackling, Buzzing, Clattering, Stomping
5—Percussion noises on metal, wood, skin, stone, clay, etc.
6—Voices of men and animals; cries, groans, howls, laughter, rales, sobs.

Conclusion: The realm of sounds must be constantly broadened and enriched. This answers a need of our sensibilities. In fact, we observe that all the contemporary composers of genius tend towards the most complicated dissonances. By moving away from pure sound, they nearly reach sound-noise. This need and this tendency can only be fully satisfied by fusion and the substitution of noise for sound.

Russolo's first futuristic instruments had evocative names: gobbler, smasher, whistler, screecher, snarler, buzzer, splitter, thunderer, croaker, crumpler, hooter. But a question arose about these instruments. Despite the virtuosity of their performers, and despite some fairly sophisticated second-generation instruments for imitating noises—such as the rumorharmonium and the enharmonic bow, which incorporated all the attributes of the earlier prototypes—could these instruments produce noises other than gurgling, clattering, and whistling? And above all, could these noises be transmuted into musical sounds? Varèse reproached them for this limitation as early as 1916, to the point where he broke away from them completely, even though he had been one of their greatest admirers. In 1917, Varèse wrote: "Why, Italian Futurists, do you slavishly reproduce the hectic bustle of our daily lives in only that which is superficial and annoying?"[69]

Schaeffer claimed to have been unaware of the Italian futurists when he invented musique concrète in 1948. However, he had surely read Claude Debussy's *Monsieur Croche*, the first edition of which dates from 1921. In this work, Debussy quoted an article from May 15, 1913, which he had published in the journal of the Société Internationale de Musique (SIM): "We will mention so-called 'futurist' music only to establish a date . . . It claims to gather varied noises from modern capitals into a total symphony, from the pistons of the locomotives to the cowbells of porcelain menders."[70]

In addition, we are all aware of the considerable success of Russolo's three performances at the Théâtre des Champs-Élysées in Paris in 1921, attended by Tzara, Stravinsky, Ravel, Honegger, de Falla, Mondrian, and Claudel, performances that were etched in every memory.[71] The fact is that Pierre Schaeffer had many conversations with Paul Claudel during the Studio d'Essai period, which is an additional argument for affirming that he surely knew about futurism through him.

Finally, it is very likely, quite simply, that Schaeffer had read the futurist manifesto of 1913, *The Art of Noises*, which had been translated into French from the beginning by its authors and was in the National Library in Paris.[72]

Let us leave to Pierre Schaeffer his declarations of ignorance of the existence of futurism even if, like Maurice Lemaître, we are not really convinced of it.[73] How could a man like Schaeffer, so inquisitive and informed about everything, so close to Claudel and Honegger and many other connoisseurs of the bruitists, have ignored such an artistic event?[74] Moreover, his actions betray him: composing a "symphony of noises" that in 1950 would ultimately be titled *Symphonie pour un homme seul* (Symphony for one man alone), and recording locomotives and their whistles at the Batignolles train station in Paris for *L'étude aux chemins de fer*, the first of the 1948 *Five Studies of Noises*. But in his defense, let's not forget that, in this period immediately after the Second World War, a strong desire to wipe the slate clean animated everyone. Many philosophical questions arose after the massacres of the war and the attitude of the Nazis and fascists. In these circumstances, was it possible to endorse futurism, even in France in the 1950s, without being considered a fascist or a "collaborator"? Because unfortunately, the partisan impulses that had led some Italian futurists to embrace Mussolini's fascism had been very real.

In any case, the aesthetic kinship with futurism, and in particular with bruitism, would be long-lasting: abstraction, montage, noises of the industrial world and of daily life, "lyrical obsession with matter" as the Italian Marinetti put it, attraction for all the noises of the modern world. Pierre Henry composed *Futuristie* in 1975, in homage to the Italian futurists, and in 1993 he wrote music to accompany the film *Man with a Movie Camera*, shot in 1929 by the Russian Dziga Vertov. Alongside Italian futurism, other futurisms had emerged, such as Russian futurism, also contributing to the artistic avant-garde of the interwar period.[75]

Russian Futurists

One does not only change countries in moving from Italian futurism to Russian futurism. The great novelty with the Russians is that they took an interest in the physical support and the techniques that are associated with it: montage, mixing, and manipulation. It was through their research into cinema, more precisely the image, but also into the relationship between image and sound, that the Russian futurists most contributed to reflection about the new arts. The filmmakers Dziga Vertov (1896–1954) and Lev Koulechov (1899–1970, the inventor of the famous "Koulechov effect"), took up some ideas from the Italian futurists, for example that of a "mechanical man."[76] However, the Russian approach to futurism, created beginning in 1911 by Maïakovski, Burliuk, Malevich, Larionov, Gontcharova, and Khlebnikov, maintained its independence from the Italians, notably from an ideological point of view.

Dziga Vertov—who is of particular interest to us because of his advances in sound matter—was not even aware of the Russian futurists when he founded his Laboratory of Hearing in 1916. He was simply at the cutting edge of the technical progress of his time, driven by a great desire to experiment. Unlike the Italian futurists, he did not use instruments that imitated sounds, but rather recorded sounds at first on acetate discs, and later on optical film tracks. He was immediately interested in the physical support. This observation is significant because this new approach permitted transformation of sound, as later on in musique concrète, unlike the Italian bruitists' instruments, each limited to a few effects played live.

Let us now return to the idea of trying to reconstitute an aesthetic kinship to musique concrète, following a chronology of events. Dziga Vertov's film *Histoire d'une bouchée de pain* was screened, and appears to have won an award, at the Exposition des Arts Décoratifs in Paris in 1925.[77] Then in 1931 Vertov made a quasi-triumphal tour of European capitals, including Paris, to present his two most recent films, *Man with a Movie Camera* (1929) and *Enthusiasm: The Symphony of Donbass* (1930). In order to achieve a "cine-perception" of the world, Vertov wanted to use the camera as a "cine-eye," more perfect than the human eye, to explore the chaos of the visual phenomena that pervade space.[78] And since Vertov conceived of cinema only in terms of editing, he was immediately interested in the correlation between sound and image. Vertov even created the term *cine-object* around 1925, which can be perfectly grasped by analogy with Schaeffer's notion of the *sound object*: an entity plucked from reality and *fixed on a support*, subsequently functioning autonomously.[79]

The German scholar Oksana Bulgakowa, from Stanford University, analyzed Vertov's film *Enthusiasm* as a "practical implementation of the ideas developed by Russolo in the Futurist Manifesto of 1913."[80] In effect, in this film, the sounds of sirens punctuating the lives of the workers serve as a trigger, at each of their on-screen appearances, to a descent into the heart of matter—in this case, molten steel in a metallurgical complex. What's more, by varying the speed at which the sound support was played, and by making successive copies, Vertov managed to tune the pitches of these sirens to one another, thus creating a sort of "symphony."[81] Added to this are the effects of synchronization between sound and image. Using all these examples, Oksana Bulgakowa demonstrated that Vertov was indeed a precursor, on many levels.

In parallel with his invention of the concept of the *cine-eye*, where using a subjective and portable camera, in conjunction with editing techniques, "shows the world as only I can see it," Vertov envisaged the *radio ear* as early as 1922. But the technology for recording and editing sound was still so rudimentary that it was not until the 1930s that his German disciple Walther Ruttmann finally put into practice the principles of sound editing imagined by the master. For example, Ruttmann's "sound film" *Weekend* (1930) was projected without images, the screen remaining black for the entire twelve minutes of the projection, in order to force viewers to create their own images, listening to the sounds (in the sense of a pictorial thought). Today we would speak of acousmatic listening, or listening without seeing (the source).

From this foray into the world of futurisms and their possible relationship with musique concrète, we maintain that the instruments built by the Italians constitute the apogee of the evolution of the tools and of the entire arsenal of ingenuity employed by bruitists, noisemakers, and other entertainers of the silent cinema era and even of theater. In contrast, Vertov stands out as the first representative of the sound artists of the new era of image-synchronized sound, known as "talking pictures," where it was now possible to mechanically *fix a sound on a support*.[82] So on this point, there is no continuity between the two futurist movements, Italian and Russian, but a rupture. One can differentiate the period before the sound recording and the period after. In his futurist manifesto of 1913, Marinetti only caught a glimpse of all these possibilities. He stated: "Almost no one today using the telegraph, telephone, gramophone, train, bicycle, motorcycle, automobile, transatlantic, airship, aeroplane, cinematograph, and the great daily newspaper (the synopsis of the planet's day) ever imagines that all of this has a decisive influence on our consciousness."[83]

On the other hand, the Russian futurists were unaware in 1916 that in Paris the French poet Guillaume Apollinaire had been advocating the use of the phonograph as a new means of poetic expression since 1914. In an article in his review *Soirées de Paris*, Apollinaire evoked the Italians:

> The new technique of words in liberty, coming from Rimbaud, Mallarmé, and symbolism in general, and the telegraphic style in particular, is quite popular in Italy, thanks to Marinetti. We even find some poets using it in France, in the form of simultaneities similar to the choruses found in opera librettos.[84]

Apollinaire even identified a precursor to all this:

> Jules Romains, who in 1909 had a poem entitled *L'Église* rehearsed for recitation during a conference of the Indépendants. This poem was to be recited by four voices, answering each other, intermingling, in authentic simultaneities unattainable by any other means than direct recitation, or reproduction by means of the phonograph.[85]

The Dadaists

Precursor of precursors, Erik Satie, by introducing typewriter noises, revolver shots, and other real-world clicking into his music, inserted himself into the future history of musique concrète. He is one of the favorite composers of electroacoustic musicians. Moreover, his connection with the Dadaists allows us to establish another link between musique concrète and the past.

French Dadaism extended from about 1915 to 1924, led predominantly by the painters Francis Picabia and Marcel Duchamp, photographer Man Ray, and poet Tristan Tzara. Zurich Dadaism was launched in 1916 at the Cabaret Voltaire, while the American faction took shape at the Armory Show in New York in 1915. Erik Satie (1866–1925), while being a contemporary of the Dadaists and linked to them, never claimed to be part of any movement. He always remained singular even as he actively took part in the artistic scene of his era. Erik Satie had abandoned his studies at the Conservatoire to turn to the bohemian milieu that revolved around the Chat-Noir. His anarchism naturally linked him to Dadaism. It was with Satie that Man Ray created his first Dada objects.

The scandal of the ballet *Parade* in 1917, with music by Erik Satie, made him famous. *Parade* was staged by Diaghilev's Ballets Russes to a scenario by Jean Cocteau, with Picasso's three-dimensional cubist scenery, curtain, and costumes.

Apollinaire, in his introduction to Satie's *Parade*, coined the word *surrealism*, which he used for the first time, but spelled it out in two words: sur-realism.[86] With this term, Apollinaire stressed the subtitle that Cocteau had chosen for *Parade*, *Ballet réaliste*, because the realism that Cocteau referred to tended sharply towards abstraction.[87] It is taken to the second degree, but certainly not in a naive or anecdotal way. The realism that Cocteau evoked turns out to be just as ambiguous as the adjective *concrete* employed later by Schaeffer to characterize his music. In both cases it is a question of concepts and not of analogy. In this spirit, Cocteau introduced elements of circus and street performance into his ballet—an actor disguised as a horse plays an important role. For the first time, the audience became an integral part of the performance, encouraged to become involved in the performance in the same way as in circus. But all this resulted in distancing the audience from the piece, not identifying with it. Noises also played a big part in *Parade*. They were initially intended to play an even greater role, but Cocteau revealed in his manifesto *Le Coq et l'Arlequin* that "practical constraints (the suppression of compressed air, among others) had deprived them of the 'trompe-l'oreille' of dynamos,* Morse telegraphs, sirens, expresses, and aeroplanes [which Satie made use of in his score] in the same lines as the trompe-l'œil of newspaper, cornice, and simulated wood, which painters use."[88] Finally, referring to Satie's work on composition, Cocteau continued: "Little by little a score came into being where Satie seems to have discovered an unknown dimension thanks to which one can simultaneously listen to the parade and the internal performance."[89]

Other examples where noise acquired its letters of nobility and approached the stature of music were the ballet *Relâche* and the film *Entr'acte* (December 4, 1924). In the ultimate Dada performance, the first "instantaneous" performance,[90] this ballet in two acts, created by Francis Picabia for the Swedish Ballets and dancer Jean Borlin with music by Erik Satie and scenery and costumes by Picabia, was paired with the film *Entr'acte*. The film, directed by René Clair, was an integral part of the ballet, but it would also have an autonomous existence on its own. Francis Picabia described the film as follows: "*L'Entr'acte* of *Relâche* is a film that translates our dreams and the unmaterialized events that take place in our brains."[91] Picabia was looking more for sensitivity than for reflection. The idea of instantaneism (or perpetual motion) contrasted with the emerging surrealism, which tended to intellectualize abstraction too much for Picabia's taste.

---

* Translator's note: *Trompe-l'oreille* literally means "deceives the ear," based on the familiar visual arts term *trompe-l'œil*.

Surrealists

We are now beginning to see more clearly a genealogy of the aesthetics of musique concrète, through futurism, Dada, Satie, and surrealism.

Unfortunately, the absolute rejection of music by André Breton, a central proponent of surrealism in his first manifesto in 1924, but also in the second one in 1929, created a deafening misunderstanding. In fact music (and in particular musique concrète) should have been the leading art form of surrealism. Breton's own definition of surrealism should be enough to convince us of this: "SURREALISM, n. pure psychic automatism by which one proposes to express, either verbally, or in writing, or in any other way, the real functioning of thought. Dictation of thought in the absence of any restraint exercised by reason, beyond any aesthetic or moral concern."[92]

From the start, surrealism aimed at being multidisciplinary, subjective, and subversive. Its values were the dream, exoticism, and irreligion. All its techniques of automatic writing and collage in literature, poetry, and painting could also have been transposed into music.

Today it is believed that several factors contributed to the exclusion of music from the surrealist sphere.[93] First, Jean Cocteau and André Breton had no sympathy for each other. However, the poet-writer Cocteau had been a historic ally of musicians since the publication of his 1918 manifesto *Le Coq et l'Arlequin* (The Rooster and the Harlequin) praising the musicians of the Group of Six.[94] Many surrealists were also suspicious of musicians, finding them too academic. It is also probable that Breton rejected music because it represented, for him, bourgeois art, while he intended to be revolutionary. Breton wanted to see "the night fall on the orchestra."

Lastly, Jean Cocteau (1889–1963) had made himself the defender of music, against Breton, by advocating a simpler, more frank, less emphatic neoclassicism than the Wagnerian style in vogue. In his 1930 work *Opium*, Cocteau evoked "the astonishing possibilities of records that have become auditory objects instead of mere photographs for the ear. . . . Speak quite softly close to the microphone, press the microphone against your neck . . . let the machine use a voice of its own, new, unknown, produced in collaboration with it."[95] But Cocteau's prestigious though singular support for music would not be enough to establish a solid place for music in the vast surrealist vortex. This ambiguity would persist forever.

The French surrealist movement, armed with André Breton's anathema towards music, stands out in history as a missed opportunity for music and surrealism—even if Breton had declared that he wanted to "do justice to the hatred of the marvelous"![96] In fact, few French composers dared to proclaim themselves surrealists, even though many of them read and even devoured all the writings, paintings, and other activities of the surrealists, and of course were inspired by them.[97]

It was not the same in Belgium, where composers like André Souris,[98] Paul Hooreman, Paul Collaer, and Mesens, members of the Brussels surrealist group Correspondance, had openly declared themselves surrealists and had even drafted a few declarations of faith, in the form of tracts, such as *Musique 1* (July 1925) and *Musique 2* (September 1925), leaflets that were both informative and analytical.

> The anxious assurance that some musicians reveal—how can we deny that it stems from this form of confidence they still possess in the resources left to them? The skill they deploy to defer the inevitable confrontation with a reality that others have already faced would be enough to deceive us forever if we did not see in it, if we did not want to discover in it, undeniable testimony of their despair. The parlor trick of a little research is not enough to deceive us any longer. Will they never cease confusing means of existence with existence itself?[99]

But even among the Belgians, things became more complicated with time, with Souris excluded from the surrealists for being too artistic and not political enough. In 1953 André Souris had organized a concert of musique concrète at the Caverne du Monde des Arts in Brussels. On the program were *Étude aux chemins de fer* by Pierre Schaeffer, excerpts from *Symphonie pour un homme seul* by Schaeffer and Henry, excerpts from *Concertino Diapason* by Jean-Jacques Grünenwald, Monique Rollin's *Étude vocale*, Olivier Messiaen's *Timbres-Durées*, and *Études I and II* by Pierre Boulez.

Following the concert, René Magritte, already celebrated as a surrealist painter, addressed a very critical comment to his friend André Souris:[100] "Your musicians seem to be unaware that the object of their art cannot be reached ... They are mistaken about their vocation: they should devote themselves to the Lépine Concours award."*

And yet Souris had taken care to add a small introductory text on the concert invitation card:

> The uncompromising stringency of some musicians can no longer easily accommodate the systems, structures and materials that half a millennium of use have exhausted. Only the most recent electro-acoustic means offer them a field of action commensurate with their most delicate and sensitive intentions. This meeting of an imperious requirement and still virgin technical possibilities defines the present state of musical creation of which this modest concert reveals the first fruits.[101]

At the GRM, direct references to surrealism appeared only among certain composers in the generation following Schaeffer. Ivo Malec composed *Oral* in 1967, based on André Breton's *Nadja*, and François Bayle's *Tremblement de terre très doux*, in 1978, was inspired by Max Ernst's *Un tremblement de terre fort doux*.

---

* Translator's note: The Lépine Concours is a famous competition for inventors of everyday objects.

"Modern" Musique Concrète, between Modality and Atonality

To complete this overview of the period before 1948–50, we must finally mention, in addition to Erik Satie, other composers who provided sources for musique concrète, namely Debussy (1862–1918), Messiaen (1908–92), and Varèse (1883–1965).

As early as 1953, in a book titled *Musique vivante*, composer and musicologist Jean-Etienne Marie (1917–89) linked the history of "modern music" to two major converging currents: a *modal* current extending, after Wagner, from Bartok to Messiaen by way of Schumann, Grieg, de Falla, Brahms, Fauré, and Mussorgsky, and a current of *tonal rupture* extending, after Wagner, from Schoenberg to Stravinsky and Milhaud by way of Berg and Webern.[102] For the proponents of musique concrète, Debussy stood at the center of these vast movements as the man of synthesis and thus the father of contemporary music.[103]

This desire to go beyond the synthesis between serial music and modal music, incorporating polymodality and polytonality, was already apparent in Messiaen's work,[104] where certain melodic lines (of birdsong) contained atonal sonorities. Musique concrète, based on recording sounds of all kinds, responded to this definition of synthesis between serial and modal currents, and even exceeded it. As Schaeffer dreamt, it was indeed music striving to be as general as possible. In his book, Jean-Etienne Marie concluded that "going beyond the semitone ... seems to be an urgent imperative, and Pierre Schaeffer offers us 'dense sound' in rupture with the traditional concept of the note."[105] Here we can even detect convergence between the spectral music that was to come and musique concrète. In passing, it should be noted that spectral music composers also saw Messiaen as a source of inspiration.

Nadia Boulanger (1887–1979), herself a composer and conductor—widely regarded as the European emissary and informant of Stravinsky, who lived in New York—was an important factor of continuity for the young generation of composers after the Second World War.[106] She had begun teaching in 1921, continuing until her death. Her piano class at the École Normale de Musique de Paris, but especially the composition classes that she gave privately in her home, became an essential step for those who wanted to enter the professional world of serious music. Among her numerous students were a number of Americans, as well as major names and future composers of musique concrète: Elliott Carter, Julio Estrada, Philip Glass, Marius Constant, Francis Dhomont, Beatriz Ferreyra, John Chowning, Pierre Schaeffer (in 1932), and Pierre Henry.

EDGAR(D) VARÈSE

The world of musique concrète has always pointed to the French composer Edgar(d) Varèse (1883–1965) (Edgard, spelled with a *d*, as he preferred at the end of his life). Varèse was personally acquainted with the Italian bruitists, especially

Russolo, and had been in contact with the Dadaist movement when he was living in Europe. Already Busoni's mention in 1900 of Cahill's Dynamophone, the first electric sound machine, had sparked Edgard Varèse's imagination, and since then he had never ceased invoking science and technology to invent new ways not only of reproducing sound, but of producing it as well. When he went into exile in the United States in 1915,[107] he quickly joined the Armory Show, where he would again encounter Francis Picabia, Marcel Duchamp, Man Ray, and Alfred Stieglitz. Picabia's wife, Gabrielle Buffet, like Varèse, had studied music with Busoni in Berlin in 1908.[108]

Varèse was a precursor in the eyes of the young composers of the musique concrète avant-garde.[109] In the original version of his piece *Amériques* (1921), Varèse "inflated the orchestra," envisioning twenty-seven woodwinds, twenty-nine brass instruments, and an impressive ensemble of percussion instruments, quite "bruitist" and quite "concrete," with sirens, steamboat whistles, cyclone whistles, and herder horns. The interacting dynamics and "sound trajectories" were innumerable. The result is an extremely lively piece, in the image of Varèse's modern and urban image of America in 1918–21 (with, however, some calmer passages). Then followed *Offrandes* (1921), *Hyperprism* (1923), *Octandre* (1924), *Intégrales* (1925), and *Arcana* (1927).[110] In *Arcana*, the treatment of the orchestra prefigures the future musique concrète—it is as if there were three mixer channels working in parallel. The bass sounds are hammered, in the midrange a sort of fanfare emerges, while in the treble the strings and sometimes the woodwinds suggest the twinkling of a starry night. The three voices intertwine and collide, as if by montage. After *Ionisation* (1931), in *Ecuatorial* (1934) Varèse included for the first time in an orchestral score the pure electronic sounds of two theremins, while awaiting the availability of two Ondes Martenot. After the composition of *Density 21.5*, a work for flute in 1936, Varèse fell into a "deep silence," which lasted a dozen years.

Varèse then managed to find his place in the American musical milieu, thanks in part to Otto Luening, alongside whom he had studied with Ferruccio Busoni in Germany from 1908 to 1910. It is worth noting that Busoni had hosted the first performance of Arnold Schoenberg's *Pierrot Lunaire* at his Berlin home in 1912, in a private concert. After that, with the help of his assistant Hermann Scherchen, only twenty years old at the time, Schoenberg presented his work during a triumphant tour of Germany—notably at the Bauhaus—and Austria. When the Russian-born composer Vladimir Ussachevsky obtained a position at Columbia University in New York in 1947, he immediately contacted his friend Otto Luening. In 1948, Varèse, a friend of Luening, would be invited to give lectures. In 1955 the Columbia music studio was created. After that, Ussachevsky and Luening, joined by Milton Babbitt, saw their work acknowledged by the foundation of the Columbia Princeton Electronic Music Center studio, where the "RCA synthesizer" was repatriated in 1959.

Following his "deep silence" as a composer, Varèse composed *Déserts* for orchestra and organized sounds, which premiered with the GRMC at the Théâtre des Champs-Élysées in Paris on December 2, 1954, provoking a scandal.

The Letterists

To complete this panorama of the aesthetic cousins of musique concrète, we must not forget the Letterist movement led by Isidore Isou starting in 1942. The Letterists approach sound in a spirit similar to musique concrète, while arriving from another path, that of poetry, voice, orality, and the visual arts. Above we have already evoked the rupture of language and the verb with the emergence of radio between 1920 and 1930, notably with the Russian futurists. This new aesthetic form spread throughout Europe. The German Dadaist artist Kurt Schwitters (1887–1948), author of *Connaissances élémentaires en peinture, comparaison avec la musique* (1927), practiced "phonetic poetry." As for the French poet Henri Chopin, he conceptualized his work as an "alliance between poetry and music."[111]

But the disruption of the war had its impact. After a period of near-disappearance, Letterism resurfaced only in the 1950s, driven by some quite distinctive personalities: Maurice Lemaître, Gil Wolman, Jean-Louis Brau, and Bernard Heidsieck.[112] Letterist International, founded in 1952, yielded a radicalized fraction—Guy Debord excluded Gil Wolman in 1957 and then founded Situationist International. Later, in 1967, the musician Pierre Henry became interested in Letterism and composed *Granulométrie*, using the texts and voice of the Letterist poet François Dufrêne.

Other Antecedents

There are other precedents, unknown to Schaeffer, but that illustrate the community of thought of the time. In his 1937 text *Credo*, the American John Cage advocated making music from twentieth-century materials. As early as 1939 Cage composed a work based on sound elements played simultaneously on two variable-speed electrophones, *Imaginary Landscape No. 1*.[113] When Cage himself carried out this experiment, he was unaware of the existence of a 1916 *New York Times* article in which Varèse had already suggested using this technique.

Today it is reasonable to question which was the first musical work on a support: *Imaginary Landscape No. 1* by John Cage? Or a fragment of sound cinema without image by Vertov or Ruttmann?

THE INVENTION OF MUSIQUE CONCRÈTE

And in the midst of all these projects, one particular radio broadcast was to become the official genesis of musique concrète. In 1948 on Chaîne Parisienne, on the 5th of October, Club d'Essai broadcast Pierre Schaeffer's *Concert de bruits*.

Was this period running up to 1948 the "one bar for nothing," beat in silence before beginning to play together just to coordinate the musicians' gestures and breathing, just to set the perfect tempo and come into perfect harmony? One essential measure giving meaning to everything that follows?

The first question is asking where the electroacoustic genre starts, and how. In other words, when does noise become abstraction? When do composers shift from sound scenery to music?

For the time being, we will respond that electroacoustic musical work begins "exactly" where creative radio work and the film soundtrack end. This ambiguity about the boundary between genres perfectly illustrates the history of the invention of musique concrète. Originally, it was born on radio while being directly linked to cinema, because the two forms of media use the same technical resources—machines for recording.

The use of recording machines—machines for inscribing memory—led to deep reflection about the notion of time. Pierre Schaeffer read the philosopher Henri Bergson extensively, becoming absorbed with his thought. Bergson's reflections date from the 1920s, coinciding with the beginning of the emergence of recording.[114] Bergson sees time as manifold. In his view, when the present passes, the past cannot follow, but can only be contemporary with it, coexist. Consequently, information is stacked up, not sequential.

Moreover, the act of the annotation of time makes it possible to establish interactivity between two a priori separate senses, hearing and touch. This exploration of "doing and listening" through direct manipulation of the medium finds its analog in cinematographic practice. But it is in the wake of reflection about the technique of the *closed groove* in radio that everything would crystallize for musique concrète. The composition of sequences obtained by repeating loops of varying lengths is the offspring of the closed-groove process. And then Pierre Schaeffer, in an effort to transcend the effortless magic of this process, encouraged us to forget the root of the sound source and to enter into the abstraction of what he called *reduced listening*, at the origin of the concept of the *sound object*, and later of the *musical object*.

In film, the technique of optical sound allowed filmmakers to work with sound as they did with images, editing, mixing, using synthesis and all manner of manipulation.[115] At the dawn of the advent of cinema in the 1930s, filmmakers even took the lead over musicians because the technical tools of film made it possible.[116]

But with the arrival of broadcast radio in the 1920s, a specifically radio-based art form developed (the German Hörspiel), akin to recorded theatrical drama. Little by little, the work on *sound scenery* gained specificity, moving towards abstraction, leaving aside the actual origin of the recorded sounds and the use of words. One can even notice induced polarization between radio drama and musique concrète. Schaeffer noted this in 1942, in his article *Esthétique des arts relais*: "These words, which we have used for thousands of years as convenient bridges to designate objects, lose their usefulness within a means of expression which is itself a bridge."[117]

The second question consists of determining at what precise moment in history musique concrète appeared. Spontaneous generation? Or not? The official history of the discovery of musique concrète involves the Studio d'Essai, the Club d'Essai, and the *Concert de bruits* in 1948, but it stumbles over one question, that of its (sometimes obscure) links with the past, notably futurism and bruitism, as well as Dada, surrealism, Satie, Debussy, Messiaen, Varèse.

It is quite curious to observe to what extent things were built on the unspoken, on the difficulty of expressing the immensity of the consequences of such a discovery! On one hand, Pierre Schaeffer hesitated to flaunt his aura as a guru, but at the same time, he seemed driven by a sort of quasi-religious conviction in an assumed sequence of unfolding events in apparent contradiction with the rudimentary resources available and a skeptical institutional hierarchy. I suggest the hypothesis that he felt secure in his technical knowledge, in his experience gained from previous work, particularly in cinema, and in his skills in theater, but fragilized by the sociopolitical context of the period: the war, the occupation, and the modest means at his disposal, in terms of both finances and collaborators. He was also weakened by his isolation within his parent institution, RDF, and by his lack of connection with the musical milieu.[118] In these circumstances, the public broadcasting of *Concert de bruits* in 1948 was a landmark, both in the eyes of the audience and in the writings of Schaeffer, who himself presented this event as the birth of musique concrète.[119] But this voluntarist proclamation on the part of its inventor should not lead us to believe that musique concrète resulted from spontaneous generation.

For François Bayle, "the breakthrough" was situated in the historical and artistic context of the immediate postwar period, which served Schaeffer as a breeding ground:

> The rupture was the result of a discovery (but also of the sensibility of an epoch that was equipped to appreciate it), the accidental discovery of the closed groove, present in the early records used at the broadcasting studios in mixes and cut edits. While Schaeffer, and later Henry, were the first to reveal this, it was also because all of Surrealism was obviously implicated in it. In analogy with the radiophonic simulacrum and the phenomenon of displacement in dream work, was the electroacoustic experience anything other than this series of truth tests marking the irruption of the existential into the musical domain, until then reserved for the pure play of abstract values?[120]

Is it possible today to establish historical continuity between the different aesthetics identifiable in earlier artistic movements and the avant-garde of musique concrète during the immediate postwar period? We can answer yes to this question, but we must qualify the answer with one aspect often forgotten today: the dates of the events. It is important to remember that Schaeffer's discoveries were made just after the war. The psychological circumstances of the time were a major factor. The intense desire to feel alive, to forget the war, to wipe the slate clean of the entire weighty past of the Vichy period, the camps, the atrocities, and the deprivation, all incited artists (and not only artists) to look for new directions, new paths for expression, while avoiding looking backwards. Abstraction made it possible to whitewash the past. It was central to Schaeffer's thinking. For him it was necessary not only to forget the war and the occupation, but also to forget his own past, namely the tragedy of the death of his wife and daughter. Abstraction is able to dispense with precise manifest reference points and discourse. This was precisely what he needed to protect himself, to reject his legacy. In his effort to forget and reconstruct, he succeeded in singularizing his work with sound and media by bringing it closer to music (a paradise for abstraction), where he believed it could find its full flowering. As he liked to say, musique concrète is the most abstract of all. This path suited Schaeffer perfectly. The son of musicians, he honored his parentage by seeing music as the supreme paradigm. It led to delving deeply into (sound) matter, to the heart of sound. It led to immanence, perhaps even to the underworld, whose entrance Schaeffer sought, like Orpheus, in search of his own Eurydice.

2

A Name—a School—a Style of Music

A NAME

What is the name of this musical corpus we are exploring? Is it musique concrète? Experimental music? Electroacoustic music? Acousmatic music? How can it be that, after more than fifty years of existence, we are still struggling with what to call it? The question of the evolution of the name is not inconsequential. Hoping not to add to the confusion, I won't take sides, believing that my contribution to the definition of the genre consists rather in describing the stages of its development. In general throughout this book, usage of these terms will reflect either the heyday of their popularity or the preferences of their advocates.

Moving from noise to sound, and to music, the genesis of electroacoustic music, placed under the constant tutelage of technological developments, is portrayed as a given (for its practitioners) and as an aberration (for many musicians from the classical school). Originating from theater and sound illustration on the radio, and revealed through techniques for *fixing* sound on a *support*, electroacoustic music was initially termed *musique concrète*, which essentially remained the exclusive term through the late 1950s. Pierre Schaeffer recounted in his journal how he set out in 1948 "in search of a musique concrete," inventing it (invention, in its etymological sense of discovery).[1]

Let us take up one of the numerous definitions of musique concrète, proposed for explanatory purposes, as was often the case at the beginning. Louis Pradalié, deputy director and administrator of the Centre d'Études Radiophoniques (CER), described musique concrète in 1950:

Contrary to traditional or abstract music which is conceived by the mind and then notated theoretically and executed instrumentally, Musique Concrète is assembled from pre-existing elements, derived from any type of sound material, whether it be noise or traditional music, and then composed experientially by direct construction, resulting in the expression of compositional intention without recourse to the aid of traditional musical notation, rendered impossible.[2]

But things got complicated in 1958 at the time of the creation of the GRM, when the inventor himself, Pierre Schaeffer, decided to rename musique concrète and call it "experimental music."[3] This choice was fraught with consequences, creating confusion even in the ranks of devotees. In a question-and-answer exercise for *Revue d'Esthétique* in 1968, Pierre Schaeffer declared, "Musique concrète does not exist. Or at best, it was only a passing moment, a period of time, an initial phase. It is dead."[4] Pierre Schaeffer further explained:

> I therefore abandoned the name *musique concrète* in 1958, not without congratulating myself on this initial stage, to which I still owe everything I have done. But it was necessary to avoid misunderstandings, tenacious as are all misunderstandings when they are both aesthetic and technical. If these first experiments had any consequences beyond particular procedures and the inspiration of a few, it is because it became possible to conceive of an experimental music that made every experimental process its own and preceded all aesthetics.[5]

So as of 1958, the term in use became *experimental music*, relinquishing the term *musique concrète* tethered to the group of the same name—the Groupe de Recherche de Musique Concrète (GRMC)—and also linked to the person of Pierre Henry and the other members who had resigned (or been dismissed) from the now-defunct GRMC. The period that followed led to the Service de la Recherche de l'ORTF (1960), in which the GRM served as a stimulator. Until 1975, "experimental" works followed one another at an intense pace. They took the form of short "études"—studies—or of music "applied" to visual arts (television, cinema, animation, fine arts) and to performing arts (dance, theater). Theoretical research, in preparation for writing *Traité des objets musicaux*, occupied a considerable chunk of the work by composers. A new avant-garde was being born, at the cusp linking visual and performing arts.

Following the creation in 1968 of the class of "Musique fondamentale et appliquée à l'audiovisuel" at the Conservatoire National Supérieur de Paris, the term *electroacoustic music* began to be used. It had appeared long before—the German Jörg Mager had founded an electro-acoustic association in Germany in 1929. The word was in circulation. We find it under the pen of Pierre Henry, in a brief 1952 text titled *La musique concrète et le XXè siècle*: "Here is music that is neither refusal nor acceptance of itself but that imposes itself in the same way as communication in its most modern form: electroacoustics."[6] Later the members of the GRM, in an effort to distance themselves from the periods of musique concrète and then experimental music, met several times in the early 1970s to decide on a new name: What to call this genre situated on the boundaries, intended to extend the field of music to the entire domain of sound as Varèse had hoped in the 1930s—music that, as Schaeffer often said, would be "the most general music possible"?[7] And was it best to put a hyphen between electro and acoustics (electro-acoustics) or better without—electroacoustics? This latter solution was finally retained and adopted by the Société des auteurs compositeurs et éditeurs de musique (SACEM), which had already recognized the genre since the mid-1950s, following the initiative of Philippe Arthuys when he was secretary general of the GRMC.

Michel Chion in 1972 had already identified four stages in the construction of the identity of electroacoustic music, drawing a parallel with the history of cinema, another art form based on using a support.

> In its early years, cinema was satisfied with the simple demonstration of its premise: imitation of reality.

For electroacoustics, this first period was found in the *Études de bruits* (1948).

> Period two: (cinema) sought its cultural letters of nobility by imitating earlier established arts vested with tradition and prestige, such as painting or theater.

In electroacoustics, this second period corresponded to works referring to traditional music: *Suite 14, Bidule en ut, Symphonie pour un homme seul*.

> Period three: this allegiance to tradition is denounced and the intention of forging cinema into a distinct art form is proclaimed.

In electroacoustics, Schaeffer and his collaborators set out to find their own language, which they experimented with in rigorously strict studies: *Études aux objets, aux allures, aux sons animés*.

Finally period four, in the cinema:

> identity having been established over time, and assured through a basic language, strong works, and confirmed authors. Concern with identity weakens, and puritanism is relaxed. New generations enjoy rediscovering the flavor of ambiguity and of the confusion of genres and references. Cinema incorporates everything, and incorporates itself into everything.[8]

It is the same in electroacoustics. But, as Michel Chion said, "that's where we are." And that was in 1972.

Around 1974 a new term came into use, coined by François Bayle: *acousmatics*. For him, electroacoustic music was concerned with the register of playing, acousmatics with the register of listening. Declension of the word led directly to the

Acousmonium—the loudspeaker orchestra—an indispensable tool for those who wanted to project their music into the space of the concert hall, in the spirit advocated by François Bayle—listening to the movement of sounds.

Without wishing to insult Michel Chion, who worked diligently in the 1980s and 1990s to establish a single precise term recognized by all, it must be admitted that the absence of a single name for this musical genre made it difficult to capitalize on and consequently to disseminate it. "Teach? But teach what?" asked Chion—who himself had returned to the original name, *musique concrète*,[9] faithful to its inventor and master, Pierre Schaeffer.

Moreover, expansion in the teaching of electroacoustics, notably at the Paris Conservatory beginning in 1968, gave the GRM the status of a school. Under these conditions, the naming of a genre was important to establishing concepts, all the more so since knowledge of the techniques and conventions of electroacoustic music had practically become essential for anyone wanting to practice music, even if it is only to make a record or broadcast on the radio.

For its part, the German school associated with Karlheinz Stockhausen identified itself from the beginning as "elektronische Musik," electronic music. In the United States, as early as the late 1920s, John Cage had begun speaking of "electric music," but this denomination did not cover any specific group of composers, nor any studio or school. In the end, the term *electronic music* would become the most common on the other side of the Atlantic, even though other names also flourished, such as sound art or sonic art.

During the 1980s and 1990s, the multiplication of mixed works—combining written instrumental sound sources performed live with concrete or electroacoustic sounds, or even electronic sounds—led to a fusion of the different genres in music in general. One would have thought that the question of naming would no longer be an issue. That was without counting on the contribution of new ways of communicating sound, the CD, and then the internet.

Since the year 2000, we have observed the arrival of more and more composers, often emerging from the underground scene, identifying themselves under the succinct designation *electro*, a term that seems to be spreading with the generalization of personal computers and the internet.

Globalization has not spared music, which finally seems to have chosen the greatest common denominator as a name—*electro*—evoking the electrical energy that is behind all these genres. But this tells us nothing about the aesthetics in play.

A SCHOOL

Is the GRM a school or a studio? Let's take the definitions from the popular abridged French dictionary, *Petit Robert*:

School: A group or string of people, writers, or artists who claim the same master or profess the same doctrines.

And:

Studio: An artist's studio . . . a place set up for recordings intended for radio, television . . . [cinema].

The different names given to the GRM throughout its history are remarkably consistent with these dictionary definitions. We might say that the GRM became a school after having first been a studio (Studio d'Essai from 1942 to 1946) and then becoming a club from 1946 to 1951.

At the Studio d'Essai, from its inception in 1942, the Beaune course revealed Schaeffer's strong interest in training. This pedagogical activity placed him in the position of a master intent on professing his doctrine. At the Club d'Essai (1946–60), the Centre d'Études de la Radio et de la Télévision—inaugurated in December 1948 by Wladimir Porché, general director of French Radio and Television—organized numerous conferences where Schaeffer was the leading figure.

The function of a school, in the sense defined by the dictionary, only appeared in 1951 when the Groupe de Recherche de Musique Concrète (GRMC) was founded. The master, Pierre Schaeffer, professed his doctrine, musique concrète. One of the first disciples was Pierre Henry. The first training workshop in 1951 gathered Pierre Boulez and André Hodeir, and then Michel Philippot, Jean Barraqué, and Olivier Messiaen. In 1958, the Paris School continued its existence under the name of GRM, despite Schaeffer's absences. During his missions abroad on behalf of Radio Television, various collaborators ensured continuation of the doctrine. In 1968 the creation, in collaboration with the GRM, of the electroacoustic class at the Conservatoire National Supérieur de Musique de Paris (CNSMP) further reinforced this role of school.

After 1975 the GRM gradually became a studio again following Schaeffer's retirement at the end of 1974. Schaeffer was replaced by two people, François Bayle at the GRM in 1975 and Guy Reibel at the conservatory in 1979. At first, the two designated successors to "the Master" fulfilled their functions well, but competition between the two ended up creating tensions. In 1982, the conservatory's "school" studio, which had been relocated to the GRM, was repatriated to the conservatory, and from that day on, links between the two institutions weakened. The conservatory's course gradually became part of its mainstream curriculum, in the 2000s becoming nothing more than a computer-assisted music (CAM) option.

On the other hand, the passing of Pierre Schaeffer in August 1995 and the retirement of François Bayle in 1997 did not particularly impact the destiny of the GRM as a school. When Daniel Teruggi took over the direction in 1997, he was able to rely on collaborators who were already trained, both from the musical point of view and in the structural workings of the Groupe. Moreover, development of the software suite GRM Tools, a pure product of the GRM ideology developed since the beginning, had consolidated the role of the GRM as a school since the 1980s. To this must be added the electroacoustic music CD-ROM published in 2000, the research seminars accessible online on the GRM website, and the various CD publications and books. Last but not least, the role of hosting composers in residence in the three 116 studios—A, B, and C—continued full force.

Schaeffer remained the undisputed master. His pervasive influence was always subliminally present, even after his retirement. Pierre Schaeffer was a brilliant man, a voice that some even called a guru—similar to his own master, Gurdjieff. No personality, even strong ones, ever seriously undermined his influence. The people of character or genius who had rubbed shoulders with him at one time or another in their careers all ended up distancing themselves, notably Pierre Boulez, Pierre Henry, Eliane Radigue, Iannis Xenakis, Beatriz Ferreyra, Luc Ferrari, Michel Chion, and Guy Reibel. The case of François Bayle is a bit unique because, in not focusing exclusivity on pedagogical, research, or musical projects, his devotion to the vital cause of the group put him in the position of director of the studio, rather than of the school, whose mission would be assumed in a diffuse way by diverse collaborators, bearers of Schaefferian doctrine.

Musically speaking, Pierre Henry, Bernard Parmegiani, Luc Ferrari, and François Bayle, but also Michel Chion, Francis Dhomont, Annette Vande Gorne, Beatriz Ferreyra, and so many others, have all perpetuated the school mindset in their own ways, continuing to create works, organize concerts, write books, and teach, based on the principles learned in the past from Schaeffer.

But in 2007, a question surfaced. Little by little, young collaborators were arriving at the GRM who had never trained under Schaeffer's strict guidance. What would happen when the "old" generation became a minority in the structure?

In Germany, unlike at the GRM, the Cologne Studio had always remained a "studio," even in its name. It functioned primarily as a home for production—and much less as a forum for promoting ideas. Stockhausen's function as a master played out elsewhere, in his summer workshops and his university lectures. But when he worked in the studio, he focused purely on personal compositions and research.

As for John Cage (1912–92) in the United States, one can justifiably speak of the New York school—Morton Feldman (1926–87), Earle Brown (1926–2002), and Christian Wolff (1934–) all considered Cage a master, at least in the beginning, if only because of their age differences.

It can also be said that each of the centers for electroacoustic production, since the 1950s, has been founded by a local master: Luciano Berio in Milan, Toshiro Mayuzumi in Tokyo, Iannis Xenakis at the Cemamu, Henri Pousseur in Liège, Gottfried Michaël Koenig in Utrecht. But unlike all these other masters, Schaeffer never presented himself as a composer. He wanted above all to be a researcher-inventor, intent on passing down concepts, a teaching method, tools, and a permanent institutional framework.

A STYLE: THE MOST GENERAL MUSIC POSSIBLE

How did musique concrète manage to become part of the musical world? Pierre Schaeffer himself admits that "we hesitated to call music" these peculiar works sparked off by modern technique. He even questioned the interest of continuing his experiments.[10] That would be the business of the second generation—Ferrari, Xenakis, and Parmegiani—who immediately followed the pioneers Schaeffer, Henry, and Arthuys.

Schaeffer had valid reasons to hesitate. Electroacoustic music had often been rejected by the musical world. The question "but is it really music?" was often pronounced by critics and music lovers, especially in the early days. The musicologist Célestin Deliège still gave voice to the objection in 2003: "Wasn't the big mistake, unfortunately, labeling this sort of act performed on sound objects as 'music'?"[11] The rejection was as much aesthetic as political; it took one back to the atmosphere of the immediate postwar period, when fledgling musique concrète was in direct competition with the avant-garde of serial music, which spared no effort in attracting the public's favor.

But Schaeffer also had his defenders, including the composer Jean-Etienne Marie, who in 1953 wrote in his book *Musique vivante*:

> Pierre Schaeffer is a discoverer: he knew how to bring together earlier isolated initiatives and produce initial foundational works, creating a current of interest for his research: a new music was born.
> Or rather no: new *musics* were born....
> But beginning with Schaeffer, music saw an immense field of research open up to it, into time itself, or in addition tackling the unknown world beyond the tempered semi-tone. Let us not be mistaken, *Symphonie des bruits*, *Suite 14*, and *Symphonie pour un homme seul* are perhaps the first sketches for a piece that will be to the music of tomorrow what *The Rite of Spring* is to the music of today.[12]

The principal reproach inhibiting recognition of electroacoustic music *as music* was that it did not accede to notation, to musical writing. The second complaint was that it was not melodious.

In response to this absence of notational writing in electroacoustics, Pierre Schaeffer had from the beginning turned to graphic listening transcriptions. As a good student trained at the conservatory, he reproduced what he had been taught. He tried to impose on himself and his collaborators the process of beginning with a score, or at least with a predictive schema, before composing the music. But this did not work well. The empiricism of the method of "doing and listening" gave much better results. The paradox was that Schaeffer was evolving in a scholarly environment, oriented towards notation, from which he himself came, but that the music he had just invented was first and foremost an oral tradition. Probably blinded by his sociocultural background, he seems to have been the first unable to see that music does not need *writing* to be played or transmitted—memory and practice alone are sufficient. Notation is always introduced after a time lag, when a corpus of works begins to form. It then contributes to stabilizing the genre and thus to legitimizing it. In electroacoustics, the interrogations about writing, based on graphic transcriptions, the first of which are more than fifty years old, made it possible to go forward with the subject. But the task was complicated by the multiplicity of musical criteria at stake, making elaboration of a universal code practically impossible. The public had also noticed this, from the start, by reproaching electroacoustics for not being melodious, thus acknowledging, in a way, that musical narrative constituted from a succession of note pitches and durations was already a thing of the past.

Nevertheless, the inclusion of the electroacoustic genre into the learned musical community has been possible through exploiting numerous other consecrated channels:

—use of music vocabulary: concrète opera, *La Symphonie pour un homme seul*, etc.;
—constitution of an instrumentarium (the potentiometric control panel designed by the engineer Jacques Poullin,[13] phonogènes and their entire progeny, up to and including *GRM Tools* software;[14]
—initiation of composers initially trained in classical music, beginning from the very inception of musique concrète: Pierre Boulez and André Hodeir in 1951, Olivier Messiaen, Michel Philippot, and Jean Barraqué in 1952, Darius Milhaud in 1954;
—in-depth description of different musical criteria—in addition to the traditional pitch and duration—in view of developing an "extended" solfeggio;
—organization of concerts, accompanied by the resulting public recognition;
—production of recordings;
—opening a course at the National Conservatory of Paris (in 1968);
—rapprochement with the Western academic musical tradition through so-called "mixed" music;
—collaboration with French National Education, notably for development of musical teaching methodology for early learning (from the 1970's); and
—teaching in musicology programs at the university.

Electroacoustic music has found its place at last in the French musical landscape, all the more easily since the old markers that delimited the field of learned music have become blurred with the generalization of computer-based software tools for composition. Since the 1990s, musical composition has left the academy. The trend has been much the same across the globe. And simultaneously, the explosion in distribution first made possible by the CD, and then by the internet, has erased the borders between genres and between musical circles. We have indeed reached Schaeffer's ideal of the "the most general music possible" in terms of the musical criteria involved and the techniques of production and distribution. The electroacoustic genre simply enlarges the realm of music. But all this still tells us nothing about the essential aesthetics at work, because we are exploring a deeper level, at the roots of the works, in the realm of communication theory. This reveals one of the justifications for the proximity of electroacoustics with communication media.

In an attempt to define the aesthetics of the music composed at the GRM, here are a few clues. The GRM style remains linked to its foundations, the musical object. Works are most often "pure" electroacoustic music, that is, without the addition of instrumental parts played live. The basic sounds come primarily from acoustic sources, recorded from close-up, to preserve the maximum of energy. They are then transformed using machines and software, in accordance with the criteria of typo-morphological analysis described by Pierre Schaeffer in his *Traité des objets musicaux*, namely, manipulation of the dynamics, colors, and shapes of sounds. According to its proponents, it is an aesthetic of "beautiful sound," which "resonates." This is in part because of the addition of the concept of the movement of sounds in space, intensified in concert by use of the Acousmonium. As far as composition and mixing are concerned, the GRM has a reputation for "meticulous" and highly "written" work, predetermined, leaving little room for improvisation.

3

Concepts—Pedagogy—Tools

In fact, I do compose at the piano and I have no complaint about it. What's more, I think that it is a thousand times better to compose in direct contact with the sound material than to compose by imagining this material.
—IGOR STRAVINSKY

Today we are aware of the difficulties that electroacoustic music has had in imposing itself in a musical milieu that was fixated on written music, especially at the end of the Second World War, when an entire youthful generation of classically trained composers was impatient to establish themselves professionally. The criticism that is still often put forward to exclude electroacoustic music from the musical sphere has almost always been the same: "electroacoustic music does not have access to notation." But by the early twenty-first century the trend had finally been reversed. Almost all musicians, even those trained in the strictest school of the written score, have progressively understood that electroacoustics favors a return to the sensory, a quality they have themselves pursued with less and less confusion following all the years of serial and postserial austerity. Since it does not depend on notation, but on the contrary on praxis, electroacoustics is renewed proof that one can approach music directly through the senses, through touch and hearing—doing and listening. In addition, as early as the 1950s it became possible to rediscover, thanks to radio and dissemination of the research work of ethnologists, that the same was true of many other musical traditions, both popular and serious, that could only be learned through praxis, aurally (and orally) communicated through a master. But at the end of the war, the primacy of the concepts of the classical tradition, still linked to the quasi-monopoly of teaching music by way of notation, remained preponderant.

Schaeffer was himself representative of these musicians educated in notation, and we often find this ambiguity in him. On one hand he encouraged experimentation and "doing," and on the other he tried to dominate musical parameters as much as possible, notably in describing and defining them. These two approaches to music are by no means incompatible. Pierre Schaeffer was situated in the art-science current that had been experiencing a successful resurgence since the beginning of the twentieth century. The notion of art-science, taken from the Renaissance, inspired Jean Cocteau in 1918 to write in his short volume *Le Coq et l'Arlequin*, "Art is science made flesh," and also, "The musician opens the cage to numbers, the artist emancipates geometry."[1] At that same time, the composer Edgar(d) Varèse, leading defender of art-science, was strongly inspired by rereading the writings of Aristoxenus of Taranto, a pupil of Aristotle, who in his work *Elementa Harmonica* (written about 350 B.C.) recommended resisting "turning one's back on sensation," relying on the ear rather than on mathematical reason.[2] But such an interest in the senses was not at all common among musicians at the beginning of the twentieth century. Even after the Second World War, the most widespread theses in the musical world were still largely linked to the classical heritage. Most often, it was still a question of the Pythagorean theory of the Harmony of the Spheres, which functioned on every level: from the world, and from nature, to man and all the way to music, expressed in the form of numbers. Closer to us and to the GRM, Robert Francès explained in his 1958 book *La perception de la musique* that notating music has provoked a hypertrophy of conceptualization in musicians, with its apogee found in serial music.[3]

One of the most extraordinary projects, led by Pierre Schaeffer, was the collective undertaking of systematic analysis of sound phenomenon, which culminated in publication of the *Traité des objets musicaux* (*TOM*) in 1966.[4] The *TOM* is the reference work of the GRM. Published by Éditions Seuil in Paris, this seven-hundred-page book is divided into seven sections encompassing thirty-seven chapters. Schaeffer explored the ideas and interrogations inherent to musique concrète, whether philosophical, scientific, technical, sociological, or musical in nature. This interdisciplinary approach constituted the originality and complexity of the book, which is both an essay and an account of research. In 2005, it was discovered that Schaeffer himself had drafted a reduced version of his work, to simplify access to his ideas.[5]

Complementing the *Traité des objets musicaux*, a series of sound examples commented by Schaeffer, *Solfège de l'objet sonore*—known as *SOS*—was published in 1967, in the form of three long-playing vinyl records accompanied by a booklet presented in three languages. Several reissues followed on audio cassettes, and finally on CD with a book in 1998.[6] The 282 sound examples of the *SOS* were executed by Guy Reibel and Beatriz Ferreyra, and the written notes were by François Bayle, Agnès Tanguy, and Jean-Louis Ducarme, all under the direction of Pierre Schaeffer. The *SOS* has become the GRM's "best-seller," now found in university libraries and conservatories across the globe.

Finally, the third "essential" work to approach musique concrète, *Guide des objets sonores*, was written a bit later by Michel Chion.[7] According to its author, the ambition of this guide was "to offer researchers, musicians, music lovers and all those who are interested in the world of sound, a working tool, ... providing them fuller knowledge and understanding of the considerable contribution of Pierre Schaeffer in this field."[8] Michel Chion's *Guide* was presented in the form of a very large, annotated index of the concepts broached by Schaeffer in *TOM*.

In his preface to *Guide des objets sonores*, Pierre Schaeffer reminded us that he considered the problematic of music "on three floors, or as linguists would say, with two articulations: sound/musical/meaning."[9] This quest to understand musical phenomenon motivated him throughout his life. His thinking was very open and dynamic, as I will endeavor to demonstrate in this chapter, whose title, "Concepts—Pedagogy—Tools," also emphasizes that this is a multidirectional dialectic. Following Pierre Schaeffer's example, it is the contrary of deductive linear thinking. With this basic premise in mind, it becomes simple to understand the theories that Pierre Schaeffer implemented at the GRM, and which are still in use today.

Unfortunately, Schaeffer also had his limitations, of which he was aware. In my opinion, the paradox is that he too often presented his thought in a binary form, while he experienced it in three dimensions. His paired criteria (permanence/variation, concrete/abstract, form/matter, etc.) are one example,[10] as are his numerous two-entry explanatory tables in *TOM*, to which he added subgroups and arrows to attempt to convey fundamental meanings, but without any real success.[11]

CONCEPTS

The Sound Object and Reduced Listening

If one were to reduce the electroacoustic approach to a single concept, perhaps one would elect "reduced listening" as its key. It is a core concept, more important in my opinion than that of the sound object, which has left its mark on people's minds, but it has also generated confusion. Often one hears the sound object evoked rather than reduced listening. To understand reduced listening, it is necessary to start with the sound object. Let us return to this definition from *TOM*: "*The sound object is the coming together of an acoustic action and a listening intention.*"[12] To clarify, Schaeffer gave the example of hearing an arpeggio: "musical listening, analogous to linguistic listening, will recognize a pitch structure, which can be broken down *into several musical objects* coinciding with the notes. Natural listening will recognize the unity of the instrumental gesture and, following the same criteria, musicianly listening, concentrating on the energetic, will discern *one single sound object.*"[13]

This differentiation between two types of listening allowed Pierre Schaeffer to shift from the concept of the sound object to what he called reduced listening. These two concepts are complementary. One cannot detect a sound object without practicing reduced listening, in other words, listening to a sound, or a sequence, while forgetting its causal origin. Reduced listening is only possible and only becomes automatic with training: it is an attitude resulting from an effort at abstraction.

The concept of sound object established by Schaeffer referred to finished objects, isolated or produced by an act of editing. It is a concept that originated with the use of the closed groove on acetate discs, corresponding to a sound entity severed from its origin. It was essentially tied to the "closed groove" format rather than to any other type of recording. When we speak of a sound object, we place ourselves as much in a mode of analysis as of production. Today, it is often quite difficult to identify a sound object in a musical flow produced by multitrack mixing. Moreover, the sound object (if only by the connotation of the word *object*) tends to send us back to the anecdote of the sound source, whereas reduced listening is a broader concept describing a listening attitude, regardless of the recording process used or the type of sound perceived. During the 1970s, the heirs of Schaeffer's thought—François Bayle, Denis Smalley, Marcel Frémiot, Guy Reibel, and Alain Savouret—sought to establish more functional definitions for composition, going beyond the concept of the object. François Bayle introduced the term *dynamic species* and Guy Reibel the term *sequence-play* to describe sound flux and its transformation. These different descriptive and functional categories also each corresponded to a distinct era and level of technological development: the sound object was born with the closed groove that determined the limits of the object (one circle of the groove). Then came the tape recorder. The concept of object already became a bit fuzzier: most often the sound object corresponded to a short cell, easily memorized and delimited by an editing marker (a clean cut made with scissors). But with multitrack tape recorders and mixing, the movements of the sound object and its metamorphoses became more relevant for listening, more identifiable, giving rise to the concepts of *dynamic species* (Bayle), *morpho-concept* (Bayle), *spectro-morphology* (Smalley), and *temporal semiotic unit* (Frémiot). The advent of synthesizers and MIDI systems in the 1980s would favor development of compositional procedures, sometimes semiautomated, and the term *sequence-play* (Reibel) was coined to characterize the sound flux.

In 2007 the term *sound object* has progressively taken the place of *reduced listening* in the vocabulary of the acousmates. Now when one evokes sound objects, it suggests that one is seeking the abstract character of what is to be heard. The sound object has progressively lost its original operative character.

The concept of reduced listening, for Schaeffer, stemmed from the influence of Husserlian phenomenology, which he often mentioned without ever quoting its sources. Husserl spoke of phenomenological reduction or *epoché*, a Greek word that means "put between parentheses."[14] Schaeffer wrote in *TOM*:

> If I stop blindly identifying with my perceptual experience, which presents me with a transcendental object, I then become capable of grasping this experience along

with the object it gives me. And then I notice that it is *in my experience* that this transcendence is *formed . . . the perceived object is no longer the cause of my perception. It is its "correlate."*[5]

This gives us the concept of "intention to hear," today known as "listening intention."

Listening Intention

Applying his principle of the *four listening modes*—represented by the four verbs *écouter* (to listen), *ouïr* (to perceive aurally), *entendre* (to hear), and *comprendre* (to understand)—Schaeffer raises the problem of the hearing intention. His reflection is capsulized in *TOM* in the famous sentence, "I unintentionally perceived (*j'ai ouï*) what you were saying, without listening (*écouter*) at the door, but I didn't understand (*je n'ai pas compris*) what I heard (*j'ai entendu*)."

Since listening is possibly multiple, it is necessary to highlight the different modes of functioning. According to Schaeffer, we can, for example, practice listening in a scientific or musical perspective. And in the case of musical listening, he identifies three typical listening patterns:

(1) the *ordinary* listener, in general drawn toward musical meaning and at the same time responsive to the conditions in which the sound is made; this first situation, in fact, relates to the other two:
(2) the *acousmatic* listener, and
(3) the *instrumentalist* who makes the sound.[16]

This concept of listening intention was subsequently taken up and developed, beginning in the 1970s, by François Delalande in his musical analysis research.

One of Schaeffer's ambitions through all his research was to go as far as possible in understanding music. He wanted to pierce its meaning. In his quest, linguistics was of major importance. He had read Saussure and Jakobson,[17] and throughout his work one finds traces of parallelism between language and music. He wanted to try to understand the phenomenon of listening, dividing sound into units comparable to the phonemes in language. At the very outset of *TOM*, he stated:

> General linguistics has been reflecting on language systems in this way for several decades. It was no longer content to explain language systems through one or several reference languages, as traditional linguists had done. From phonetic material to phonological functional units there are correlations that explain each other. Of course, doubt can be cast on any close parallelism between language systems and music because of the arbitrariness attached to the choice of meaning and the free relationship between signifier and signified, which makes the word into a sign, whereas the musical note has always appeared to impose itself independently of any arbitrariness, like a given from the physical world to which we seem to respond. This statement contradicts the previous one: that the musical is deduced from sound.[18]

Further on, Schaeffer continued:

> What we have learned from linguists helps us no longer to confuse the sign with a physically preexisting reality: even if the definition of pertinent features, or values, appears to be relevant in a given musical system, we will stop trying to interpret other systems in terms of the pertinent features of our own. There is no doubt that in contemporary musics, even orchestral, a study of this kind gives us other constituent objects in addition to those indicated by a no longer adequate notation—provided that this study of structures is done after the event and is not confused with the a priori schemas of composers.[19]

The entire *TOM* was thus permeated with words and concepts borrowed from linguistics. Let's conclude with these famous final sentences from the book:

> Sound objects and musical structures, when they are authentic, have no informative mission: they turn away from the descriptive world with a sort of reticence in order to speak all the better about it to the senses, the heart and mind, to the whole being, ultimately about himself. This is how languages take on a sort of symmetry. They are man, described to man, in the language of things.[20]

Real/Living, Natural/Cultural, Concrete/Abstract

The "isolation" (between parentheses) of the sound object, *fixed on a support* in its temporal development, revived the substantive debate on the notion of reality. Can we consider the recorded object as Real?

At the beginning of the development of recording techniques, the documentary mindset predominated. One recorded the voice of a famous poet, Guillaume Apollinaire, reading his *Sous le pont Mirabeau* in 1914, or one recorded a musician interpreting a piece of music. But rapidly, often as a result of errors in manipulation, the tools would be diverted from their primary use as a medium for memory, giving rise to more creative and abstract production.[21]

Roland Barthes, in his book *La Chambre claire*, provided an explanation for this change in register. Referencing phenomenology, he explained how, in photography (also an art of support), immobility due to fixation on the support "is somehow the result of the Real and the Live: by attesting that the object has been real, the photograph surreptitiously induces belief that it is alive, because of that delusion which makes us attribute to Reality an absolutely superior, somehow eternal value; but by shifting this reality to the past ('this-has-been'), the photograph suggests that it is already dead."[22]

In musique concrète this same immobility due to fixation on a support, described by Barthes, also evokes the same confusion between real and living. Since the real of the recorded sound sometimes reproduces a long segment of time, at the first listening one can confuse it with the living. This first listening leads us, as in photography, to "that which has been" and thus also to the already dead. But a simple repetition of the recorded sound (in the form of a loop for example) induces distancing, creating a new space/time tied to the support; listening to this new object leads directly to onirism, to a shift from the real to an image of the real. The sound object will then be spotted by listeners thanks to their education in perception. Schaeffer classified this last debate under the term of natural/cultural. *Natural* referred to the (almost animalistic) basic perception given to each individual from birth, oriented essentially towards survival and self-defense; *cultural* referred to the codes learned during the course of life.

At the GRM, but also earlier at the Club and at the Studio d'Essai, this debate was naturally exacerbated because of the lure created by the recorded "concrete" sounds. We use the term *concrete* here in its double sense: both sounds with recognizable sources, and sounds used "concretely" for themselves, without any preconceived notion of their origin. The ambiguity due to this double meaning even led Schaeffer to rename musique concrète in 1958, calling it "experimental." For Schaeffer, this was necessarily for the recognition of this new music by the milieu of classically trained musicians. The abstraction necessary to practice musique concrète seemed to him to be sufficient to justify its place beside other musical genres.

The dialectic between the concrete and the abstract had appeared in Schaeffer's thinking as early as 1948. He had advocated musique concrète as a reaction to the abstraction of notes used in music by the serial avant-garde of the time. But soon the most conservative musicians equated the use of sounds of concrete origin with the absence of serious musical thought, in an attempt to denigrate this nascent genre. Schaeffer, on the contrary, directed all his efforts towards a search for abstraction. He went so far as to rename *Étude aux casseroles*, his work using the sounds of pots and pans, rendering it as *Étude pathétique*.

For Schaeffer the word *concrete* meant manipulating sounds concretely, directly in a to-and-fro interplay between making and listening, without passing through the filter of a coded form of music writing. This experimental method later became, circa 1960, the watchword of the Research Department of the Office de Radio Télévision Française (ORTF), when Pierre Schaeffer undertook to impose his *Traité des objets musicaux* (*TOM*), completed by the *Solfège de l'objet sonore* (*SOS*).

In *Traité*, Schaeffer explained:

> When in 1948 I suggested the term *musique concrète*, I intended, by this adjective, to express a *reversal* of the way musical work is done. Instead of notating musical ideas using the symbols of music theory, and leaving it to known instruments to realize them, the aim was to gather concrete sound, wherever it came from, and to abstract the musical values it potentially contained.[23]

Morphological Description of Sounds

Morphological description of sounds, adopted today by the majority of composers of music on support, computer-assisted music, and music using tools of synthesis and transformation of concrete and instrumental sounds, was summarized in the form of a table in *TOM*.[24] Since it would be impossible to present here a detailed explanation of this essential element in grasping sound material, I refer readers to the two major works on the subject: first of all, Pierre Schaeffer's *TOM*, but also the book penned by Schaeffer's "pilot fish," Michel Chion, *Guide des objets sonores*. Let us simply quote Michel Chion's words: "This description will consist essentially in the detailed differentiation of sound objects, in their 'contexture,' based on traits known as *criteria*, the number of which is limited to seven: these seven morphological criteria—*mass, harmonic timbre, dynamic, grain, allure, melodic profile, mass profile*—will be examined one by one, defining different classes for each of them."[25]

It is from the basic duology form/matter that the morphological description of sounds functioned and took us away from elementary anecdotal description, by placing all the musical criteria on the same level. The abstract and the figurative collided, opening an immense space to the imaginary.

The idea of describing the morphology of sounds resulted from experimental research carried out systematically during the years 1950–60 by Schaeffer and his collaborators, in order to deconstruct musical phenomenon, and then to be able to recast it and propose "a universal solfeggio."[26] For Schaeffer, it consisted in fundamental research carried out with all the rigor of scientific research. He himself had a scientific background, an engineer like most of the research leaders he surrounded himself with. But he was not a mathematician—unlike Xenakis, for example, who failed to make his point of view heard and ended up leaving the Groupe in 1963. The researchers of the GRM were instead often physical scientists, well versed in experimental methods, centered on Cartesian logic.

It is quite common to say that the electroacoustic genre is a natural child of phenomenology. In fact, at the core and from the outset there were Cartesian ideas. One example is this quotation Schaeffer attributed to Descartes, which Schaeffer was fond of using: "The ear is a loose woman and bad counsel."* Schaeffer also took from Descartes the nature/culture opposition, terms that according to him delimited the two extremities of music. Finally, the experimental method advocated by Schaeffer was inherited directly from Descartes, who based his *Method* on observation, division into simple units, reconstruction in order, and finally, verification—when necessary, through experimentation. Moreover,

---

* Translator's note: The original French phrase, "L'oreille est bonne fille et mauvaise conseillère," contrasts "good girl" and "bad counselor." Schaeffer expresses the ideas that listening is seductive, but not necessarily accurate.

Descartes was the first to consider sound from a physical point of view. Brigitte Van Wymeersch reminds us that "the first formulation of a physical definition of sound is found in a lengthy and significant letter dated 18 December 1629 from Descartes to Mersenne. Sound is conceived as 'a beating' that is made by several turns and returns ... returns that continually undulate the air that will strike the ear."[27] Descartes had even drawn a difference between sound and noise, based on a principle of "just proportion between the object and the senses."

Timbre and Color

Even if all the vocabulary developed by Schaeffer in his morphological description of sounds did not become part of the everyday language of musicians, it is striking that in addition to the problematic itself, many concepts did succeed in imposing themselves, for example that of what has been labeled "expanded timbre" or "enriched timbre."

The word *timbre* in electroacoustics has been strongly augmented in comparison with the word *timbre* commonly used in traditional tonal music, designating the sound of an instrument—the timbre of the flute, the timbre of the violin. Even if the word *timbre* was explored more systematically by Schoenberg, Webern, and Messiaen in their essays on *Klangfarbenmelodie* (in the first half of the twentieth century), it is only with experimental music research (in the second half of the twentieth century) that it took on the full dimension that we recognize today. Indeed, Schaeffer and his team had first of all succeeded in transcending anecdotal description of the timbre of an instrument by applying it to all sound objects. Then—going beyond the physicist's intuition that timbre would be characterized by its harmonic spectrum alone—they added the criteria of form. After numerous listening experiments, they discovered for example that the attack of a sound determines its timbre—if one removes the attack from the sound of a bell, one thinks one hears an oboe. Similarly, a break in the structural shape of a sound changes its color or timbre. (Note that while the word *color* is often used as a synonym for the word *timbre*, with Schaeffer, the concept was broadened to encompass harmonic timbre).

Lastly, perceived timbre is dependent on the dynamics of the sound. In brief, Schaeffer observed, "perceived timbre is a synthesis of the variations in harmonic content and dynamic development; in particular, it exists immediately from the first moment of the attack whenever the rest of the sound flows directly from this."[28] The concept of timbre is thus correlated with the concepts of the sound object and morphology.

The Dynamics and Movement of Sounds

Another criterion of morphological description has entered everyday musical language: dynamics. As Michel Chion remarked in his *Guide des objets sonores*, "By definition, it is a criterion that exists only in time; it is therefore one of the most important criteria relating to the form of the sound."[29] To schematize, the energy contained in a sound gives it a shape, or a profile, that evolves over time. Dynamics therefore no longer only account for just the intensity of sounds but also for their general form, in which the attack of the sound is often decisive. Schaeffer was therefore induced to describe all possible forms of attack as well as all possible dynamic profiles.[30] The concept of dynamics then gives way to the concept of movement when a sound develops over a long span of time. Gradually one shifts towards the question of spatialization, because to speak of the movement of sounds is to speak of their spatial deployment over the course of time.

Schaeffer did not focus his attention on this final point. He was satisfied with defining a typology of sound objects from the most "suitable" to the least suitable for musical use. We will see in the following chapter how the thematic of spatialization would evolve within the GRM.

Acousmatics

It was during a radio program in 1955, presented by the poet Jérôme Peignot, that the word *acousmatics* made its entry into the world of musique concrète.[31] Peignot had discovered this word in the Larousse dictionary: "'Acousmatic noise refers to a sound that one hears without detecting its causes' . . . Well! here is the very definition of the sound object, the basic element of musique concrète, the most general music possible," he exclaimed. The acousmatic approach—still according to the dictionary—had been inaugurated by Pythagoras, who would place himself behind a curtain to force his disciples to listen to him without seeing him, so as not to be distracted by their eyes, devoting themselves completely to the master's discourse. Schaeffer seized on this word and even intended, for a time, to title his *Traité des objets musicaux*, instead, "*Traité d'acousmatique*." This word then fell into a state of dormancy until 1974, when François Bayle exhumed it to impose it instead of *electroacoustics*. Bayle went so far as to create variations on the word. Besides *acousmatic* music, he also coined *acousmathèque* for the entire GRM archives, *Acousmonium* for his loudspeaker orchestra, and *Acousmographe* for the software program that assisted in graphic transcription. More recently in 2001, the GRM's computerized database was baptized *Acousmaline*, even though François Bayle had not been at the head of the GRM for four years, proof enough that the root *acousma* had successfully imposed itself.[32]

PEDAGOGY

Here are some differing definitions of *pedagogy* that I will apply here:

— In the *Littré* dictionary we find: pedagogy is something other than education;

—but it is also: science of the child considered as a being whose reactions and development obey biological pathways;
—in Greek: slave in charge of directing children;
—figuratively: education, especially moral education.

Alongside *TOM*, which illuminates the results of research into sound and musical phenomena, the pedagogy of "doing and listening" is presented as a key—if not *the* key—to understanding these phenomena. This pedagogy, advocated since the beginnings of musique concrète, compensates for the difficulties of access to this music for neophyte composers and researchers who are curious about the new genre, but who are often ill-prepared by their overly codified, scholarly, and classical initial training. The experimental pedagogy of "doing and listening" is applied on a daily basis at the GRM, in research activities as well as in production and creation, also influencing the group's working methods.

The Experimental Method

The experimental method of "doing and listening" is at the source of Pierre Schaeffer's discoveries at the Studio d'Essai, and then at the Club, and later at the GRMC and the GRM. This method came from his scientific training as an engineer and was perfectly adapted to the professional obligation to innovate in which he found himself, from the beginning of his career, faced with tools and techniques in full expansion. Perhaps Schaeffer had also been influenced by his reading of Paul Valéry, who advocated combining artistic and scientific activities. "Doing and listening" was completely in line with the art-science approach in vogue at the time. One of the consequences of the use of this method is that it distanced itself from all magisterial teaching, inciting practitioners to experiment by themselves with the world of sounds and its tools, each putting their own auditory perception directly to the test. At the GRM, the method of "doing and listening" was never advocated, it was always accepted as a given. Pierre Schaeffer and his collaborators organized collective listening sessions, notably in the Service de la Recherche during the preparation of *TOM*. Sounds, sequences, and études were listened to and then described and commented upon with the constant intent of identifying the criteria of perception. All the sound documents presented for listening had been previously produced in the isolation of the studio, most of the time complying with a precise set of preestablished guidelines. In this approach, everyone was on an equal footing; the hierarchy between composer and assistant or technician was blurred. Nevertheless, Schaeffer placed himself in the role of director of research.

Gradually the pedagogy of doing and listening as practiced at the GRM became accessible to the general public, especially since the late 1990s, with the democratization of creative tools, thanks to the advent of personal computers. The number of people who had tried their hand at simple sound manipulation, and even composition, suddenly increased exponentially. Simultaneously, there was an upsurge in the volume of audiences and, above all, in their quality.

SONIC/MUSICAL/SENSORY

In this musical problematic seeking to master the act of composition, one observes that active perception, practiced during the production of sounds for both the purposes of research and musical creation, brings into play the body, gesture, touch, and listening—in other words, the senses. One avenue of research emerging clearly today, notably in the wake of more recent work such as that of the German theorist Hans-Robert Jauss,[33] involves exploring doing and listening as if they were one and the same act, where the power (of listening) is experienced in the action (of doing). In this scenario, the passage from the sonorous to the musical, and reciprocally from the musical to the sonorous, finds resolution. Schaeffer had already placed himself partially in this perspective in 1975, when he declared: "in fact, as soon as sound was more thoroughly investigated, we began to understand that the noise component of musical sounds was not the least exquisite, that it was this aspect that gave the virtuoso and the performer their irreplaceable originality. Which is why learning about noise, a game for the ear, is also very effective musical pedagogy."[34]

Jauss went as far as to think that "the most elevated forms of construction, or of poetic force, are not painting nor sculpture nor poetry, mimetic arts, but architecture and music, which can produce works free of any mimetic constraint with respect to the cosmos, to nature, or to the Idea."[35]

Teaching Electroacoustic Music

The electroacoustics course created in 1968 at the Conservatoire National Supérieur de Musique de Paris, under the name *Musique Fondamentale et Appliquée à l'Audiovisuel*, was the first to offer a training course for composers in electroacoustic music.[36]

At the outset, its structure—spread out over two years—was modeled on the workshops previously organized at the GRM to serve as an antechamber to entry into the Groupe. For the first four sets of students—1968 to 1971—Pierre Schaeffer was in charge of teaching experimental solfeggio. Three specializations were also available to the students:

—training in execution at the microphone (taught by Bernard Parmegiani and Albert Laracine);
—fundamental music (taught by Pierre Schaeffer and Daniel Charles); and
—experimental music workshop (taught by François Bayle and Ivo Malec).

Beginning with the 1972 reorganization, four workshops were created to allow students to deepen their knowledge:

— "electroacoustic music," led by Guy Reibel;
— "traditional music," led by Jean Schwarz (also a CNRS engineer at the Musée de l'Homme);
— "new computer resources," led by Francis Régnier; and
— a workshop, led by François Delalande, on theoretical research, listening, and psychoacoustics.

Despite everything, practical training remained the priority. Each week, for Schaeffer's critical listening session on Wednesday evenings, students were assigned exercises to be performed on tape, applying the musical criteria described in TOM—grooved sounds, delta sounds, profiles, reinjections, filtering, etc. They had at their disposal two specially equipped studios, maintained by the GRM, where the hardware consisted of four tape recorders (Belin tube tape recorders at the beginning and then Studer), a console, microphones, stereo monitors, and sound processing devices—filters, equalizers, speed controllers, and reverberation.

According to Guy Reibel, "the *Treatise*, a veritable listening Bible, did not pay enough attention to the creative side of music."[37] As much as the concepts of form and sound material as defined by morphology were well received by the students, and even inspiring, typology and other classification criteria were not equally well received, blocking the imagination of students. Based on these challenges, Guy Reibel came up with the idea of developing pedagogy centered on "musical invention" based on playing.* This approach did not convince Pierre Schaeffer, nor later on, François Bayle.

A letter from Schaeffer in response to Reibel, dated March 29, 1979, outlines the scope of their differences:

> I think I also understand student demands—imagine that!—for direct access to composition. While they see me as stubbornly insisting on methodology. This explains your success, since through playing, you offer them a way in. That's how our two attitudes could complement each other. While for students they are in conflict. I think this conflict is real, because I feel like you're basing not so much the pedagogy, but the music itself, on playing, and I still don't understand what that means. . . . Because it is clear to me that this debate is just a continuation of others and reminds me of my quarrels with the GRM. Do I have to repeat myself? It is not a question of reproaching composers for groping their way forward in musical invention, it is a question of maintaining, as the apple of the eye, the vocations of researchers who are

---

* Translator's note: As with the English word *play*, the French word *jeu* has two meanings. The first is "playing music"; *jeu* is the act of playing music. But also "playing," as in child's play, or as in "fun and games." In fact, the most obvious translation for *game* is the same word as for a musician's playing, *jeu*. This creates a sort of poetic ambiguity in the meaning of *jeu*. Here, when referring to Reibel's approach consisting of *jeu*, it can simultaneously refer (perhaps dismissively) to "fun and games," but also to Schaffer's "doing," as in "playing" or "playing with" the instrument to explore or compose.

not necessarily enslaved to the needs of the previous ones, or at least to immediate needs. However, your instrument-making is guarded by two sphinxes—with which you can do no more than dissemble unless you resort to fundamental research—synthesizers that allow tinkering and computers that only accept firm and duly circumscribed commands. I willingly accept the wager (for whatever future) that musical progress will obey this obligatory passage.[38]

Following these discussions, Guy Reibel continued orienting his teaching towards musical playing—a way, he said, "to escape the impasse altogether, circumventing the two sphinxes so well foreseen by Schaeffer: synthesizers and computers, keyboards and numbers. A way to access the idea directly, blending intuition and gesture while eliminating the crude fabrications of the studio."[39]

Early Musical Education

Since the early 1970s, a converging current had developed between enthusiasts of electroacoustic music and pedagogues, in particular those working with children. Education, under the pressure of the new "active" teaching methods advocated since the 1920s by pioneers such as Freinet, Montessori, Rogers, and Neill, was looking for ways to encourage creativity in general, a source of enrichment in every sense of the term for both the individual and society. Electroacoustic music, with its experiential method of approaching the world of sound, was a perfect model for this approach to learning. From the 1970s onwards, numerous experiments in early musical education—"musical awakening"—flourished in collaboration with the national education system in kindergarten and elementary classes, and sometimes even in high schools. Certain energetic teachers also emerged in conservatories and associations. With hindsight, we can see this as the beginning of a great groundswell questioning traditional teaching of music and democratizing access to the art—even more so, since the National Conservatory in Paris (CNSMP) had itself already been exploring the question with its electroacoustic composition course that had opened in September 1968.

François Delalande, in charge of research at the GRM and a promoter of this openness towards early learning pedagogies, crystallized his ideas in a famous article, "Three Key Ideas for Early Learning Music Pedagogy":

—First idea: To understand sound phenomena, children spontaneously make noise music.
—Second idea: Music is not always about rhythms and melodies.
—Third idea: Being a "musician" is not "knowing music."[40]

These three key ideas kicked off the research activity in music pedagogy, both within the GRM and with the teaching public.

In the first *Cahier recherche/musique* (1976), François Delalande's article was complemented by Claire Renard's very practical and essential article, "27 games—exercises."[41] These games were developed by Claire Renard during experiments in nursery schools of the French national education system in the early 1970s. Claire Renard was soon joined by Wiska Radkiewicz. The two musicians worked in tandem until 1979, organizing numerous training sessions for teachers.[42] Exercises included, for example:

—For 3–4 year-olds, imitation and recognition games, and advice on making sounds to accompany stories told by the teacher.
—For 4–5 year-olds, the games proposed are a little more complicated, because at this age, one can tackle symbolic games: the job game, the blind man in the city, imitation of rhythms, and a directly musical approach is possible, duration game, game without silence.
—For 5–6 year-olds, who are beginning to master their bodies and senses, musical games related to other subjects, such as mathematics, become possible—classification by gestures, classification by sound character.

The influence of the GRM on musical education has been enormous, both in France and abroad, and it has helped to renew pedagogy. Its influence was reinforced by:

—weekly educational radio programs in the 1980s under the direction of Guy Reibel and François Delalande, *Éveil à la musique*, then *L'oreille en colimaçon*, hosted for twelve years by Monique Frapat;
—numerous training courses for teachers in the Écoles Normales, conferences, lectures, and seminars;[43]
—book publications: *Cahier recherche/musique* on early childhood musical education (1976), *L'Enfant du sonore au musical* (1982), *La Musique est un jeu d'enfant* (1984)—also translated into Spanish (1995) and Italian (2001), the collection *Portraits polychromes* (2001), with fifteen issues published through 2009;[44]
—the CD-ROM *Les Musiques électroacoustiques* (2000);
—collaboration in preparing pedagogical material for the music exams of the French national secondary school diploma (2001, 2005); and
—development of the Acousmographe software in partnership with French national education (2005).

From nursery school to high school and university, the philosophy remained the same: start from listening or create based on listening.

The Group Functioning

The word *Groupe* as well as its operating model in the Groupe de Recherches Musicales had been borrowed by Schaeffer from Gurdjieff, who had organized his teachings in the form of "groups," each headed by a trusted disciple, whom he had trained,[45] surrounded by numerous candidates in training. In the same way, at the GRM researchers could only join as trainees, as interns. There were no composers as such, even if researchers could devote themselves to the activity of musical creation in their spare time. The concept of trainee was always central to the various institutional arrangements conceived by Schaeffer. For him, it was the only method for gaining access to understanding and learning the new professions of radio, music, and audiovisual. The basic model of the workshop came from the first one he had organized in 1942 with Jacques Copeau in Beaune.[46] From the very beginning, Schaeffer opted for polyvalence and practical exercises. The first GRMC workshop in 1951 placed the composers Pierre Boulez, Michel Philippot, Jean Barraqué, and André Hodeir in direct contact with technicians to learn about studio work, as did the second workshop in 1954 with Darius Milhaud and Hermann Scherchen. The studio/school at Maison-Laffitte, near Paris, founded by Schaeffer in 1955 to train the personnel of French Overseas Radio, had functioned on the same principle. In 1958, as soon as the GRM was created, new trainees were admitted. The first ones, Luc Ferrari and François-Bernard Mâche, followed soon after by François Bayle, sometimes had to train newcomers themselves, just as Pierre Henry or Philippe Arthuys had done before them.

All these trainees were nonstatutory personnel of the RTF, and later of the ORTF. The internship, which lasted for a variable period of two to three months, was designed to initiate beginners to the subject and matter of musique concrète. The first "grand" training session bringing together several people over a long period—one to two years—took place in 1961. Thereafter "grand trainings" became the standard. Beginning in 1968, participants were screened and selected before admission. The two-year course at the conservatory bestowed official student status on participants, but also resulted in lowering the average age of applicants. However, it was still possible to attend the seminars as an auditor until about 1974.

For official internships, in addition to the lecture seminars, training consisted of research and musical creation exercises using the machines in the studio, following the directives of Schaeffer's assistants. In addition, trainees had to render minor services consisting of technical material tasks, secretarial work, or documentation. During the internship, compositional activity took the prescribed form of short *études*. After the two-year internship, collaborators were sometimes offered contracts prolonging their research and participation in the life of the group. When they had the time, the right to compose in one of the GRM's studios was recognized and paid in the form of a commission, in stark contrast with the remuneration for "functional" work. In concentric circles, the trainees (volunteers), then the associate members (sometimes paid "on a fee basis" after a few years) contributed to musique concrète research, while being initiated into the social functioning of the group.[47] Apart from the hard core of the GRM's salaried collaborators—about ten administrative and technical staff—all the other collaborators were engaged in "precarious employment" (to apply the term used today). Some of them have still not forgotten the anxiety they felt every year in September when contracts were renewed—or not. Under these conditions, it was always difficult to distinguish between voluntary and involuntary departures. Moreover, there were many part-time jobs, enabling their holders to carry out other professional activities in parallel such as teaching and creative work. But not everyone could, or wished to, remain in the group. So by centrifugal force, the most dynamic people often reclaimed their freedom, while continuing to occasionally collaborate. This system of cooptation was very efficient.

Beginning in the mid-1980s, training at the conservatory became increasingly isolated from the GRM, eventually ceasing to be called a "stage" (workshop) and reverting to the name of *classe*, for at least two reasons:

—The discord between François Bayle and Guy Reibel resulted in a split: on one side the GRM directed by François Bayle, and on the other hand the conservatory class directed by Guy Reibel, after Schaeffer's retirement.

—From 1978, the status of the GRM's nonstatutory members changed with the implementation of the "collective bargaining agreement for the audiovisual sector." Annual contracts were transformed into open-ended contracts. As a result, the number of small contracts for occasional collaboration for interns and other artistic advisors slowed down abruptly. The conservatory class continued to submit its best students to the GRM, in view of obtaining a commission for a first concert work, or collaboration in some form of research work. But synergy between the GRM and conservatory trainees gradually diminished.

On the other hand, since the 1990s, a new category of "trainees" had emerged: students from universities or engineering schools as well as technicians training in specialized schools. All these young people, having acquired musical knowledge, very often supplemented by self-taught training in electroacoustic music, came to the GRM for an internship to discover the professional world and as part of obtaining their diploma elsewhere.[48] Upon their arrival, they necessarily had to adopt reflexes of autonomy in relationship to the tools, themselves adopting the "doing and listening" method to progress in their research.

In 2005, the GRM galaxy constituted a network of about twenty-five hundred to three thousand people, ranging from the most involved, those with a research contract or a musical commission, to "veteran" visitors who had sometimes only participated a short time in the group's studios.

INTERDISCIPLINARITY

Traité des objets musicaux (1966) was subtitled "an essay across disciplines," indicating the importance that Schaeffer gave to this dimension. He referred indiscriminately to mathematics, physics, sociology, linguistics, and philosophy. In a 1955 document from the Center of Radiophonic Studies, Pierre Schaeffer justified the necessity of interdisciplinarity in musical research:

> Original research has the particularity that it is usually situated on the margins of officially recognized and catalogued disciplines. While phonetics, acoustics and electronics are all specialized fields of investigation, and on the other hand, conservatories scrupulously practice traditional musical empiricism, the gap remains open between the field of experimental sciences and that of aesthetic experience.[49]

Further on in the same text, Pierre Schaeffer added:

> Once the study of the sound object has reached maturity, Musique Concrète will be able to extend its research into other disciplines . . .
>
> (1) research of a medical nature, concerning the action of sounds or noises on man;
> (2) research concerning animal cries: structure and content;
> (3) ethnological research: deciphering exotic musical forms, unintelligible in Western terms, responsive to objective analysis of the morphology of the sound object;
> (4) experimental psychology research related to the audio-visual complex and its applications (pedagogy, filmology, etc.); and
> (5) application to musical pedagogy and medical therapeutics of the analytical examination of sound samples obtained of the voice or the playing of instruments.[50]

If interdisciplinarity was at the center of the Schaefferian problematic from the very first experimentation until the end of the Research Department in 1975, it should be noted that this no longer remained the case when the activity of musical creation became predominant. What did remain was the relationship between researchers, composers, and technicians, and thus between research, creation, and production.

Within the framework of the Research Department, Pierre Schaeffer had rendered functional the triangular relationship between research, creation, and production. In concrete terms, it was a question of getting representatives of these three spheres to work together, either on common projects or simply by having them work side by side, encouraging exchange. The example most often cited is that of the development of sound processing tools: engineers strove to respond to the needs of composers, while also suggesting new functionalities. The element of active production—radio, concerts, conferences, lectures—ensured that the research remained dynamic. The imperatives of production, embodied in the dates of broadcasts sometimes scheduled many months in advance, created an obligation of result. Radio in particular, with its constant rhythm of production and direct confrontation with a very large public, required major efforts at explanation, leading to a healthy clarification of ideas. The synergy of the research/creation/production interaction also ensured the structural durability of the institutional edifice, stimulating productivity among the diverse players and harmonizing it. Ideas circulated more quickly and materialized in concrete production. Even if at the outset the logic of a radio producer is not the same as that of a researcher, the first operating in urgency and the second over the long term, by taking up the challenge of working together, they both experienced great satisfaction constructing a common body of work.

The GRM's functional model proved so effective that Schaeffer reproduced it identically in the Groupe de Recherche Image (GRI, Group for Visual Research), the Groupe de Recherche Technique (GRT, Technological Research Group), the Groupe d'Études Critiques, (GEC, Critical Studies Group), and other groups,[51] all within the framework of the RTF Research Department from 1960 to 1974. In addition, copying the GRM model, many associations and cultural structures were created in France and abroad: GMEA (Groupe de Musique Expérimentale d'Albi), GMEB (Groupe de Musique Expérimentale de Bourges), GMVL (Groupe de Musique Vivante de Lyon), the phonology studio in Milan, the NHK studio in Japan, and others.

The model is sustainable; the longevity of the GRM is the proof. A limited number of permanent staff members in charge of administration, technique, and research and production programs (about fifteen in 2007) are supplemented with external contributors, interns, and artists who come to work for the duration of a production, a research project, or a musical commission.

TOOLS

The Relationship with Source Sounds

I will first review the historical evolution of tools and compositional devices developed at the GRM over the course of fifty years, each time highlighting their originality and how they remained faithful to the basic principles of the pedagogy of "doing and listening." From the first phonogène (1951) to the final "universal" phonogène (1961), from the first Syter (1981) to the latest version of GRM Tools, these implements are all based on direct manipulation by the composer, in a predominantly intuitive mode. We will also examine the role of gesture in relation to the senses of touch and sight, alongside the omnipresent sense of hearing, particularly in the context of the advent of computer technology.

But GRM tools are also the fruit of the intersection between praxis and concept. We will not lose sight of the fact that behind hearing, understanding is always lurking, directly linked to the principle of musical research that always underpins the activity of composition: composers are also researchers, contributing to the development of the tools through their critical remarks and their practical experimentation. This focus on the importance of research activity will also help us to understand the apparent exception represented by Studio 123, the research project between 1978 and 1988 into nonreal-time sound processing, as compared to the real-time project known as Syter (SYstème TEmps Réel). In reality, the specificity of the GRM in its approach to algorithmic composition lies not in technical criteria, but in its particular relationship to source sounds. Yann Geslin, a member of the 123 research team since 1976, evoked this crucial concept:

> Any transformation, however powerful, is never equal or superior to a mechanism of synthesis if it fails to maintain a causal relationship between the sound resulting from the transformation and the source sound. This is notably accomplished by respecting and even revealing the spectro-morphodynamic movement whose natural coherence forms the most remarkable and exciting property of sounds of acoustic origin.[52]

Even if sound synthesis has long reduced the gap in both spectral and morphological richness with the concrete original sound since the very beginning of electronic music, composers from the concrete school always maintain a secret attraction for sounds of natural origin, even when they are transformed. They carry the idea that being able to perceive traces of the source sounds increases the pleasure of listening.

François Bayle, taking his experience as a composer as an example, advanced the term *morpho-concept* to describe "the circular link between idea-gesture-tool designating processes aimed at creating new musical objects, and new connections, in interaction with the corresponding sound tools, aimed at modifying and transforming perceptual effects."[53]

The description of these processes helped him invent the concept of sound image, or "i-son." Indeed, Bayle was perfectly aware of the influence of tools on aesthetics, giving as an example his own compositional work, which he parsed into three historical stages:

— that of stereophonic editing/mixing on tape, from 1963 through 1994 (*L'Oiseau chanteur, L'Expérience acoustique, Grande polyphonie, Camera oscura, Érosphère*);
— that using programmed treatments, with the first digital audio tools in deferred time beginning in 1980, then with Syter in 1988, and with MIDI Formers in 1994 (*Eros, Son Vitesse Lumière, Théâtre d'ombres, Fabulae*); and

—that which definitively adopted dynamic support and multiprocessor digital editing beginning in 1995 with GRM Tools used within Pro Tools (*La main vide, Morceaux de ciel, Si loin si proche*).[54]

In the same work, Bayle also described the specificity of the processes for creating musical objects. In his view, the processes for creating objects invariably involve:

—temporal order (editing: cut, invert, fragment/accumulate) or anamorphosis; and

—spatial order (mixing: reunite, thicken, deform/transform) or metamorphosis.

But in electroacoustic music, these processes become "dynamized" by the tools, making it possible to establish (dynamic) pairs: looping/freezing, fragmentation/breaking, stretching/compression, reinjection/accumulation, filtering/resonance, delay/harmonization, morphing/resynthesis...

And quoting Gilbert Durand, Bayle concluded, "All extracted material and every instrument are the vestige of gesture; all gestures summon their material and seek out their tool."[55]

From Early In-House Experiments to Analog Synthesis

FIRST ATTEMPTS AT SOUND MANIPULATION

Since his first works at the Studio d'Essai in 1942, Pierre Schaeffer had put forward the idea that cinema and creative radio were "bridge arts," situated between the concrete and the abstract, because they made it possible to express through form what could not be said through (spoken) language.

It all began in the radio studio with simple manipulations using 78 rpm acetate records—playing them backwards, shifting the speed from 78 rpm to 33 rpm, transposing pitch and lengthening duration, and especially looping them, using a closed groove. Mixing was also used, copying and superimposing several sounds or sequences. The other major technique for sound transformation came from the microphone, used in every conceivable way—close-up, far away, in motion—in all the spaces where sound can travel, to capture it in its every aspect: reverberated, filtered, masked, or simply direct. As can be appreciated, from the very beginning this extreme use of the tools of sound capture and recording facilitated their detournement from their original function, transforming them into the first instruments of acousmatic music practitioners.

Two things should be noted. The record player also became a tool for transforming sound, by way of the mechanical actions of the user. In the same way, working with the microphone directly solicited "virtuoso" gestures from the individual manipulating the sound body. Schaeffer analyzed this shift from using studio tools for purposes of radio production towards the concept of their use as musical instruments. He explained in his *Traité*:

> Instrumental activity, the first and visible cause of every musical phenomenon, has the peculiarity that above all else it tends to cancel itself out as material cause. And this in two ways: The repetition of the same causal phenomenon, through saturation of the signal . . . The variation of something perceptible within the causal repetition.[56]

Schaeffer here expressed the principle of the permanence/variation dialectic defining instruments as the source of all music. So from the moment one voluntarily creates a closed groove on a record and spins it several times in a row, the record player is no longer a simple tool—it becomes a rudimentary musical instrument. The repeating sound loop of the groove is thus heard as *ritornello*.

Schaeffer went even further. He tried to pierce the meaning of music because in the *homo faber* who was a musician, he saw *homo sapiens* in the making. While Leroi-Gourhan claimed, "philosophy has distinguished two successive forms of humanity, *homo sapiens*, ourselves, and *homo faber*, a creature whose only human characteristic was theoretically the possession of tools,"[57] Schaeffer argued that "our *homo faber* would not have been able to get by without the vision which made him look for meaning in what he did."[58] Schaeffer condensed this thinking a few years later in his book *Faber et Sapiens* in the sentence, "I listen, therefore I know. I hear, therefore I am."[59]

This first period of discovery, exploiting rather rudimentary means, lasted until the arrival of tape recorders and magnetic tape in 1950. Then everything accelerated.

TAPE RECORDERS

The first monophonic tape recorders date back to 1935, but it was not until the 1950s that they were actually used in the RTF studios. Next to the first tape recorder, which spun at a speed of 76cm/sec. (30 ips), the engineer Jacques Poullin installed a "potentiometric control panel" in 1952, which allowed for controlling sound diffusion in the space of the room.[60]

Right away, all the manipulations already used with turntables were transposed to the tape recorder and magnetic tape—acceleration, deceleration, inverting playback direction, looping, and reinjection (by sending a single tape through three or four aligned tape recorders, the first ones in playback mode and the final one recording). Editing became possible and mixing was greatly simplified. But the tape recorder was still limited to its primary function of recording, or memorizing.

Schaeffer was perfectly aware of the duality of using tape recorders.

> In the sense of *making* or even analyzing sound, the tape recorder is a laboratory or instrument-making tool. It works at the basic level, let us say the level of objects. In

the sense of *hearing*, the tape recorder becomes a tool to prepare the ear, to provide a screen for it, to shock it, to remove masks from it. The tape recorder, but no more than any other acoustic device, cannot exempt us from a thorough study of listening, but it prepares the way for this through new contexts.[61]

RTF, and later the ORTF, had a policy of buying French products (Tolana, Sareg, Bourdereau, Belin, Schlumberger), but also of constructing their own prototype devices. The GRMC technicians developed a three-channel tape recorder used in 1952 for Olivier Messiaen's piece *Timbres-Durées*. Divided into left, right, and center channels, it made it possible to give harmonic depth to the piece and also to control spatialization during broadcast or sound projection.[62] This three-track tape recorder, rather cumbersome and complicated to handle, would remain in the studio until 1958—without ever being used again. Admittedly, it produced a quite consequential level of background noise!

In fact, until stereo became widespread in 1958, work at the GRM was either full-track mono- or dual-track (a separate signal on each track).[63] Almost all of Pierre Henry's early works were bi-track. Iannis Xenakis was the first, in 1958, to record in stereophony, for his composition *Bohor*. Then came multitrack tape recorders beginning in 1961. They would later become the norm in the electroacoustic music studio, as in all studios. Arriving first at the GRM was a 1-inch 4-track Schlumberger, then a Studer 2-inch 8-track in 1978, and then a 4-inch 16-track analog in 1987. Since 1999, analog equipment has been on the decline in studios, faced with the rapid rise of audio sequencers like the Pro Tools system, which itself went from 4 to 8, then 16 virtual tracks, and finally to dozens of tracks.

The passage from monophony to multiphony sounded the death knell for the tape recorder used as a tool for sound transformation. The multitrack tape recorder once again became a simple recorder, to which more and more sound transformation devices were added. In contrast, the increasing number of tracks made it possible to develop harmonic work, as well as the spatialization of music. The virtuosity of recording at the microphone was a thing of the past; now everything could be calibrated and corrected track by track.

We must not forget the revolutionary introduction of the Nagra III Kudelski portable news recorder in 1955. The word *revolutionary* is not too strong, because this portable recorder liberated the sound recording operator, suddenly able to leave the studio, recording in total autonomy the sounds of nature and people, in stereo, captured on the spot.

Phonogènes and the Morphophone: Organology

Development of the phonogènes extended over ten years. Several prototypes were designed under the direction of Jacques Poullin, an exceptionally talented engineer who worked with Schaeffer until 1975. The first two phonogènes, with keyboard and slide controls (Schaeffer/Poullin patent no. 561539, in 1951, France), then the morphophone (Poullin in 1954), and the universal phonogène (Poullin in 1961) all shared in common being sound processing instruments.

These devices required gestural skill by the manipulator. After the initial phase of recording sound material, composers used them as sound processing devices to obtain greater variation, as well as to better control sound color and, when possible, to lend continuity to morphological variations in the sound material.

All these devices invented in the Research Department were prototypes designed in the workshops of RTF and then fabricated by the private firm Tolana, which already provided RTF with other technical equipment. Today the various phonogènes and the morphophone have all vanished, destroyed because they were considered too cumbersome by the Museum of Radio France, which was their depository. Only a few photos remain.

Digital implementation of the universal phonogène was effectuated in 1979 in Studio 123 of the GRM on Bénédict Mailliard's software instrument Etir. Its operation was in delayed "nonreal" time. Later the first commercial digital machines performing the same function were designed by the American company Eventide, which trademarked the name Harmonizer in 1982.

THE KEYBOARD PHONOGÈNE

The first use of this newly invented *keyboard phonogène* was in 1951 for Pierre Schaeffer and Pierre Henry's work *Orphée 51 ou Toute la lyre*. The prototypes produced at the GRM would later be produced in series by the Tolana company, which constructed a *chromatic phonogène* also operated by keyboard, controlling transposition by semitones over one or two octaves.

Pierre Schaeffer explained:

> With the phonogène you could run the scale with any sound. While the sound was a bit disparate, at the same time as raising pitch the rhythm also accelerated. Since the effect was tiresome and not what I was looking for, I instead sought homogeneous sounds on the spinning tape reels, producing timbres—be they rich or poor—without asperity, and without introducing the rhythmic element. This is how I came across a new approach, which was not calibrating the sounds as a traditional instrument would have done, but instead creating sound montages obeying other laws than the law of pitches and the law of rhythm. But what laws? And that is precisely where we encountered the problematic of music.[64]

In a 1954 article, Jacques Poullin described the technical aspect of the keyboard phonogène:

> The keyboard phonogène (patented by Schaeffer, manufactured by Tolana) essentially consisted of a drive system for a closed loop on magnetic tape, with twelve wheels of suitable diameters producing twelve different speeds, the keyboard serving

to engage one or another of them. The relationship of the twelve speeds corresponded with the ratio of frequencies of Bach's tempered chromatic scale. The drive motor, with two speeds, doubled the possibilities of immediate transposition, thus covering two octaves. There were actually no mechanical limits to transposition—with an ordinary tape recorder to make intermediary copies, it was possible to record a loop of sound material running at slower or faster speed, and a new loop could then be recorded, transposing by another two octaves in the chosen direction, and so on.[65]

THE SLIDE PHONOGÈNE

Like the keyboard phonogène, this new device was used for the first time in the work *Orphée 51 ou Toute la lyre*, by Pierre Schaeffer and Pierre Henry. The slide phonogène enabled glissando by stepless control of speed variation. In addition, it could be fed closed loops of variable length, using standard tape reels. It was fabricated by the S.A.R.E.G. company. The best-known example of the use of this device is found in the piece *Voile d'Orphée*, where Pierre Henry used the slide phonogène to "tear" the veil of Orpheus.

THE MORPHOPHONE

Created in 1954 by the engineer Jacques Poullin, according to specifications defined among others by Abraham Moles, the morphophone was constructed by the S.A.R.E.G. company. It was the ancestor of modern digital delay, enabling modulation of dynamics, and secondarily of the spectrum. A signal was recorded on a tape loop placed on a drum, then played back in turn by ten playback heads set at adjustable time intervals. Each head was connected to an individual preamp whose gain and response curve could be modified by filters. The signals were then mixed at the output. Unfortunately, the position of the heads could not be modified during playback to achieve variations in delay. Each time a change was made, the heads had to be readjusted, while bringing the heads into contact with the tape could result in damage.

THE UNIVERSAL PHONOGÈNE

In 1961, at the Groupe de Recherches Techniques (GRT) directed by Francis Coupigny, the engineer Jacques Poullin developed the universal phonogène. Contrary to previous slide and keyboard phonogènes (patented by Pierre Schaeffer and Jacques Poullin in 1951), this last device made it possible to independently vary duration and pitch; or, as it was also said at the time, "separate the elements of form and the elements of matter contained in the sound object."[66] And so the analog studio never ceased refining sound processing tools, revolving around work habits linked to recording—sound recording (generally in close-up) and the

FIGURE 2. François Bayle at the morphophone. © INA/Laszlo Ruszka (1962).

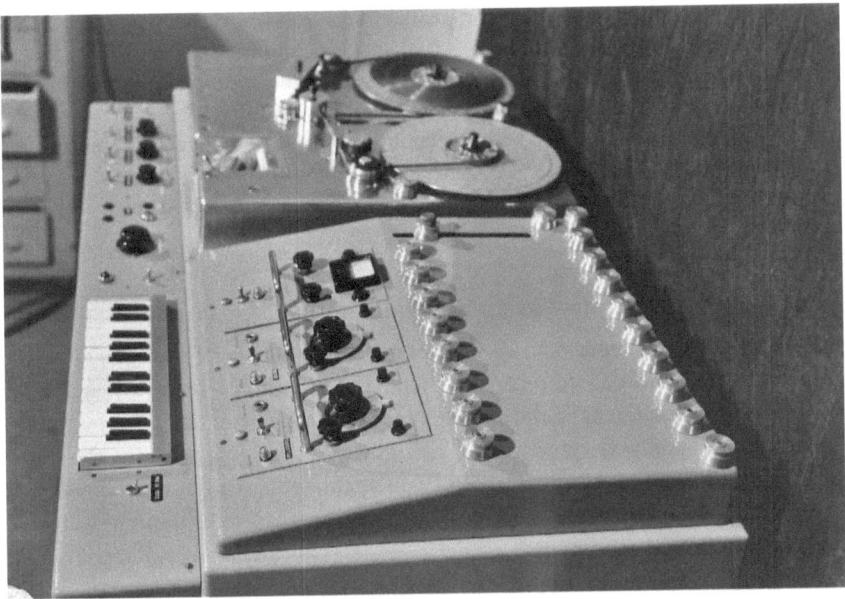

FIGURE 3. The universal phonogène. © INA/Laszlo Ruszka (1965).

successive basic forms of manipulation: speed variation, filtering, reversing playback, accumulation, looping, editing, mixing, spatialization.

Here is a detailed description of the universal phonogène, found in the GRM archives:

> It is a device making it possible to carry out transpositions on sounds—temporal, harmonic, and global.
>
> It works by varying the playback speed of a magnetic tape, using a rotating playback head. In the case of global transposition, the magnetic tape is played back on a fixed magnetic reading head at a speed different from its original recording speed.
>
> Playback speed can be controlled over a wide range, continuously, by a mechanical speed regulator and electronics.
>
> In the case of harmonic transposition, the tape is maintained at its original recording speed (i.e., no time transposition). The playback head rotates, and depending on its direction of rotation, its peripheral speed is subtracted or added to the tape speed. The speed of the head can be controlled either by a two-octave keyboard (one octave in the low register, one octave in the high register), tuned to the tempered scale, or by an electronic speed variator whose action supplements that of the keyboard—enabling either continuous variation or tuning to a different reference standard.
>
> In the case of temporal transposition, the tape spins, as in the case of global transposition, at a speed different from its original recording speed, and the head is

assigned a peripheral speed that compensates for the difference between playback speed and the original recording speed, thus recreating the original pitch of the sound. Controlling this process is identical to the one used for total transposition, with the exception that pitch correction is automatic.

These three operations can be carried out simultaneously whatever the character of the desired transpositions.

For example, it is possible to simultaneously obtain a faster tempo and a deeper sound.

ANALOG SOUND SYNTHESIS—STUDIO 54 AND THE COUPIGNY SYNTHESIZER

Even if the first electronic sounds incorporated into a piece of musique concrète realized at the GRM date to 1956 with Pierre Henry's *Haut voltage*, it was only in the mid-1960s that research and development of tools for synthesis, led by Enrico Chiarucci, became widespread.[67] The first entirely electronic work in the GRM repertoire, *L'Instant mobile* by Bernard Parmegiani, did not appear until 1966. For comparison, Karlheinz Stockhausen in Cologne inaugurated combining electronic and concrète sound sources in *Gesang der Jünglinge* as early as 1954, and composed the first of his versions of *Kontakte* entirely with electronic sounds in 1960.

In fact, between the beginning of the Service de la Recherche in 1960 and the 1970s, the GRM was primarily focused on intense audiovisual production activity and on the research undertaken by Schaeffer for classifying sounds in the context of *Traité des objets musicaux*, which appeared in 1966, followed by *Solfège de l'objet sonore* in 1967. Once *TOM* was finished and the troubles of May 1968 were over, new research projects flourished again in the early 1970s.[68]

The GRM finally acquired an electronic studio in 1970, Studio 54, designed by Enrico Chiarucci. Its most striking attraction was a synthesizer made up of about twenty generators—the Coupigny modular synthesizer—where the generators could control each other through a matrix of pin connectors. Guy Reibel presented the synthesizer:

> The modules include LF and VLF generators,[69] amplitude modulators, a ring modulator, vibrators and multi-vibrators to obtain varied waveforms, correctors, and amplifiers. All the connections between the modules are made by means of a "matrix" panel, which also handles mixing. A system of different colored pins allows connecting the desired types of output and input using the same bridge on the panel. Finally, there is a pin programmer making it possible to "draw" a waveform by setting, at twenty successive points, the duration of each segment and the dynamic slope of this segment along a non-linear scale. This programmer can also be used to control other modules.[70]

The studio was also equipped with filters and a Moog synthesizer (with VCA) beginning in 1964, all in conjunction with a sophisticated mixing console designed by the engineers and technicians of the GRT directed by Francis Coupigny.[71] But it

FIGURE 4. The Coupigny "Cube" synthesizer still in use at the GRM. © Évelyne Gayou.

should be emphasized that the composers' use of the new studio involved, as always, a "concrète" approach, exploiting the limits of tools to produce unexpected and innovative sonorities. Practically all the great works of the 1970s, which made the fame of the GRM, were created "in the 54" on this console in tandem with its synthesizer: *L'Expérience acoustique* by François Bayle in 1972, *La Divine comédie* by François Bayle and Bernard Parmegiani in 1973, *Requiem* by Michel Chion in 1973, *Triptyque électroacoustique* by Guy Reibel in 1974, and *De natura sonorum* by Bernard Parmegiani in 1975. The Studio 54 console remained in active use until it was replaced, in 1992, by a new analog console from Studer. But starting in the 1980s, the synthesizer, although still in working order, no longer interested composers who had switched to commercial synthesizers. The first synthesizer from the outside, the AKS, arrived in 1976, packaged in a suitcase, small and therefore transportable. It immediately found numerous fans, despite its limitations and its instability in sound production. A reduced, "portable" version of the Coupigny, built with ten synthesis modules and christened "The Cube," is still occasionally used in creation at the GRM, following its 2005 restoration by Jacques Darnis. The Coupigny Cube is prized by those nostalgic for the "good old days" of experimental music and analog sound. The Japanese composer Ryoji Ikeda, who was commissioned to create a new work at the GRM, used The Cube in its premiere March 9, 2006, at the "Présences *électronique*" festival.

MIDI, MACSOUTILS (MIDI FORMERS)

In 1983, with the appearance of MIDI (Musical Instrument Digital Interface), a standard for digital description of musical events that enabled integration of gesture sensors, synthesizers, samplers, expenders, and computers of various brands, the GRM acquired a Yamaha DX7 digital synthesizer using the MIDI standard in order to further diversify its sound sources and diffusion methods. The DX7 was made available to composers working in Studio 116. The MIDI standard—the result of an understanding between the companies Sequential Circuits, Roland, Yamaha, Korg, and Kawai—led to considerable expansion in live electronic music, particularly using digital keyboards, which the GRM did not escape. From 1988 to 1995, a "MIDI class," led by Philippe Mion, was offered by the Parisian cultural program Adac, with support from the GRM. Instruction was in the form of weekly sessions spread over a semester to groups of five or six "amateur" composers.

In addition to the adoption of MIDI tools in production, the GRM launched a research program, running from 1988 to 1998, entrusted to Serge de Laubier with his dual expertise as composer and engineer, who had trained at the prestigious national Louis Lumière school. MacSoutiLs, which he developed, also known as MIDI Formers, was a set of tools intended for the generation and flux control of events. Serge de Laubier also worked on Mars Tools—which would expand significantly, later on in the 2000s, with the creation of a community of users around MSP.

In 1992, Serge de Laubier described MacSoutiLs in its user manual, version 1.0:

> MacSoutiLs reappropriates MIDI keyboards to generate not notes but fluxes of notes. Reminiscent of the musicality of musique concrète, both in the complexity of the sound objects it generates, and by certain familiar acoustic models the sounds resemble: rebounds, echoes, wheels of fortune.
>
> The MacSoutiLs Bounce, Random, Wheel, Grain-glide, and Metro process MIDI data from one or more MIDI keyboards and continuous controllers. All of this data must be merged before being sent to one of the MacSoutiLs. At output, they control one or more MIDI synthesizers.
>
> To work properly, they require Max 2.2 software or a later version, or MAXPlay version 1.0 originally developed by Miller Puckette, sold by OPCODE SYSTEMS.
>
> Hardware requirements:
>
> —a fast Macintosh (68020 processor at 12MHz minimum, Mac II type);
> —a MIDI interface on the modem port;
> —a MIDI keyboard;
> —at least one synthesizer, if possible multitimbral; and

—optionally, a series of MIDI controllers, emitting control change values, such as modulation wheel, volume pedal, breath control, MIDI potentiometer boxes, etc.

The system developments made by Serge de Laubier, under the name of MacSoutiLs and then MIDI Formers (distributed by Opcode System), were awarded a prize in Bourges in 1993. Unfortunately, as MIDI Formers could not run on the new Mac OS X platform, the entire program would have had to be recoded. Due to lack of financial means, in 1998 the GRM finally decided to terminate research activities related to MIDI tools, and Serge de Laubier's annual contract was not renewed. But he continued his research and software development with his non-profit organization Puce Muse, extending it to visual imagery. His work took the form of performances where he projected—often on giant screens—images that "changed" the sound. In other words, in real time he transformed sounds, based on manipulating images of virtual sound objects.

Research in Computer Music

BIRTH OF STUDIO 123

One should not conflate the early computer research at GRM, initiated around 1970, with equipping GRM production studios, beginning in the 1980s, with machines containing digital technology purchased on the open market. Acquisition of these machines led to the GRM's first "all digital" studio in 1999 (Studio 116A),[72] computer research perpetuating the GRM's R&D tradition and leading in turn to creating the in-house software GRM Tools and Acousmographe.

Studio 123, dedicated since 1976 to computer music research, was the first tangible manifestation of a new direction taken at the GRM, from 1970 onwards, to remain in step with the technological evolution of the time. For the GRM to launch itself into veritable research in computer music, and not just computer-assisted music, was a fundamental and courageous act, almost as bold as the commitment to the saga of musique concrète in the 1950s.[73] Tribute must be paid to François Bayle, then director of the GRM, for having been able to both detect the right people for the job and find the material means to carry out this new research, all the more so since means were severely limited at the Institut national de l'audiovisuel (INA, which neither encouraged nor disavowed the project), and since he was also confronted with a very powerful newcomer on the scene: the Institut de recherche et coordination acoustique/musique (IRCAM).

In 1970, the challenge consisted in searching for a new direction in computer music, in accordance with the concrete tradition of "doing and listening," giving perception its rightful place without "giving in" to the temptation of adopting the already open breach of music composed by computer. Francis Régnier, the first person in charge of the GRM's computer workshop from 1970 to 1973, quickly understood the full scope of the work and thinking of the American Max Mathews, author of the Music V program, on which the GRM was to build.

In a public lecture at the International Festival of Sound in 1971, Francis Régnier clearly outlined the internal debate within the GRM:

> It is not enough that the sounds that can be synthesized by Music V are theoretically rich, it is also necessary to know how to specify the sounds that you want. The computer needs a complete description of the structure of the sound to be synthesized. We cannot, a priori, provide it with psychological information, expecting it to know how to interpret what is meant by a brassy sound, by a stiff attack, by a clear timbre, etc. Unless, of course, one has already been able to translate such notions into physical parameters....
>
> A first step in this direction was made by the Groupe de Recherches Musicales, in particular by Guy Reibel and Enrico Chiarucci, who made possible Pierre Schaeffer's work of consolidation in the form of a "Solfège des objets musicaux."[74]

Francis Régnier also asked himself, using the specific example of the attack of a sound, how to develop terminology for the phenomenological description of sounds. He realized that the gradation—null attack, sforzando, soft, flat, elastic, steep, abrupt—was not very effective. In the end, Francis Régnier joined Max Mathews's viewpoint, declaring:

> The author of the Music V program has shown, in our opinion, astonishing perspicacity when he opted for the concept of the instrument as a fundamental concept. It is due to this idea that Music V will owe its efficiency; it is due to this idea, we hope, that the music created will be intelligible.[75]

It was thus this philosophy of *the instrument*, applied to computer tools, that would guide the research work at Studio 123, initially in deferred time due to technical limitations, and shortly after using real time with the Syter project from 1985 onwards. This new way of producing sounds, from virtual instruments, would (temporarily) sound the death knell of the musical gesture in the analog studio. The composer would no longer create sounds by hand—rubbing, banging, striking, and tearing the sound material. The composer would no longer work standing up, in direct contact with the sound material *fixed* on the magnetic tape. With computers, the composer worked sitting down in front of a screen.

APOGEE OF "DEFERRED TIME" STUDIO 123, 1978–1988

As soon as he received the first DEC (Digital Equipment Corporation) PDP-11/60 computer in 1978, Bénédict Mailliard started transferring the Music V program to it and started developing applications.[76] The basic functional tool was Fourier theory, and its associated concept, the spectrum. During the summer of 1978 (in July, to be precise) the work *Erosphère* by François Bayle was the first to exploit this new tool. The presence of an engineer at the composer's side quickly became indispensable. Jean-François Allouis, the second engineer of the studio, unwilling to perpetually assist composers, had the idea to develop, in a few days, two or three small "black boxes," very simple to use, to "hand back control" to the composer. Thus were born RAL, FLT, and REV. One month later, in August 1978, Bénédict Mailliard systematized this idea of black boxes, deciding to impose it as the basic architecture for a digital research/composition studio, even as he continued development of Music V. Faithful to the GRM sound processing philosophy, Bénédict Mailliard transposed all the techniques used in analog to a digital mode. The black box for ritardando, RAL, would be supplemented for example with BRAGE (shuffling), itself entirely rewritten a little later by the computer engineer Alain Dumay. FLT (filter) would be tweaked in 1981 by Yann Geslin, who remained attached to the further developments of 123 Delayed Time going forward. REV (reverberation) was also enhanced. In addition to Alain Dumay, it is also necessary to mention the engineers Denis Valette, Jean-Pierre Toulier, and Jean-Yves Bernier, all of whom contributed to the research and the development of 123. Jean-Pierre Toulier notably rewrote the entire Music V program by hand. In 1981, Bénédict Mailliard listed the tools that had been developed:

> We have a sound editor (editing and micro-editing in semi-real time), a blending program (but not mixing), reverberation, speed variator, ring modulator, phase modulator, amplitude and shape modulator, formant modulators with continuously variable filters, resonant filter banks, filters for lengthening and shortening of materials, general vocoders (analytical and modulating sources) with filters and linear prediction, material shuffling.[77]

Bénédict Mailliard described the studio:

> It is a computer studio for working on sound (now understood as an arbitrary set of purely functional elements of audible events), of the same size as a traditional electroacoustic composition studio. The spirit governing its conception is similar, allowing for the same working attitude. As in the early days, it is largely experimental, in every sense....
> The approach to sound situated between simple periodic sounds and noise, the most important for music, is still largely open in its theory. This is why, although the existing tools are rich and give solid reference points, no general process that is likely to concern all sounds is scientific or exact, and must therefore be seen as actual musical production.[78]

And so we reencounter in the software development of the "deferred time" Studio 123—first on the DEC PDP-11/60 computer and then on the Real Time System—transposition of the different sound manipulation techniques from the analog studio. These techniques were now called "instruments": the speed variation instrument, the reverb instrument, and others. We also note that Mailliard had little interest in sound generation.

According to Daniel Teruggi, in his doctoral thesis on the Syter system, "the Studio [123 deferred time] experienced its period of glory between 1983 and 1986, in terms of the numbers of productions hosted and people trained. Out of approximately 109 works composed by 60 composers—out of 124 who trained there—some of them would become part of the "great classics of GRM musical production."[79] In addition to François Bayle's *Erosphère*, we should mention Jean Schwarz's *Les quatre saisons*, Bernard Parmegiani's *Lumière noire*, Denis Smalley's *Tides*, and above all *Germinal*, a collaborative work from 1985, all emblematic of Studio 123.[80] *Germinal* was the sum of fifteen different works composed according to the common principle of starting from a very short musical nucleus—a few seconds long—and developing it over three to five minutes.

But examination of the work composed in 1987 at Studio 123 shows a general slowdown in activity, tied to the Syter. It must also be recognized that despite the wish of the majority of composers to maintain gestural access to the tools, computer music—in general, and at the GRM in particular—offered only visual access through the intermediary of the computer monitor, in a static, seated work position. The 123 "deferred time" project even took on the aura of a demi-fiasco, following the virtual rejection by composers of its musical approach, judged insufficiently intuitive. Even if centered on hearing (listening), it still lacked the doing. In these circumstances, the phrase "death of gesture" seemed justified. It appeared necessary to seek out access modes to sound processing that would involve greater emphasis on sensory feedback. This concern would become central in the future work of the engineer Jean-François Allouis, guiding spirit in development of the Syter, and then with Hugues Vinet and Emmanuel Favreau with GRM Tools. Today we could say this was a winning hand. Ergonomic transparency of the user interface has become a signature of GRM Tools.

APOGEE OF "REAL TIME" STUDIO 123: SYTER, 1987–1995

Peak musical production with Syter was between 1987 and 1995. Each year six to eight pieces were composed predominantly using Syter. We note a peak in 1991 with fourteen pieces, but after 1995 use of Syter decreased inexorably.

During the period of its construction, between 1976 and 1985, Syter benefited from the searing technological progress of its day. Microprocessors and diverse components made it possible to provide composers with a real-time sound processing tool. Syter's philosophy was an extension of that of the 123 time-delay computer developed by Bénédict Mailliard: building software instruments capable of reproducing the basic operations of sound manipulation already discovered previously in the analog studio, while adding capabilities specific to computers—

namely programming, but also power and speed. It was literally the practical implementation of one of Schaeffer's greatest ideas: building instruments capable of exploring one by one the criteria of sound in order to transform it.

But within the GRM, Syter was the object of fierce debate (from which it would emerge victorious) on the appropriateness of working in real time, in contrast to working in the deferred time as defined by using the PDP-11 computer. Working in real time led composers to rediscovering the reflexes of instrumental improvisation. Working in deferred time required greater abstraction and control over the global musical stratagem. The main criticism addressed at defenders of real time was that they often took the "easy way," favoring facility and even just "fooling around," using Syter as a diabolical machine capable of generating random sound processes. The real-time attitude appeared backwards to a group that had just spent twenty years analyzing each and every sound phenomenon in the slightest detail, with the clear objective of mastering and understanding them—which had been given concrete form with *Traité des objets musicaux* and uncountable musical works in the form of études.

What was not apparent in the work undertaken at 123, at first sight, was the depth of the computer developments underlying these new tools. Many composers lacked the scientific expertise that would have allowed them to form an educated opinion. Today the quarrel between real time and deferred time seems obsolete. It appears clearly that the Syter (real time) and 123 (delayed time) projects were complementary. The instruments developed in the one merged into the other and, later, into GRM Tools.

The principal innovations contributing to the glory of Syter were on several levels (I'm referring here to the most fully developed 1986 version, with SYG version 5 software):

—First of all, the possibility of real-time calculations. The processor connected to the analog inputs and outputs could control up to sixteen audio channels simultaneously, at a sampling rate of 10 to 50 kHz in 16 bits.

—The SYG sound processing software permitted creating algorithms with a programming language, configuration of instruments and modules, playback and recording of sounds, visualization and editing, and performance in real time.

—Combining all these functions into a single program was a great innovation, supplemented with the capacity to store settings and control several parameters at the same time.

—This final improvement, memorization of states in the variables, was one of the major original advances of Syter proposed by Jean-François Allouis, under the names *interpolator*, *interpolation screen*, or *ball interpolator*.

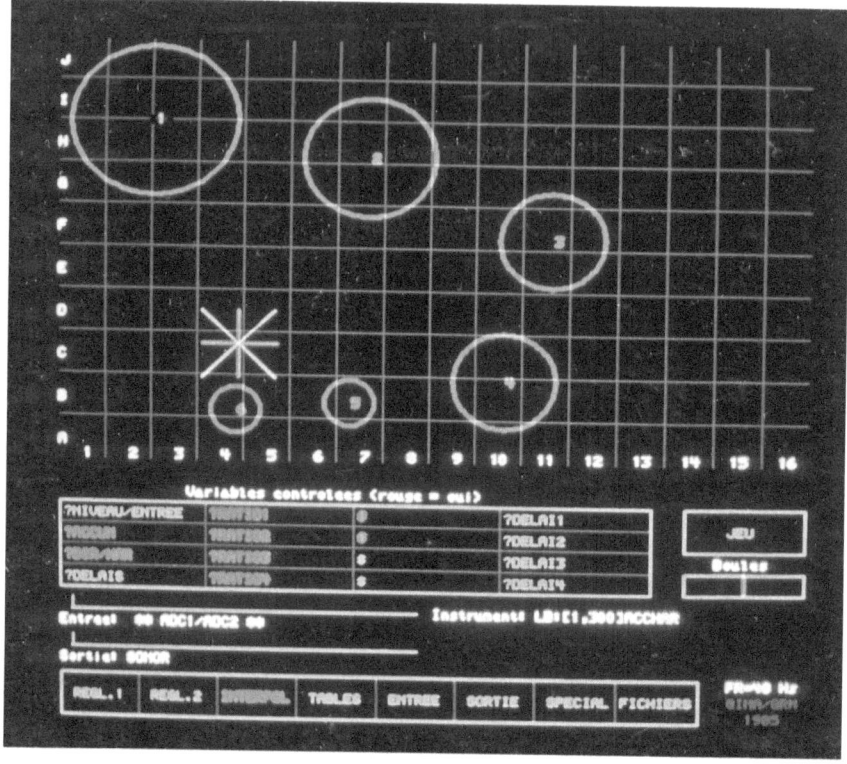

FIGURE 5. The famous Syter ball interpolator. © INA (1986).

Interpolation allowed for passage from one processing effect to another without any audible rupture. This function was adopted in subsequent developments at the GRM, such as MIDI Formers and GRM Tools.

Altogether, Syter proposed a little more than thirty instruments, obviously including all the sound manipulations already conceived—including montage and micromontage, filtering, speed variation, stretching/contraction, looping, playback direction inversion, phase inversion, reinjection, Doppler effect, ring modulation, frequency modulation, and granular synthesis.

Out of the ten Syters built and marketed, only one was still working at the Paris Conservatory when this book went to press. But since the reorganization of their teaching program in 2000, Syter is no longer included.

GRM TOOLS

With the appearance of personal computer systems (the first Apple Macintosh came out in 1984),[81] adapting Syter to personal workstations became a feasible

FIGURE 6. Emmanuel Favreau and the trophy awarded for the new GRM Tools ST. © Évelyne Gayou (2004).

challenge. Hugues Vinet, a young Telecom engineer trained at ENST, was at the origin of this project that would eventually take the name GRM Tools.[82]

Hugues Vinet presented DSP Station, the ancestor of GRM Tools, at the International Computer Music Conference (ICMC) in Montreal in 1991.[83] It consisted of tools associated with Digidesign's DSP 56000 accelerator card. At that time, audio plug-ins did not yet exist and personal microcomputers were just beginning to provide the computing power necessary for real-time sound processing. Consequently, transferring the sound processing programs developed for Syter would only become possible if Hugues Vinet could streamline them considerably. GRM Tools 1.0 was first marketed in 1992, consisting of thirteen algorithms working with a Sound Tools II card. In 1994, GRM Tools 1.5 became compatible with the Pro Tools environment. The first plug-in versions date to 1995–96 on Pro Tools TDM. A version for the Steinberg VST environment saw the light of day in 1999— GRM Tools could finally run on Windows. In 2001-2, Emmanuel Favreau, the engineer in charge of GRM Tools development who had taken over from Hugues Vinet in 1994, proposed four new treatments based on vocoder phases: Contrast, Equalize, Freq Warp, and Freq Shift.

In 2007, there were three possible GRM Tools configurations: a VST plug-in for Macintosh or PC, and an RTAS plug-in as well as a TDM plug-in, both reserved for Pro Tools.

In addition to Contrast, Equalize, Freq Warp, and Freq Shift, the original eight treatments offered were Band Pass, Comb Filter, Delay, Doppler, Freezing, Pitch Accum, Reson, and Shuffling. In 2008, Emmanuel Favreau presented Evolution to the other members of the GRM, a new plug-in producing continuous timbral evolution by frequency sampling of the input signal.

From 1994 to 2009, commercialization of GRM Tools was entrusted to an American distributor, the Electronic Music Foundation (EMF). Since 2009, INA, now with its own online store, has taken over marketing. GRM Tools is a commercial success, both with professionals and the general public. The success with professionals is in part due to the GRM having maintained a privileged relationship with Digidesign, owner of Pro Tools, since the very beginning of the GRM Tools venture. There is no commercial contract, but rather an NDA (nondisclosure agreement), yielding a mutually beneficial partnership, the GRM developing software and Digidesign granting access to its hardware system. Thanks to Pro Tools being widely used in professional sound processing studios, especially in cinema, GRM Tools has acquired an international reputation. There are countless testimonials and glowing press articles. The fifteen algorithms owe their strength to their precision, but also to their ergonomics. Their manual utilization is remarkably user-friendly. They are self-contained tools offering great flexibility in practice. For example, the principle of interpolation between presets, adopted from Syter and from Serge de Laubier's MIDI Formers, has been retained in GRM Tools.[84] Just as with all the other tools preceding them throughout GRM's history, GRM Tools aims at rendering composers autonomous and intuitive, privileging the creative work with the material, through transformation of the different sound criteria.

Limited Instruments versus Open-Ended Instruments. The example of interpolation potentiometers between "presets" (series of parameter presets, aggregated on a single button, which vary the ensemble of settings by changing only one or two parameters) requires clarifying the concept of a self-contained (or "closed") instrument. At first sight, these presets may give the appearance of limiting the composer's creative possibilities. But on closer inspection, there is an advantage in the sense that these limitations offer a framework within which the composer can act with complete independence, without the risk of getting lost in aberrant or inaudible sound effects. It has even been observed that Paris Conservatory composition students prefer self-contained instruments. The closed instrument favors a performing-musician mindset, which often appeals to young composers who have also been trained to play a traditional acoustic instrument. On the other hand, so-called open instruments, the Max software being the best example, sometimes create a feeling of vertigo, since the possibilities seem unlimited.[85] As Schaeffer had already remarked in 1966: "It is at the elementary level that we, who are focused on the material, are seeking to use machines—synthesizers as much as calculating machines in association with them."[86]

THE ACOUSMOGRAPHE

I will briefly outline the Acousmographe here, to then clarify the logical continuity of its technical development. More detailed elements of the genesis of the tool can be found in chapter 5.

The Acousmographe, software developed at the GRM since 1988, is the counterpart to GRM Tools. While GRM Tools is dedicated to *doing*, the Acousmographe is a tool for *listening*. It is designed to aid in the analysis and graphic transcription of sound and musical works (of all genres) by way of listening to them. Even if it enables production of multimedia, of interactive visual documents, it should not be confused with a machine for drawing music, like the UPIC designed by Iannis Xenakis in 1977.[87] The functioning of the Acousmographe is very simple: the music one has chosen to analyze is first screened using a sonogram, which appears in a window with a time scale and a cursor. Some parameters can be adjusted, such as the colors of the spectrum, the scale, the time window, and the height of the tessituras that one wishes to emphasize visually on the control screen. Then the musicologist, or any other user who wishes to annotate their listening, places markers on the screen one by one, or other symbols drawn with the aid of a palette of drawing tools, highlighting their chosen musical elements.

The Acousmographe assists in esthesic analysis.[88] But listening must come first, only to be followed later by graphic transcription. People sometimes mistakenly come to believe that the Acousmographe has automatically produced finished graphic transcriptions, or that it was used to make the music they hear behind the graphic transcriptions they are viewing.

This misunderstanding concerning graphic transcription perfectly illustrates the fundamental question of the relationship between listening and doing. In effect, successful graphic transcription sometimes reaches a level of concordance between sound and imagery that conceals the preliminary effort of analysis. When users of the transcription tool understand that it is their responsibility to analyze their listening, second by second, and to transcribe it by hand, they are often terribly disappointed. After this initial setback, the true skill of the teacher is to inspire hope, by making apprentice analysts understand their initial impression was not entirely off-mark, that there can indeed be a correlation between music and representation. But it is necessary to search for it in listening, constructing it visually, bringing it to light. The expression "do you see what I hear?" then takes on an entirely new dimension, returning us to the concept of "music in reverse" formulated by Schaeffer.

Several theoretical debates accompanied the development of the Acousmographe software:

—Is it necessary for musicians to use visuals to talk about sound?
—What exactly is transcribed after listening?
—What models for transcription can we consider proposing (to users)?
—By focusing on graphic representation, doesn't the Acousmographe distance us from the concept of musical gesture?
—Can we expect the Acousmographe to help us elaborate a code for writing sound, or even for writing music?
—Should we banish the search for "beautiful" transcriptions, contrasting it with "serious" didactic analytical work?

This final point about the contradiction between (ludic and analytical) approaches brings us back to the recurrent debate concerning two extremes in experimental music: one oriented towards musical creation and the other towards musical analysis.

The Acousmographe is not a tool for doing music, but that may soon change. During an in-house seminar at the GRM in July 2005, it was suggested for the first time that the next developments of the Acousmographe could consist in using it as a controller for certain treatments that could be integrated into the Acousmographe. The point would not be to synthesize sound from scratch, but to use visual representation of sound as a tool for modifying the sounds. "Doing and listening" once again demonstrated that they were not mutually exclusive, but on the contrary able to mutually contribute to the musical act. I will address this in greater detail in the chapter dealing with music writing.

Studio Ergonomics

TOWARDS AN "ALL-DIGITAL" VIRTUAL STUDIO

The prototype for radio production studios was developed in 1943 at the Studio d'Essai, then extended to the entire RTF: a horseshoe-shaped console with two turntables on the right and two on the left.[89] Around 1950, tape recorders began to replace the turntables, but the general layout remained unchanged. The tech booth was separated from the recording studio by a double bay window enabling visual communication between technicians and performers while isolating the studio from the background noise of the machines. A third room, a natural reverberation room, completed this arrangement, which architects drew upon to design and build the Maison de la Radio studios in 1963. At the GRM in the 1970s and 1980s, studios were still conceived in this traditional manner, consisting of three or four tape recorders for playback and one for recording, a 4-track tape recorder (gradually replaced by 8 tracks, then 16), a mixing console with linear potentiometers (before 1960 they had been rotary knobs), along with various effects boxes such as filters and reverb. The ergonomics of the music composition studio, inherited from that of the radio production studio, did not change much until the advent of digital technology in the 1990s. One minor change was that composers started to work sitting at the mixing console, as soon as it had been enlarged and enriched with memories, remote controls, and groups. The only real change was that the natural echo chambers were replaced, first by plate reverberation boxes around 1970, parallelepipeds 2 meters long, 1.5 meters high, and 0.5 meters wide. They were still bulky, but so much less so than an entire thirty-square-meter room. Then came electronic echo chambers (the GRM had acquired the famous EMT 250 in the early 1980s), and finally around 1995 reverberation software became commonplace. The natural reverberation and special effects rooms were thus transformed into production booths. At the GRM this was the case for the reverberation chamber of Studio 116, a room with a surface area of about 30 m², which was renamed Studio 116C when it was converted in 1975.

With the introduction of digital equipment, workstations shrunk in size as the processing devices became virtual, along with both the consoles and recording equipment. Today a small, soundproof room with a computer and good speakers is all that is necessary, composers seated at the computer, manipulating their original sounds stored on hard drives.

THE TRIUMPH OF THE VISUAL

This working posture, seated in front of the computer, introduced two new elements. Composers were no longer standing, no longer physically active while composing. They hardly ever used their sense of touch anymore, which was one of the reasons why Michel Chion refused to compose using computers. In contrast, the presence of the monitor screen, along with composition being facilitated by the visual representation of waveforms, shifted the sensory activity of the composer towards sight. The visual interface created a bridge between music and visual arts.

This sensory dislocation profoundly changed the relationship of composers to production tools. Some composers, nostalgic for robust instrumental gesture, have sought to develop new modes of access based on touch or gesture. Serge de Laubier introduced the Méta-instrument in 1989, a device operated by the arms, wrists, hands, and feet, enabling thirty-two continuous variables to be controlled simultaneously. Michel Waisvisz presented The Hands (1984), a gesture-sensing glove developed at the STEIM Institute in Amsterdam, which is dedicated to interactive music and installations. Faced with the same problematics, in 1980 the American researcher Max Mathews developed the Sequential Drum / Radio Baton for controlling "intelligent" instruments, instruments that have knowledge of the music to be played.[90] This gesture-capture device, built and programmed by

Mathews, inspired works by Jon Appleton, Richard Boulanger, Andy Schloss, David Jaffe, and Chris Chafe. At the ACROE in Grenoble, Claude Cadoz and Annie Luciani went even further. In 1994 they presented a prototype of an interactive "gestural transducer," a tool for synthesis of tactile sensations.[91]

At the GRM in 2005, François Donato and Philippe Dao launched a research project on the issue of gestural interfaces, to overcome the GRM's lag in the field.

The Acousmonium

The conceptual dimension of the original Acousmonium is explored in chapter 4. Some additional general information is covered here.

The Acousmonium, or loudspeaker orchestra, was conceived by François Bayle in 1974, initially to renew the concert experience. The Acousmonium quickly proved to be essential in revealing the contours of works in the context of the specific acoustics of concert halls. In general, composers themselves performed their pieces, because with their intimate knowledge of the works, they were well placed to give the best rendition. But in recent years (since about 1995), a new tendency has emerged, consisting in having the pieces performed by people highly experienced on the Acousmonium—or on any other similar diffusion device. This trend is further favored by the fact that the works regularly circulate from one continent to another without the presence of their composers. It therefore becomes necessary to have someone competent in spatialized performance execute the sound projection.

The technique for "playing" the Acousmonium is directly linked to traditional studio work by musique concrète composers. The work—although already recorded with all its nuances, spatial interplay, and colors—must not be perceived by the listener as static and distant, but rather as alive, giving life to the moment of the concert. The best way to give life to recorded sounds is to incorporate them into the performance space, creating the illusion that they emanate from it.

Jacques Darnis, the GRM technician in charge of the Acousmonium from its creation in 1974 until his retirement in 2007, described the set of forty speakers: "The spheres are sort of our trademark, there are six of them, six JBL sound system 'monsters,' then there are four JBL 4343s ('big blue ones'), eight JBL 4311s, two 'stars,' ten JBL 4425 'little butts,' four very large JBL 4435s, two JBL TC Boomers, four 32-speaker trees and four 16-speaker trees." In 2005, in addition to the new console, the Acousmonium was given four new speakers and a subwoofer, inspired by the world of pop music.

All of this equipment is stored in flight cases that can be transported to the concert venue, since the GRM does not have a permanent concert hall. Even at the Maison de la Radio, each concert requires a team of transporters who carry the Acousmonium from its storage location in the basement of Studio 116 to Olivier Messiaen Hall, where concerts take place. From 1977 to 1985, the GRM even owned a Berliet truck to facilitate transporting its technical team and equipment. Today there are other systems similar to the Acousmonium all over Europe, but for the GRM the Acousmonium remains a fundamental element of group cohesion, centered on the aesthetics of "beautiful sound" in concert.

From -thèques to Acousmaline—the Relationship between Heritage and Creation

Since the origin of musique concrète, recording—on records, tape, CD, hard disk—has been at the heart of composers' questioning. Recording—or *fixation on a support*—has a twofold importance, being the object of creation and at the same time its subject. But when they are in the process of creating, composers do not worry about *conservation* of their work. The only question is *how* to record. Only once the work is finished do they ask themselves how to preserve their heritage. The libraries, the different *-thèques*, are answers to these questions.

THE SONOTHÈQUE—THE SOUND LIBRARY

The sonothèque was inaugurated by Pierre Schaeffer at the Service de la Recherche in 1958 to classify sounds to aid in writing *Traité des objets musicaux* and especially *Solfège de l'objet sonore*. Mireille Chamass-Kyrou was in charge, assisted by Luc Ferrari. From 1963 until 1965, the baton was taken up by the "Solfège group" composed of Beatriz Ferreyra (a musician Bernard Baschet had noticed at the Conservatory), Enrico Chiarucci (a physicist), Guy Reibel (an engineer), and Simone Rist (a singer). This small group directed by Bernard Baschet spent two years confirming, through listening, the hypotheses put forward by Schaeffer in *TOM*. They either listened to and described "simple" sounds (in mono) that were provided to them or created sounds according to certain predetermined criteria. This was how the GRM sonothèque of sound examples was constituted. Afterwards, during the research campaigns, each composer (always under a temporary two- or three-month contract) would enrich the sound cabinet with indexed recordings classified by headings: grooved, granular, smooth, short attacks, steep, and so forth. This "didactic sound library" was primarily used for conferences and teaching sessions at the conservatory. But Schaeffer's big idea (which would never be achieved) was to make it a general tool shared by all members of the group.

A second sonothèque, "the dramatic," had been entrusted to Bernard Parmegiani in 1962. This sound library, intended for radio production and for music for film and television, contained sounds and atmospheres from animals, industry, airplanes, forests, and such. It was also stored in the same cabinet that served as a repository for directors of dramatic works and music for film and television. In fact, the majority of the sounds in the library had been made by Parmegiani for his own television productions. But it had long been understood, even before 1958 when

FIGURE 7. Geneviève Mâche and the music archives at the Centre Bourdan. © INA/Laszlo Ruszka (1967).

Pierre Henry was still part of the team, that it was not easy for an author to make one's own sounds available to the community. Sounds that one records oneself are too distinctive and personalized, and even too intimate, to be distributed freely. In fact, when Pierre Henry left the GRM in 1958, he took all his sounds with him, to the great despair of Schaeffer, who from his point of view as a researcher did not understand this attitude of "authorship." This concept of "signature of a sound" made its way among musicians, becoming central for some: one now commonly speaks of the "sound" of a group or a composer to define their aesthetic identity.[92]

Today the *sonothèque dramatique* no longer exists, because the last interesting sounds it contained were all recovered by their composers. Only the sound examples of the didactic sound library remain; they were used for the realization of the boxed record set *Solfège de l'objet sonore*. But the still-remaining documents from that time, in particular the filing cards corresponding to each sound, testify to the hundreds of hours spent by each individual listening to, classifying, comparing, and publicly presenting all these sounds: a painstaking undertaking of remarkable finesse.

THE PHONOTHÈQUE—THE AUDIO LIBRARY

When the young Geneviève Mâche joined the GRM in December 1962 as an intern, she did not yet know that she would become its sound librarian for more than thirty years.[93] When she was asked to help Philippe Carson tidy up the archives that were stored in the cellar of rue de l'Université, no. 37, Paris she could not believe her eyes.[94] In the damp dirt basement, the tapes of the first *Études de bruits* were lying there, in a broken cabinet, among other unidentified boxes. The first sound library of original tapes of musical works had been created in 1964 by François Bayle, as soon as the GRM moved to the Bourdan Center. Geneviève Mâche devoted her entire professional life until 1995 to the phonothèque, in collaboration with various colleagues including Jean Schwarz, who since 1964 had divided his time as an engineer between the GRM and the ethnomusicology research division of the Musée de l'Homme. From 1975 to 1999, Schwarz was responsible for overseeing the conservation of the archives, a behind-the-scenes yet extremely important task consisting of copying the original works onto new media, in particular CDs, as new technologies emerged. Geneviève Mâche recalls the difficulty she encountered in explaining her work to the uninitiated; she would explain that "a phonothécaire is like a librarian, but for sound tapes."[95] She was also proud to tell us that the GRM was the first major historical studio to establish a sound library: "In Cologne, they had none."[96] Works were first catalogued, followed by the associated written archives, photographs, program notes, scores, concert reviews, films. As documents accumulated, it became necessary to find ways to streamline the collection. François Bayle had special cardboard boxes purchased to visually differentiate musical works from radio broadcasts, lectures, and

TABLE 1 Standards for indexing analog and digital audio documents held at the GRM on physical media

Code	Channels	Width (in.)	Speed (in./sec.)
KUR	Mono	¼	7½
KSUR	Stereo	¼	7½
KXUR	4-track	¼	7½
KWUR	4-track	1	7½
LUR	Mono	¼	15
SUR	Stereo	¼	15
XUR	4-track	¼	15
VUR	4-track	½	15
WUR	4-track	1	15
WWUR	8-track	1	15
W16UR	16-track	2	15
DUR	Stereo	Original DAT	
DSUR or SDUR	Stereo	Backup DAT	
CDUR	Stereo	Original CD	
CDR	Stereo	CD copy	

other audio documents. The indexing system for recorded media, which was in use until the beginning of digitization at the GRM in 2001, was inspired by the ORTF system in place since 1955, but François Bayle had to adapt it to the eclecticism of techniques and media used at the Service de la Recherche. The indexes had three letters and a number, the first letter following the nomenclature already adopted by ORTF for its own documents—K, L, M.[97] This was followed by two letters to identify the department: UR, standing for "Université Recherche" for the Service de la Recherche, and finally a number in chronological order.[98] A fourth letter later appeared with the advent of stereo (S). For recording standards specific to the Service de la Recherche, such as 4-track or ½-inch, François Bayle, as secretary general of the GRM, personally created codes: the letters W and X (graphically defined by four dots) stood for 1-inch four-track. And "V was half of W," in other words, ½-inch four-track.

In 1977, the first breach in this organization occurred with the adoption of the Radio France standards for radio broadcasts. The numbering of works and other sound documents remained unchanged. But symbols specific to digital technology appeared in 1998: DUR and CDUR.[99] In 2002, coded indexing was permanently discontinued. A simple serial number, entered into the Acousmaline database in digital form, was introduced for all GRM sound and music documents.

THE ACOUSMATHÈQUE

Established by François Bayle in 1984, the Acousmathèque, inaugurated with great pomp and ceremony in 1993 in the presence of the secretary of state for communication, Jean-Noël Jeanneney, and the president of the INA, Georges Fillioud, only formalized the work undertaken since the 1960s phonothèque, while simultaneously clarifying the responsibility of the INA administration. In 2005 the GRM phonothèque contained about fifteen hundred works preserved on different media (tape, DAT, or CD), two thousand program notes, four hundred fifty biographies of composers and performers, three thousand photos, a file of three thousand addresses of frequent audience members, a library of scholarly articles, and a library of books on music, all of which have been digitized.

Following Jean Schwarz's retirement, his successor Diego Losa was asked to begin digitizing the archive of analog works, starting with the oldest, Schaeffer's first *Études de bruits* (1948). Digitization began in 1998 on CD media in 24/96 — 24 bit and 96 kHz. Beginning in March 1999, the storage medium was the new digital DAT format DTS 2 (Digital Data Storage 2), then, in 2003, it changed again to DVD. In 2007, the first eight hundred works in the GRM catalog were digitized, the last three hundred fifty analog works from 1978 to 1992 still remaining to be done. The next step consisted in using the Acousmaline database, a 500 GB digital server, expandable at will, dedicated to the conservation of the GRM archives and their dematerialization.

THE ACOUSMALINE DATABASE, A TOOL FOR CONSERVATION—
AND PRODUCTION?

In 2005 the Acousmaline project was extended to production, thanks to a second phase carried out by the INA's IT Services Department, the DSI, at a cost of €40,000.[100] The ambition was to refine earlier developments of the server, at first making simple archive-based productions possible and, why not one day, creating music directly from the server. As of 2024, transforming the Acousmaline into a tool for production or even creation is not yet a reality. But the idea is still worth exploring.

As in different stages of developing other tools, we observe two successive phases. At first the tool serves primarily for preserving, memorizing, and transmitting. But afterwards other applications and hijack are discovered, oriented directly towards production, or even creation. It is this process of inversion in the use of the tool that François Delalande described as a "revolution" in the case of the advent of recording.[101] At the GRM, this was first the case for the record, then for the tape recorder, and finally for digital storage. The tool gradually becomes an instrument —which can be played—thus opening the way to new creation. The machine can be diverted from its initial intended use, like any machine, making it possible to explore the limits of its capabilities, with an experimental attitude, transforming it into an instrument There is every reason to believe that the database will one day follow the same evolution, considering that it exploits all sorts of sources. Its power lies in the number of documents it contains and the computer links it can make.

Historically, starting from complete disinterest in the conservation of sound archives, with a sonothèque dedicated only to research until 1966, followed by a coherent system for classification of the cultural legacy, in 2001 a dedicated database was developed, culminating in a project for modeled production. In 2024, after many technical improvements the Acousmaline database is part of a bigger server at INA. At this final stage, with all the conserved documents now dematerialized, it is the conservation tool itself that should become the focus of all attention to follow the cycle of inversion in the use of the tool.

Could the Acousmaline finally be "the most general [meaning universal] instrument possible" that Schaeffer thought he had already invented in 1948? At that time his device was "a 'turntable piano' linking twelve gramophones to a controlling device that enabled them to be 'played.'"[102]

4

Space—Concert—Audience

Composing is recomposing one's own space, trying to understand what is happening to us.
—PIERRE BERNARD

One of Pierre Schaeffer's favorite phrases was, "Music is a mediated phenomenon, it is made to be heard." He knew that a work fixed on a support is only potentially music; it will actually come into existence only in the act of being presented for listening. Recording media is not enough to bring it to life, at most serving to preserve it. The same is true of the classical score on paper, which leads to music only through the interpreter's performance.

To reach an audience, electroacoustic music sought out listening situations resembling traditional concerts, even while developing new formulas for representation. But how do you hold the attention of the audience when there is no performer on stage, no conductor, no score, no libretto to follow—and sometimes very "unusual" sounds to listen to? One solution for the electroacoustic concert was to *inhabit* the space of the hall, in the visual sense of the term. This work on space in electroacoustic music led to a renewal of the concert form. The new challenge was to make the space "sound," and to create an exceptional moment, sometimes bordering on the sacred.[1]

In 2004, a survey based on interviews with composers invited to compose at the GRM revealed that 100 percent of them declared their intention to explore (among other things) the concept of sound space in their next composition.[2] Of course, since the GRM possessed its own diffusion device, the Acousmonium, made up of a hundred loudspeakers in fifty speaker enclosures, it was obvious that no invited composers would want to deprive themselves of using such a setup to spatialize their work in the concert hall.

Since the earliest use of electricity in music, different systems have been experimented with, and not only at the GRM: stereophony, triphony (with three sound channels), quadraphony, surround, 5+1, 7+1, multichannel, wall of sound, ambisonic, sphere of sound, etc. The most famous precursor was Varèse's 450 loudspeakers installed in the Philips Pavilion at the Brussels International Exhibition in 1958, for the premiere of his work *Poème électronique*. Karlheinz Stockhausen had experimented with spatialized diffusion in 1970, using fifty loudspeakers in a sphere 28 meters in diameter, in the German Pavilion of the World Fair of Osaka in Japan. In France, the Gmebaphone, a diffusion device resembling the Acousmonium, was inaugurated by the Groupe de Musique Expérimentale de Bourges (GMEB) in 1973.

Concerning the recent past of humanity, I will evoke a single example as a reminder and just to put things in perspective. Already during the Renaissance, Andrea and Giovanni Gabrieli had experimented with "responsorial" arrangements using two choirs—proof that the problematic of space predated modern technology. Not to mention prehistoric men and their resonant caves. So when the second half of the twentieth century saw the explosion of the conquest of space—Apollo 11 took off for the Moon on July 16, 1969—it's not surprising that the music world also embraced the subject of space.

In electroacoustics, work on the concept of space has correlated with the available tools. Preparing themselves for mediation with the audience, composers construct and *fix* their space on the support, at the internal level of the composition. Later during the concert, at the external level, the space of the performance hall becomes the privileged site for a form of celebration, reinforced by the technical installation, aimed at captivating audiences.

In order to emerge from the vagueness created by the multiplicity in usages of the concept of *space* by composers, I will begin below with an inventory of different conceptual approaches used in the musical milieu. We will then be able to follow the evolution of the relationship between composers and the audience, by observing the different transformations of the concert form in electroacoustics.

SOUND SPACE

Occupying Space as a Reference to Visual Arts

In electroacoustics, because composers are in direct contact with the recorded sound material, they can transform it by manipulating it in a variety of ways, until they obtain the desired morphologies and effects in the internal space of their work. Originally sound treatment was carried out in a very direct physical relationship with the recording medium, hence the term *manipulation*. The turntable was slowed by hand, transposing the sound to a lower pitch; magnetic tape was tugged, accelerating its playback, transposing the sound to a higher register; recorded media were played backwards, inverting dynamics; tape was sliced with a razor blade, creating dramatic breaks. But then, through technological evolution, these procedures became virtual, thanks to the use of software. The original way of working was once comparable to that of the visual artist, painter, sculptor, photographer, or filmmaker, who also worked directly upon the material. Electroacoustic music composers often use vocabulary from the visual artists, linked both to the visual and to touch (color, reverberation, matter, support, etc.), but also constantly evoking movement, bringing them nearer to filmmakers.

This focus on movement is explained by the fact that music is a temporal object. Its perception corresponds to its unfolding in time. Musicians speak of their sounds being mixed, copied, cut, sculpted, erased, accelerated, slowed down, stretched, stopped, frozen . . . Musicians also "project" their music in space during electroacoustic concerts, in the same way as a film is projected onto a screen in a cinema. All this terminology is linked to using temporal objects.

As Jean-Claude Risset pointed out, "electricity has disrupted the possibilities of sound, to the point of calling into question our prevailing views. Recording sound applies time onto space." Bringing together this remark by Risset with François Bayle's quote from Jean Piaget, "space is a snapshot of time and time is space in movement," one has gathered together all the ingredients of our current musical problematic about space.[3] This is valid concerning the procedures for constructing space during composition, but also during the restitution of the work in concert.

This aesthetic problematic finds an echo in the philosophers and scientists of the twentieth century who nourished Schaeffer, and whom he frequently quoted, notably Paul Valéry and Henri Bergson. The latter wrote in his work *La pensée et le mouvant*:

> Radical is the difference between evolution where continuous phases interpenetrate in a kind of interior growth and development where distinct parts are juxtaposed. . . . But true evolution, whether accelerated or slowed down, is modified as a whole from the inside. Its acceleration or its slowing down is precisely this internal modification. Its content becomes one with its duration.[4]

It is obvious that the problem of space is not specific to music on a support, but in a certain manner, acousmatic music presents itself as a textbook case. The fact that the composer and the performer are one and the same person, and that composers are their own first listeners (as soon as the work has been completed in the studio), crystallizes even further into a single point (of view) the question of the position in space of the individual composing, transmitting, and listening. The ideal site from which one could best listen is precisely the place where the transmitter (the composer/performer) is located, either in the studio when they compose or at the console when they direct the performance in concert. And since they tell us intimate, secret things, the ideal would surely be to be "in their skin." This extreme situation of acousmatic music places us in a kind of oxymoron reinforcing the ritual aspect of the concert, which in certain cases may even border on the sacred. Karlheinz Stockhausen claimed that his music was "a rapid vessel to the divine,"[5] leading himself and his listeners to absolute illumination.[6] As Michel Rigoni described his perception of Stockhausen's music, "the listener is enveloped in a pervading sensation superimposing the unchanging state born of the infinite slowness of the great cycle and the rapid events that emerge."[7] But are listeners always able to sense the composer's consciousness-raising devices, notably the symbol of the ascending spiral leading to transcendence? To this question, Karlheinz Stockhausen invariably replied that it didn't matter: "if you don't understand it today, you will one day."[8]

Besides the standard configuration of electroacoustic concerts (composers performing their music alone at the console), other hybrid settings, with other musical genres, other installations, or other arts, have also been created: mixed works, multimedia works, installations. But the treatment of space is always taken into account.

However, today we observe the term *space* becoming a catch-all. Below we will explore some definitions of the term *space*, illustrating the broad variety of theoretical approaches that have emerged in and around the GRM.

Two Types of Space—a Few Conceptual Points of Reference

ABRAHAM MOLES: WANDERING AND ROOTEDNESS

In his book *Psychology of Space*, Abraham Moles (who worked closely with the GRM between 1951 and 1956) differentiated two extreme attitudes to space. In one case, the wandering man (for example the nomad, but also the camper or the tenant) does not appropriate space, "he makes use of it." For the wandering man, "the world is uniform and unlimited," it has no center. On the contrary, man in our world "seeks to be rooted in space, to be anchored in the ground . . . by believing that the real is necessarily located in a place in space." Occupying space takes the place of appropriating this space. The place of rootedness is then demarcated and personified. Walls or any other limit are in this case only the "crystallization of distance." At that moment in time, "place begins to exist—insofar as it is a source of aesthetic perception."

From these two types of appropriation of space, Moles derived two philosophies. Wandering produces a philosophy of *expanse*, characterized by "Cartesian coordinates where the origin of the axis is arbitrary." In this unlimited expanse without predefined center, "coexistence is essential, with spots of variable density, and free movement." Beings are situated in volumes. "Space is material to be shaped."[9] In the case of enrootedness, space is egocentric, its center is attained through combat.

DELEUZE AND GUATTARI: SMOOTH SPACE/STRIATED SPACE

Gilles Deleuze and Felix Guattari speak of smooth and striated space in their joint work *Mille Plateaux*.[10] We could draw an analogy between smooth space and wandering, as well as between striated space and rootedness or centered space. Musically speaking, two works composed by Parmegiani can serve as examples of these concepts of space. The first, composed in 1970, *L'Œil écoute*, with its strong repetitive nature, can be assimilated to smooth or wandering space. The second from 1972, *Pour en finir avec le pouvoir d'Orphée*, where Parmegiani succeeded in producing a more dramatic work after struggling to detach himself from the repetitive style, instead evokes centered or striated space.

PETER SZENDY: CENTRIPETAL/CENTRIFUGAL

In a 1994 article, where he took stock of the theory of space in contemporary music, Peter Szendy reviewed research by various authors demonstrating how musical spatiality, initiated by Wagner (according to Szendy), gradually freed itself from time. Peter Szendy evoked *dis-location*: "In the classical music ideal, a work has no space, no outside, it is pure interiority."[11] But implementing the concept of the concert gave rise to a dialectic of inside/outside, the internality of the work in contrast with its external representation in a spatial setting. Rather than inside/outside, Peter Szendy referred to centripetal/centrifugal, while Algirdas-Julien Greimas spoke of engagement/disengagement. Numerous other questions then arise concerning the limits or framework of space.

MICHEL CHION: INTERNAL SPACE/EXTERNAL SPACE

Let's go back to 1951. In a photo taken in the studios of the rue de l'Université, we see the young composer-performer Pierre Henry (or sometimes Pierre Schaeffer) moving an electric coil between metallic hoops with his hand.[12] He is directing sound projection in spatial relief of *Orphée 51 ou Toute la lyre*, creating spatial effects by altering the balance between loudspeakers, and thus facilitating the mediation of "giving to hear" by way of "giving to see."[13] What is his goal? Egocentric sound projection? Or sounds that wander in space? In fact, both approaches to space already coexist in this mode of spatial projection. But this evokes only the "external space" of music, according to Elsa Justel's terminology adopted from Michel Chion—the space of interpreter "is constructed at the moment of performance."[14] To this must be added the music's internal space, that which is *fixed* on tape or other media, interwoven with the sounds. Internal space can also satisfy one or the other of the philosophies of space described by Abraham Moles, wandering or rootedness, or move in succession from one to the other, going from wandering to the egocentric "me-I-here-now."

Michel Chion's vision of space raises the question of framing, as described by Peter Szendy and Jean-Louis Déotte in another article from the work cited above:

> From the moment [the works] are seen primarily "for themselves," from the moment when they tend to belong to what we could call the sphere of art, the role of *framing* is affirmed. And this role of framing is characteristic of showing, of pointing out.

Before the appearance of museums, framing would have been a kind of museum in deed alone, preceding its own invention.[15]

And Szendy continued further on in referring to Walter Benjamin, "the ornament, the aura, or the frame of a thing is all the more visible when it is in motion." This brings us to the concept of interaction as elaborated by Jean-Claude Risset.

JEAN-CLAUDE RISSET: SPACE AND INTERACTION

In his 2004 article in *Portraits polychromes* no. 7, dedicated to John Chowning, Jean-Claude Risset reminds us that concepts

> about the conception and representation of physical space ... permeate the most recent mathematical physics. Certain of its developments suggest replacing space with an algebra of operators. This obliterates the concept of space seen by Kant as an *a priori* framework, and even Einstein's space-time, which the presence of masses only curves. Primacy is transferred to reciprocal interactions, to observables—physiology rules over anatomy, function over structure.[16]

FRANÇOIS BAYLE: SPACE AND ACOUSTIC IMAGE

It is well known that François Bayle was fascinated by the question of space in music.[17] As a complement to the approaches presented above, I will quote an excerpt from his book *Musique acousmatique*, in which he posed the question of the limits or framework of the perception of space in an acousmatic concert—in other words, a concert where there is theoretically nothing to see. But from his viewpoint, vision is never really out of the picture.

> "Nothing is happening," claim those who absolutely need in music the sensory *redundancy* of the visible accent of seeing a gesture, of seeing instrumental causality.... In the acousmatic concert the entire history for the eye has become a story of sounds between themselves.... Nevertheless what remains of the real is the space of the "listening eye." The listening eye provides acoustic space with context, a frame, a scale. *The acoustic image must occupy the space provided*, otherwise the effect will be artificial, lacking credibility.
>
> And this impression—that nothing is happening—is justified if the deployment of the acoustic image *does not correspond* to the framework of scale provided by the visible clues.[18]

FINDING BALANCE

Today we are well aware of how to obtain spatial effects using a variety of sound processing techniques—including filtering, reverberation, masking, Doppler, and phase shifts—to create the sensation of ascending or descending shapes, trajectories, spinning, dispersion, or depth. These same techniques can be applied indifferently to both the internal and external space of music.

But while internal space results from prior construction where the composer takes the time to select and implement such and such effect quietly during the process of composing the piece of music, external space results from after-the-fact fashioning, which must take into account the physical characteristics of the performance space and of the sound equipment used. In electroacoustic music—but one can also extrapolate to all other musical genres—each individual sound projection and concert is unique to a given site and a given interpreter. During the performance, the composer-interpreter must immediately ascertain the relationship between their perception of the site and the music to be played. Understandably, the inspiration of the moment (in concert) is not always enough to achieve agreement between the internal space created in the composition and the external space that must later be constructed live. This is the challenge the composer faces in concert. Only effective rehearsal can mitigate the risk of errors in judgment.

The GRM's First Attempts at Staging Space

This question of "space" in music has always been part of the theoretical debate at the GRM, as well as in technical and software development, focusing on better rendering the spatial effects desired by composers.

FROM STEREOPHONY TO THE POTENTIOMETRIC CONTROL PANEL

The earliest productions simulating spatiality at the GRM date back to the Club d'Essai. Jean Tardieu—director of both the Club d'Essai and the Centre d'Études Radiophoniques (CER)—recalled the events of 1949 when he opened the ninth session of the CER:[19]

> Four years ago, right here at the Centre d'Études Radiophoniques, while just upstairs Pierre Schaeffer was inventing "musique concrète," two other researchers, José Bernhart and Jean-Wilfrid Garrett, studying ways of adding the illusion of depth that had already been obtained for dramatic and musical work by the artifice of radio, set to exploring another spatial dimension: the impression of lateral movement or lateral placement, as it arises under normal conditions of biauricular hearing.[20]

Everyone is aware of the success of this technique of *sound relief*, or *stereo* as it came to be known, and the numerous utilizations that were then made of it at the Club d'Essai, both on the airwaves and outside of French national radio. René Clair's adaptation of Théophile Gautier's unpublished play *Une Larme du diable* would be the first work broadcast.

But the issues of relief and sound projection could also be explored from other angles. On July 6, 1951, Pierre Schaeffer and Pierre Henry performed two works in sound projection and in spatial relief using a potentiometric control panel, *Symphonie pour un homme seul* and *Orphée 51 ou Toute la lyre*. The spatialization

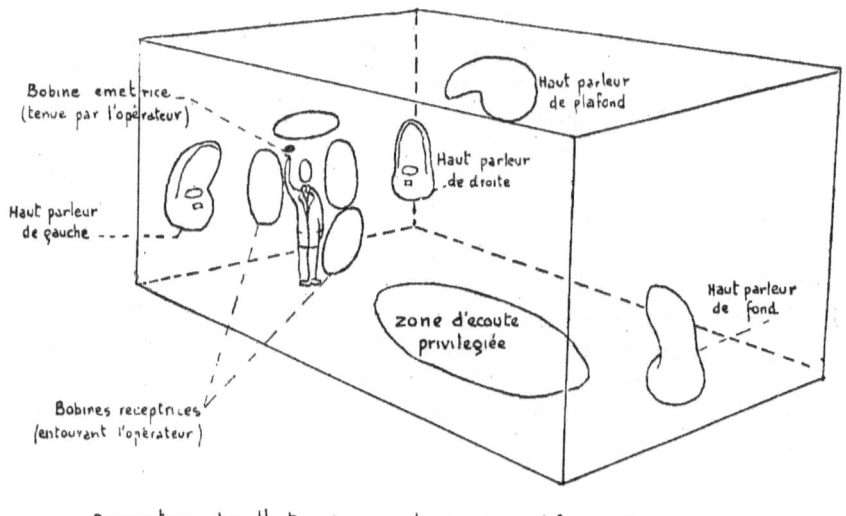

FIGURE 8. The "spatial projection" device devised by Jacques Poullin. © INA (1951).

console, the *pupitre d'espace* (Schaeffer patent no. 605.467, February 26, 1951, France), had been developed by the engineer Jacques Poullin. In its original version, potentiometers were connected to rings sending sound to four loudspeakers hung on a frame—one each at the right, left, top, and front—with the audience placed in the center. But this installation was not visually very alluring. It was rendered more spectacular in 1951, at the Salle de l'Ancien Conservatoire, with four large rings—variable voltage receiving coils—between which Pierre Henry moved a transmitting coil powered by alternating current, shifting the modulation from one channel to another, independently adjusting each amplification channel.[21] In technical notes for the July 6, 1951, concert at Théâtre de l'Empire in Paris, Pierre Schaeffer specified:

> Sound projection:
> An implicit concept in radio broadcasting, sound projection has emerged increasingly as a decisive factor in direct listening. To the extent that Musique Concrète is played in concert, loudspeakers constitute its instrumentarium, whether virtuoso or clumsily improvised, depending on the case.
> The system developed by Jacques Poullin, with aid from the specialist constructor M. Léon, made it possible for the first time to fill a large volume like that of the Theatre of the Empire with satisfactory dissemination of the sound masses and with appropriate dynamics.

Sound relief:*
This is not a matter of stereophonic restitution, pointless since the works only rarely include references to ordinary sound sources. On the contrary, the domain of the "unheard of" in Musique Concrète needs to be liberated from the pointillist prison of a single loudspeaker. Associating musical form with spatial form, either static or kinematic, is the objective of this first attempt at complete relief projection, projection in three dimensions. Using four acoustic channels, sounds are projected following perspectives or trajectories that are expressed in height, width, and depth. The potentiometric relief control panel . . . for lack of specialized potentiometers, still allows only rather crude localization.[22]

On May 21, 1952, Olivier Messiaen premiered *Timbres-Durées*, with spatialization by Pierre Henry, using three channels—left, middle, and right—on the six-reel tape recorder (making it possible to simultaneously play back three tapes), plus a so-called kinematic channel on a normal mono recorder that was synchronized with the three-channel machine. The sounds had already been fixed on tape, but Pierre Henry and Jacques Poullin added "kinematic relief" at the moment of performance.

On May 23, 1952, Pierre Henry, on stage, presented *Antiphonie* in spatial relief by field effect, or kinematic relief.

In 1953, the engineer Jacques Poullin, member of the GRMC team, described the state of his research in an article titled "Sound and Space." He started by evoking the sound phenomena emitted by an orchestra and "generators of musical emotion," then explaining that

> in creating an entirely reconfigured sound space, we can . . . extrapolate our usual practices to musical listening. . . . While it is a priori inconceivable to imagine the layout of instruments within a volume where the listener would occupy the center, it is not inadmissible to conceive of spatial music that would draw our attention towards all the surrounding directions.[23]

At the time, Jacques Poullin argued that quadraphony was the minimal system necessary for spatial projection of sound, essential for reconstruction of three-dimensional space, with four loudspeakers arranged at the vertices of a tetrahedron, at the center of which is placed the audience.

On December 2, 1954, Edgar(d) Varèse presented his piece *Déserts* (for organized sounds) in dual track spatialization, alternating tape with sections played by the orchestra in the classical manner.[24] In addition, the work was broadcast live and in stereophony. From a technical point of view, it was a first.

---

* Translator's note: In the early days, the GRM used the term *relief*—sound relief—in the same sense as relief in sculpture. Later they would more often speak of *l'espace du son*—the sound space.

From October 5 to 10, at the Philips Pavilion of the 1958 Brussels World's Fair, Edgar Varèse presented *Poème électronique*, and Iannis Xenakis *Concret PH*, on 450 loudspeakers arranged throughout the site in groups. "For the first time, I heard my music literally projected into space," said Varèse. The sound diffusion installation at the Philips Pavilion amazed everyone. A description from the event's presentation dossier:

> Four hundred and fifty loudspeakers have been fixed to the walls of the Pavilion for the performance of the score by Edgar Varèse. Behind the balustrade there are also twenty-five 20-watt loudspeakers, especially designed for their bass reproduction. Their positioning—and the "sound route" they follow—was essentially defined by the shape of the Pavilion. A special acoustic recording of the composition, divided between three sound tracks and a control machine, provoke a multiplicity of impressions in listeners. Three sound tracks were used not only for recording the acoustic action, but are also used for routing sounds to the different fixed locations or along certain paths, "sound routes." The individual loudspeaker groups are fed by 20 control amplifiers (plus 20 in reserve) and 20 power amplifiers of 120 watts each (plus 20 in reserve), while a 1000-watt amplifier is provided for the bass loudspeakers. Between sessions, the main control unit is locked, and the tapes are rewound to their starting positions. During this period, a distinct installation, also automatic, is played through the loudspeakers, a sound game by Yannis Xenakis, while cinematographic equipment projects an explanatory text concerning *Poème électronique* on the wall, in three languages.

Following this pioneering period, over the years every possible nuance and improvement was devised by a multitude of individuals, facilitating optimization of equipment for both production and broadcasting, and in the early days, even their invention. The *artificial recording head* was developed in 1958, under the name "couple ORTF," and fabricated by the firm Schœps. This artificial recording head consisted of two microphones, connected by a rigid 170 mm rod, with a 110° angle between the two microphones. This spacing simulated the natural separation between the ears of a human head. During the same period, ORTF engineers and technicians also designed various prototypes of sound direction interfaces such as "db keys," allowing the simulation of space by adjusting the sound levels assigned to each key on a two-octave keyboard. This entire period, until the early 1970s, was one of deep exploration of the basic processes of spatialization. But although numerous works from this period registered in the GRM catalog were conceived in quadraphony, the often-precarious technical conditions for dissemination obliged composers to also provide a "simple" stereo version.

François Bayle's Acousmonium

François Bayle, with his loudspeaker orchestra christened Acousmonium, earned the GRM the reputation as a maker of beautiful sound and consequently of beauti-

FIGURE 9. The Acousmonium on the stage of the Grand Auditorium of Radio France. © Évelyne Gayou (2005).

ful concerts, where sound spatialization was one of the main preoccupations of composers. The Acousmonium was inaugurated February 24, 1974, at the Espace Cardin in Paris, one month after an initial trial run at the Saint-Séverin church in Paris, on the occasion of the premiere of Bayle's *L'Expérience Acoustique*.

Following the events of May 1968, François Bayle, realizing that electroacoustic concerts had lost their appeal, asked himself how to renew audience interest. Little by little due to lack of funding, the sophisticated spatial projection experiments inaugurated in the 1950s had been sidelined. Until May 1968, simple straightforward playback of "experimental" musical passages on a few loudspeakers in a poorly equipped hall could still be considered an avant-garde concert. After May 1968, this picture of counterculture and experimentation completely disappeared. It was necessary to reinvent the concert, to reconnect with an audience.

In the words of François Bayle, the Acousmonium consisted in "substituting the outdated traditional 'sound system,' which transmits sound from the periphery to the center of a room, with a set of sound projectors that construct 'orchestration' of the acoustic image, exploiting the most effective perspectives for acoustic propagation. For example, in a theater, from the stage, or in an amphitheater, by occupying the central space, etc."[25]

Based on the model of the symphony orchestra organized in the manner of Haydn, François Bayle envisioned installing loudspeakers on the stage of the concert hall, in the place of the performers—a hundred or so in general, depending on the concert—creating varied effects of sound levels and colors depending on the choice of the characteristics of each loudspeaker: highs, lows, or mids, dry or bright, weak or strong. The composer, like a conductor, would direct the piece from a control panel, generally located in the center of the audience, facing the stage. He is illuminated by a projector, while the rest of the room remains in darkness. Thus he interprets his recorded work live, playing on intensity, dynamics, and most importantly, on the spatial distribution of different sound events. He sends the sound to the sound planes that he considers optimal based on the acoustics of the venue and his personal tastes. The Acousmonium is not a static mechanism, it is an instrument that can be played. Certain composers, such as Pierre Henry, would sometimes move about or execute bold gestures while directing the sound from the console, adding to the ritualistic feeling. When the event was enhanced with visual effects, such as lighting or lasers highlighting the sound effects, the concert became genuinely original, rendering the audience's effort to attend completely worthwhile. For those who have experienced it, it is undeniable that this concept added a spatial dimension to the music, extremely enjoyable and enriching from the point of view of the concert experience, a powerful spatio-temporal experience.

It should be noted that another system similar to the Acousmonium had already been developed a year earlier in 1973 by Christian Clozier, Pierre Boeswillwald, and Jean-Claude Le Duc, of the Groupe de Musique Expérimentale de Bourges, the GMEB.[26] While the two groups could be described as rivals, the GRM benefited from important advantages: its establishment in Paris, substantial financial means, and its roots in the field of broadcasting. Armed with the anteriority of his system, and critical of the Acousmonium, Christian Clozier underlined that "one of the ultimate objectives of the Gmebaphone is to create an acoustic space that is alive, evolving and encompassing, and not just a response pathway of lines and points."[27]

From its inception in 1974, the GRM Acousmonium, like the Gmebaphone, contributed to revitalization of the concert. Since then, numerous other loudspeaker orchestras have been created and the use of elaborate systems has become widespread. Little by little, a culture of "acousmatic" transmission has been constructed. Today there are specialists in interpretation,[28] as well as different schools: on one hand those promoting "straight" diffusion consisting of a sound work completely under control from its moment of composition, and on the other hand, those favoring the moment of interpretation, adapting their effects live based on the venue and the audience. Once again, we encounter the same two dominant tendencies: (predictive) *writing* on one hand, and (procedural) improvisation and play on the other.

François Bayle became the committed promoter of a new formula for the diffusion of works at the GRM. And if the Acousmonium immediately won over both audiences and composers, the theoretical debate that should have accompanied it never materialized, certainly because of François Bayle's stature and position. As he himself stated, "in an institution it is always difficult to appropriate an invention when it emanates from a hierarchical superior."[29] François Delalande in particular, who was in charge of theoretical research at the GRM under François Bayle's direction at the time, always gave the impression that he was not particularly enthusiastic about this research topic, which was not his own. On the other hand, it could be added that François Bayle may not have felt pressed to launch and orchestrate debate, leaving the vague impression that this was his purely personal project.

It is also true that, at this same moment, another important member of the GRM, Guy Reibel, was trying to orient research towards consideration of musical gesture.

Then after 1975, GRM composers progressively adopted the various tools for assisting spatialization and sound diffusion that were becoming available on the market, whenever they proved useful for musical projects: joysticks for controlling sound directly at the console, projection consoles, specialized loudspeakers, Dolby encoders. Since 2000, other software applications for controlling sound have become available: the Holophon, developed in 2002 by the engineer Laurent Pottier at the Groupe de Musique Expérimentale de Marseille (GMEM), which the GRM acquired, as well as the Ambisonic system developed by Michaël Gerzon in England, and various other proprietary systems.

The Two Extremes of Space in Electroacoustics

In 1997–98, debate concerning space reemerged at the GRM. The retirement of François Bayle and the need to modernize the Acousmonium's equipment coincided with renewed curiosity for spatialization equipment, which had recently been launched on the consumer market—Dolby surround, 5+1, and 7+1. Dolby Surround consisted of encoding sound based on phases and counterphases using five loudspeakers, three in front and two behind. The 5+1 system consisted of five loudspeakers, two in front and two behind plus a subwoofer; 7+1 was similar to 5+1, but with two additional loudspeakers on the sides. Given this context, what spatial setup should be adopted? Should the same system be used for creating the work and for its diffusion? Or could the composition be in stereo at first, and then potentially extended to other more elaborate formats? Or is it better to conceive the piece for a precise type of spatialization from the beginning? And although stereo diffused with a loudspeaker orchestra, as well as "8 channel," held on to its supporters, the fundamental debate was not closed:

—What importance should be given to the spatial projection of sound?
—Should one privilege the work of space at the time of the composition of the piece?

—Or (at the extreme) should we compose a "straight" piece, with little or no spatial effect on the multitrack medium, waiting until the concert to give the piece its full spatial scope within the performance space?

The new millennium has seen the various competing systems of diffusion now in widespread home use multiply the available aesthetic choices.

If we exaggerate just a little in order to identify and understand the extreme tendencies in the treatment of space in electroacoustics, we discover that they reflect the entire aesthetic history of the genre, at first theatrical with radio, before becoming musical. If we also bring together Michel Chion's internal space / external space, Abraham Moles's rootedness/errancy, Peter Szendy's centrifugal/centripetal, and Jean-Claude Risset's interaction, we are able to discern two basic conceptions of space among the authors, naturally intertwined and functioning dialectically.

FIRST TENDENCY

There are those who prefer to compose in stereo, creating works akin to program music intended for broadcasting on the radio. These works focus on structuring the sequence of sound morphologies and colors through the use of effects of dramatic construction obtained through montage, privileging a narrative model, with relatively few sound layers. For this work, radio is an appropriate vehicle. The spatial interplay is already entirely inscribed on the medium (internal space) and can be broadcast as it is.[30] If the piece is given in concert, on the Acousmonium, then the spatial effects, already recorded on the support, are merely enhanced by the Acousmonium, or at best, embellished by a well-conducted interpretation by the performer "projecting" the sound. But frequently composers are disappointed by the concert, no longer recognizing their piece. Its rather dramatic nature, created in the intimacy of the studio using overly uniform sound layers, no longer makes sense in the large reverberant space of a concert hall.

SECOND TENDENCY

On the opposite side are composers who give paramount importance to the concepts of dynamics and spatial movement, both internal and external. Led by François Bayle, and then François Donato and Philippe Dao (of course most composers do not hesitate to explore both tendencies), from the outset they are intent on the quality of spatial rendering, already anticipating the privileged moment of the concert (external space). Their entire approach consists in mastering effects and movement aimed at the future concert.

But well aware that they may only have limited time available for rehearsal before the concert (primarily because of financial costs), increasingly often they prefer applying their spatial effects directly on the original media, leading to the popularity of multichannel audio. Eight-track appears to be a good compromise, because it enables all the basic effects to be realized in complete freedom, in other words in all axes, both in the studio and in concert. But 8-track systems are much less widely available than 7+1, and especially 5+1, which has become the standard on home players because of DVD cinema. Unfortunately, 5+1 is rather restrictive for musicians. Its central front speaker, monophonic, incites plagiarism of the cinema sound system, which places voices in the center and ambiences at the periphery. Abandoning this center risks creating a vacuum (a vacuum in musical interest), while occupying it is equivalent to imposing a sonic "hero," risking sinking into Abraham Moles's philosophical scheme of "me-I-here-now."[31] In contrast, it is logical that this process is progressively adopted by those engaged in works of a dramatic nature.

Experience reveals that composers are increasingly resorting to composing multiple versions of the same work, in function of the ultimate target—a stereo version for radio, a 5+1 version for concerts with limited means and for a possible CD release for home listening, or a 7+1 version that can be reduced to 5+1. In the end, 8-channel versions are produced exclusively for large concerts with good sound projection facilities and sufficient rehearsal time.

THE ELECTROACOUSTIC CONCERT

Let's take a concrete example, that of perception of the overall form of a composition. Focusing during both composition and performance on the qualities of space sometimes causes one to lose sight of the general temporal development of the piece. One criticism often made of electroacoustic music, that it is unable to cope with large-scale form, brings us back to the visual arts. Music can be perceived as sculpture. Listeners speak of forms unfolding before them in the space of the hall. Appreciation of the present moment, during listening, sometimes leads the listener to evaluate only the plastic qualities of the composition: sound relief, movement in space, planes of presence, evolution in the morphology of sound. Spatialization in music acts to slow down time, sometimes to the point, according to some listeners, of making it seem as though it has come to a complete stop. But this does not signify the absence of form or narrative. It simply highlights that the act of listening can transport the listener to a number of places in their consciousness, far from detached perception of the work's overall global form.

The moment of the concert has the particularity of establishing mediation of the music between the composer and the audience. But since not everyone perceives or listens in the same way, consensus must be found in concert so that all are satisfied. The GRM has sought out its place in the world of music by gradually defining ritual for the acousmatic concert, restoring a place to the interpreter.

Audiences were not always receptive. It was necessary to accept criticism and evolve.

The Electroacoustic Ritual

Let us here separate science from technology, as the mathematician René Thom urged, believing in short that science is multifaceted but that technology serves as its bond.[32] During the electroacoustic concert, composers, through their technical activity at the console, interpret and direct the spatial projection of the piece in an act of linkage. The composer/interpreter becomes simultaneously a technician at work and the demiurge officiating, introducing a form of sacralization and of communion, and therefore a form of religious experience. The concert is experienced as a parcel of utopia. In the early days of concerts with technology, at their conclusion one regularly witnessed crowds of curious and admiring spectators gathering around the technology and material installed in the hall.[33]

But as the ethnomusicologist Bernard Lortat-Jacob remarked, "music is nothing but the acoustic form of memory."[34] In the case of electroacoustics, this memory is built on in-depth listening to sound material. Without this praxis, can one really understand music—in the sense of Schaeffer's four modes of listening: *écouter* (to listen), *ouïr* (to perceive aurally), *entendre* (to hear), and *comprendre* (to understand)?[35] Yet "acousmatic concerts" where there is "nothing to see" have also become a GRM specialty, given in complete or partial darkness, leading the audience to fully concentrate on the sound. Most of the time, listeners acknowledge that this attentive listening increased their musical pleasure.

I will advance a few hypotheses, while providing some conceptual reference points, to better evaluate how the modes of organization of electroacoustic concerts at the GRM have recently evolved. The relationship with audiences has evolved over time. There have been periods of splendor, and other periods that have been more ambivalent. Four periods of time stand out as particularly positive, 1948 to 1958, 1962 to 1968, 1975 to 1981, and 2000 and after.

A Few Conceptual Points of Reference

MUSICAL SPACE AS A PLACE FOR AGREEMENT

In his book *Musiques en fête*, the ethnomusicologist Bernard Lortat-Jacob observed that in the traditional popular musical practices of Morocco, "the music really gets going" the moment when the animated and sometimes violent exchanges are exhausted.[36] But music cannot be practiced without consensus. Accordingly, Bernard Lortat-Jacob deduced:

> If music has a function, asserting itself when it is of a group nature, it is indeed to bring into harmony those who devote themselves to it (and one could equally well say to consecrate those who harmonize themselves with it)....
>
> This idea takes on even more force if we admit not only that music reveals entente, but also that it is the principal device that people use to express it.

After also mentioning his research in Sardinia and Romania, Bernard Lortat-Jacob concluded his book with this general reflection concerning traditional music, which could be extended to all music:

> Shared musical practice is always a form of commemoration, in the sense of appealing together to memory. Music is nothing more than the audible form of this memory: it carries experience and arouses feelings that are rooted in our intimate interpersonal life, largely escaping scientific discourse.[37]

GUIDED LINES OF FORCE

In their analysis in *Psychology of Space*, Abraham Moles and Elisabeth Rohmer concluded that the sacred comes into play when the rational is no longer operational. Furthermore, the sacred has its topography: "the sacred is concentrated on guided lines of force" in the case of unlimited space, while in limited space it follows a vertical axis. In music, this would apply as much to internal space as to external space. This problematic deserves to be further explored in the specific context of music.

I would like to draw a connection between this concept of oriented lines of force and the account brought back from Australian aborigines by the writer-traveler Bruce Chatwin:

> It was during his time as a school-teacher that Arkady learned of the labyrinth of invisible pathways which meander all over Australia and are known to Europeans as 'Dreaming-tracks' or 'Songlines'; to the Aboriginals as the 'Footprints of the Ancestors' or the 'Way of the Law'.
>
> Aboriginal Creation myths tell of the legendary totemic beings who had wandered over the continent in the Dreamtime, singing out the name of everything that crossed their path—birds, animals, plants, rocks, waterholes—and so singing the world into existence.[38]

A SPACE OF CELEBRATION

Sociologist Michel Maffesoli described contemporary religiosity as "essentially syncretistic."[39] He elaborated:

> One could say that the megalopolis is constituted of a series of "high places," in the religious sense of the term, where various cults with a high aesthetic-ethical coefficient are celebrated. These are the cults of the body, sex, image, friendship, food, sports ... the list is, in this respect, endless. The common denominator being the space where this cult is carried out ... space becoming bond.

Maffesoli uses a formulation from Rilke to summarize his argument, "the space of celebration"—celebration that confers to the religious its original dimension of interconnection:

> It could be a technical celebration (La Villette museum, Videotheque), artistic (Beaubourg), ludico-erotic (Le Palace), consumption (Les Halles), sport (Parc des Princes, Roland-Garros), musical (Bercy concert hall), religious (Notre-Dame de Paris), intellectual (the great amphitheater of the Sorbonne), political (Versailles), commemorative (the Arche de la Défense), etc. Spaces of celebration, made by and for the initiated: in the etymological sense of the term, spaces where *mysteries* are celebrated. One gathers, one recognizes the Other, thus recognizing oneself.[40]

We could continue Maffesoli's list by declaring radio-television a space of politico-cultural celebration, and the internet a space of communicational celebration. And then, wouldn't the concert hall with its Acousmonium also become a space of musical celebration, just like Bercy? And wouldn't listening at home, on "home cinema" equipment, become a private space reserved for ludico-artistic celebration?

CONNECTING WITH AUDIENCES

1948 to 1958—"Presentations"

The first public presentations of musique concrète took place on radio. On June 20, 1948, at Radio Paris, Jean Toscane presented Pierre Schaeffer's *Concert de bruits*. These "essays" or "études"—always short, less than five minutes—were recorded on 78 rpm records and presented with commentary. There were regular broadcasts until about 1951. Pierre Henry, like many other listeners, discovered musique concrète on the airwaves. "Hearings" outside of radio—one cannot yet speak of concerts—took place in universities (Sorbonne, École Normale de Musique) or cultural centers. Starting in 1952, concerts in specialized halls became more frequent: Salle de l'Ancien Conservatoire, Théâtre des Champs-Élysées, Salle Gaveau.

Concerning the social act of communication—the rituals of the concert or of interpretation—Schaeffer had been the first to seek new formulas to occupy the stage space deprived of performers. As early as 1952, he had suggested turning the electroacoustic concert into a spectacular event by having Pierre Henry climb on stage to direct the sound with the help of the spatializer.

1958 to 1974—Experimental Music

The experimental music period saw the enlargement of audiences, in proportion with the diversification of forms of expression: radio, concerts, festivals, dance, mime, puppet shows, theater, music applied to the image, cinema, television, and visual arts. Multidisciplinarity was not only a word. It was a reality. Experimental music associated with the Service de la Recherche of the ORTF became widely known through famous television productions such as *Les Shadoks*. But the intention was not primarily to expand audiences. Inaugurating the first Research Festival in 1960, Schaeffer confessed, "juxtaposing collecting stickers and Tachism, Tino Rossi and Lettrism, we are badly placed to create popular art."[41] In the prolific output of the 1958–74 period, "pure" musical works were less numerous (about 200) than "applied" works (about 350).[42] Noteworthy pieces included Bernard Parmegiani's *L'instant mobile* and François Bayle's *Espaces inhabitables*, both first performed at the Exposition des Musiques Expérimentales in 1966 and 1967, respectively.

1975 to 1999—"Traditional" Concerts

After 1968, the effervescence of avant-garde cultural circles slowly faded, reaching a low mark around 1975. Attempts were made to reach out to the general public by fusion with more popular musical genres such as rock music. Pierre Henry composed *Ceremony* (1969) with Spooky Tooth, and François Bayle worked with Robert Wyatt, the drummer of Soft Machine, for *I Did It Again* (1972). These collaborations were ephemeral because sociologically, electroacoustic music remained in the sphere of scholarly music, even if its approach to sound investigation could be compared to commercial music techniques, or even to amateur musical practices from the oral tradition. Electroacoustic music "professionals," like all classically trained musicians associated with so-called postmodern contemporary music, appeared as an elite cut off from the general public. Today it is instructive to compare this assessment with the observation made by Guy Reibel in 1977: "Although electroacoustic music halls are emptying, on the other hand the number of requests for introductory courses in electroacoustics, at all levels, is mushrooming, as if the public confusedly senses that before *listening*, it needs to *do*."[43] Audiences slowly dwindled, forcing François Bayle, then director of the GRM, to seek out new ways of disseminating music. In 1974 he inaugurated the Acousmonium, a system that he continued to develop until the end of the 1990s. In 1979, the first season of concerts coproduced with Radio France under the title *Cycle acousmatique* marked the beginning of an outstanding period for GRM concert attendance, with between two hundred and four hundred people at each concert. All the works performed in concert were then broadcast on radio. Until 1981, the public was assiduous. But because of an aging audience, in conjunction with the political change in 1981 with the arrival of the Left to power, the avant-garde and even the countercultural aura that had been associated with the GRM's output since its inception faded during the 1980s and 1990s. The GRM's concerts blended into the general landscape of "contemporary music" with a slight decline in attendance, until the arrival of the Techno wave in 1997–98.

One could easily assert that the Acousmonium concerts in the late 1990s constituted an apogee in electroacoustic celebration. With a composer-interpreter as master of ceremonies, backed by well-organized production and distribution structures (commissions, radio, record publishing), with the help of institutions (INA, Radio France, Ministry of Culture) the system functioned perfectly. There were on average about fifteen concerts per year, with premieres of two to three new works. The majority of the composers programmed belonged to the worldwide network of large and long-standing research studios.

2000 and Beyond—Diversification

Since the year 2000, social polarization of the modes of presentation of electroacoustic works has emerged.[44] François Delalande, introducing the in-house seminar of the GRM in March 2004, addressed the issue from the perspective of the verticality in the relationship between the composer and the public, and the horizontality of exchanges between amateurs.[45] Top-down verticality is encapsulated in a composer creating works that are performed for an audience whose only role is to applaud. This vertical model, linked to show business, the star system, and mass media, is increasingly driven by the laws of commerce. On the other hand, the horizontal model is linked to marginal productions distributed by small labels for fragmented audiences, based on the principles of sharing and exchange.

In recent times—2000 and beyond—the diversity of forms of music distribution (small labels, mail order, internet, network of collectors) well serves the function of multiplicity linked to the baroque resurgence of our time, as described by Michel Maffesoli.[46] The exchange of information on the internet constitutes a new circuit outside the vertical top-down model, even while the "GRM-style" concert (at a set time, in a concert hall, seated, with frontal projection) still remains a *high place* for musical celebration in the vertical model. The significant participation by young audiences is evidence of the potential for electroacoustic music to move towards the horizontal model, encountering enthusiasts and future active participants. In this scenario, perhaps abandoning its name—instead labeling itself simply Electro.

We less and less often see crowds of admirers at the end of concerts swarming around the equipment. We are moving away from the vertical model of the composer presenting music to an applauding audience. Now that hardware and software are more commonly accessible, concertgoers are looking for a more horizontal relationship with composers, even more so because they are often composers themselves. It should also not be underestimated that GRM concerts are usually free, maintaining them in this horizontal universe.

The experience of encountering sound matter during the concert becomes the focus of social sharing. Electroacoustic music is a means of expressing and communicating the intimate emotions of the composer, rather than their sentiments, echoing Damasio's argument that tends to reveal that "emotions are manifested in the theatre of the body; sentiments in that of the mind."[47] Following this logic, the success of the spatial interplay during sound projection would signify that audience approval of the spatial figures proposed by the composer is related more to emotion than to sentiment.[48]

One Million Composers

And then in 2002, it was discovered that, according to a survey by the Ministry of Culture, at least one million people in France had purchased computer equipment for the purpose of composing musical works—without attending a conservatory or receiving any academic training.[49] Another surprise was that these self-taught people (50 percent of whom are women) circulated their works outside of formal channels, whether privately or through nonprofits, or semiprofessionally, through local radio stations, small self-produced labels, or the internet. These amateur composers complained about being isolated from the professional world, but they were perfectly organized in networks among themselves. In short, the entire scholarly musical system of teaching and dissemination seemed to be hermetically sealed off from this movement, and consequently unable to tap into it.

We now better understand the gap created between the music lovers who are exclusively followers of the written classical repertoire and the training provided in conservatories, and all the other music lovers who have been entering the world of sound by side roads for more than fifty years now. Because in France, contrary to many other countries, there is no alternative to the musical education provided by the conservatories: no sacred musical tradition as in American churches for example, no family musical tradition, and finally only early music education at school, which we will characterize as "insufficient."[50] This final point deserves a little nuance, because the GRM is actually in contact with the most dynamic and innovative teachers in the French National Education system in order to promote the use of new pedagogical tools, such as the CD-ROM *Les musiques électroacoustiques* (2000) or the Acousmographe (2005).

But in contrast with the deficiencies in teaching, we note that since the early 2000s, electroacoustic concert halls are full, certainly in large part with all these new enthusiasts at their personal computers manipulating sound.

The Temptation of Techno

It is in this context that since about 1999–2000, we have witnessed a convergence between electroacoustic music and techno. Pierre Henry—celebrated by the DJs themselves as their "grandfather"—has taken advantage of this, promoting his music among this wider and younger circle of listeners. He has even reintroduced into his music, sometimes outrageously, pulsating beats inciting dance in the functional techno style. Most often from one movement to another of a piece, as in his *Une tour de Babel* (1999), he alternates two opposing aspects—dance (the party) and meditation. On one side, the body, on the other, the mind—without altering in the least his sound ritual.

In 2003, Christian Zanési, composer and GRM member, organized an encounter with some of the stars of techno, the Austrian Christian Fennesz and the Russian Mika Vaïnio, in an event titled *GRM Expérience*. For his part, beginning in 2002, Luc Ferrari has given concerts with DJ Olive, DJ Scanner, and eRikm. Denis Dufour organized Acousma-Raves from 1997 to 2000, mixing techno and acousmatic music.

All these connections with techno can be in part explained by the sharing of closely related techniques and compositional tools, linked with the technique of the studio, but also by a reciprocal desire for encounter—for techno to build serious musical legitimacy and for electroacoustic music to renew and enlarge its audience, while confronting itself with the live concert, and certainly also in search of the expression of pure emotion.

TECHNO CELEBRATION—ELECTROACOUSTIC CELEBRATION

It is important to stress that this connection was essentially made with "serious" techno, the movement in techno moving away from *dance* towards a more minimalist style. But even so, deep differences between the two genres rapidly became visible. Techno and electroacoustic music soon realized that they were not sisters, but cousins. So why did the linkup take place at all?

My hypothesis is that techno music is to electroacoustic music what trance is to rapture.[51] Or as some of the DJs put it so well, "electroacoustic music is techno without rhythm"—meaning without the famous 4/4 "boom boom" techno beat.

If we refer to the sacred, we can say that electroacoustic music is on the side of rapture, which brings one out of oneself and is attained in the solitude of the work on composition in the studio, while techno is on the side of trance, which is reached in company, in movement, even in Dionysian euphoria. According to Gilbert Rouget, "ecstasy and trance must be seen as constituting a continuum of which they each form one end."[52] The Dionysian aspect of trance can be seen in participants surpassing their limits, while community is born from the intoxication of the moment, in a form of possession—recalling the techno of the dance floor, where the act of being together is what counts.

In contrast, this Dionysian aspect is mostly absent from the electroacoustic concert, or perhaps barely glimpsed in the interpersonal relationships created by the shared interest in technology. The electroacoustic concert valorizes the cult of the aesthetics of the sound, of its form, characteristics no longer Dionysian but Apollonian, in the Nietzschean sense of the terms—Dionysus, creator of intoxication and oblivion, Apollo, creator of form.[53] Continuing with Nietzsche's terminology, which explains how the "duality between the Dionysian and the Apollonian" is at the source of the development of art, between the authentic natural truth that one seeks to reach and the lies of civilization,[54] we can also conclude that the disgust following the return to the banality of everyday life can find its resolution in art.

From this perspective, it is intriguing to compare the electroacoustic music concert with the techno concert, because they share many points in common while representing precisely the dualism described by Nietzsche between Dionysian and Apollonian, or the two extremes of ecstasy and trance that Gilbert Rouget evoked. Meanwhile Michel Maffesoli clearly differentiates the techno gathering, situated in the present, in the qualitative, and therefore focused on presentation, contrasting it with (classical) representation, "as much as today there is something intense and nomadic in the *being together* of the techno gathering, at a time when everything was based precisely on representation, there were institutions that were static and immobilized."[55]

SIMILARITIES BETWEEN ELECTROACOUSTIC MUSIC AND TECHNO

Many common points link electroacoustic music and techno: a singular site where listeners gather, performances well defined from the temporal point of view with a beginning and an end, music realized from processed sounds fixed on a support, diffusion by means of a loudspeaker system, few or no interpreters, partially identical audiences, especially among the young generations of DJs who have the same intuitive way of approaching and transforming sound matter as electroacoustic adepts, and finally aesthetics that are highly dependent on repetition and exploring timbre, but relatively disinterested in melody.

And not only do techno lovers claim Pierre Henry, Luc Ferrari, and Bernard Parmegiani as their spiritual forefathers, but they also identify shared lineage with the German electronic music epitomized by the electro-rock group Kraftwerk, itself influenced by Stockhausen.

DIFFERENCES BETWEEN ELECTROACOUSTIC MUSIC AND TECHNO

The primary difference between electroacoustic and techno music is the absence of a pulsating rhythm (in electroacoustic music) inciting dance (in techno). Techno is above all functional music, meant to make people dance, and this dancing, backed by 120 bpm—one hundred and twenty beats per minute, simulating rapid beating of the heart—easily leads to trance and oblivion. This orgiastic character contrasts with the effort of intellectual concentration that electroacoustic music demands, to reach refined auditory perception. Another difference appears. In electroacoustic music the concepts of authorship and the work of art are central, while in techno there is "art without a work" where the artists remain anonymous, or simply cosign their performances in the form of credits.[56]

This contrast is translated by physically visible differences between the two types of concerts. In the techno concert, the rave party in England, the "free partie" in France, the public is in a standing position, dancing to the rhythm of the music, in a state of celebration, striving if possible for a state of trance. This atmosphere of transgression often leads participants to using drugs and organizers to adopting extremely high sound levels to inflame the audience through a kinesthetic effect. Due to the risk of disruption and unrest, techno concerts have sometimes been banned, outlawed by the police. As a result, they have often been organized clandestinely. This notion of the illicit only increased the attraction for younger generations, especially in the early 1990s. Later in the 2000s, techno concerts were gradually "socially recuperated," no longer necessitating their being declared illegal.

In contrast, with electroacoustic concerts, the venues and programs are announced in advance, the audience is seated, the atmosphere is serene and calm. Concentration reigns.

But after the enunciation of these differences, it is important to recall the plurality of techno music with its numerous hybrids since its very beginnings in Detroit in 1988. Some techno music resembles electroacoustic music to a fault, and vice versa. The reason lies not in aesthetics, but in the social network to which the authors belong. Techno artists originating from rock will remain associated with their rock milieu. In the same way, techno artists educated at the conservatory will remain a part of their serious music milieu, since funding and distribution networks are often impermeable.

As for the GRM, it has undertaken its share of rapprochement by programming artists from the techno scene in its concerts, sometimes commissioning works from them. For instance, faithful to his techno rock origins, at a GRM concert Arnaud Rebotini titled his March 13, 2004, premiere *Grind*,[57] evoking the term Grind-Core, a style of radical rock, whose aesthetic signature is: fast, loud, incisive.

What about Installations?

With the advent of the internet and multimedia, the different media arts have come closer together and interart encounters have multiplied. GRM composers have never been attracted to installations nor to the extreme freedom—grounded in chance operations—of the early American happenings of the 1950s, with the well-known exception of Luc Ferrari, who was always fascinated by American artistic productions. While installations generally do without the presence of the composer-performer at the moment of presentation, on the contrary at the GRM we have always valued the sound quality of the concert, directed by the composers themselves, who are regarded as the best possible interpreters of their own music and the focal point of the presentation. It must "sound"! It must be remembered as a unique and exceptional moment. A successful concert must not only gather a large audience, but also attain sound rendering of high quality, in harmony with the site and its audience, under the patronage of the author projecting their sounds.

Conversely, an installation is spread out over time. It resembles a pictorial exhibition. Access is free and sound projection is usually automated. One does not experience the singular character inherent to the concert. In addition, in installations, sound played in a loop is often associated with images or lighting. It is treated as a visual work of art exposed to the eye. So it is not the audio quality that distinguishes the concert from the installation (it can be just as good or bad), it is the ritual of the occasion. In one there is a master of ceremonies, in the other there is none. The electroacoustic "celebration" contrasts with the "self-service" installation.

And yet thanks to multimedia and the internet, the genres are succeeding in coming together. For example, the Australian artist Ros Bandt has taken it upon herself to collect and put online all existing installation experiences.[58] In doing so, by fixing these installations on a multimedia support, she is revisiting the practice of music fixed on a support.

CONCLUSION: REINVENT THE ELECTROACOUSTIC CONCERT?

Since the early 2000s, there has been overproduction in electroacoustic musical works. New channels need to be found. All these enlightened amateurs, new tool users, estimated at one million people, are both a breeding ground and an audience for future professionals. More than an aesthetic problem, we are witnessing the emergence of a social problem in musical creation. The question today is to determine who can (legally) access the status of composer. The obvious answer should be, "in a democracy, everyone!" However, faced with the influx of works inevitably clogging up traditional channels (concerts, radio, records), instead we hear the answer, "who to choose? And how?" It is clear that the distribution channels are no longer adapted to creative production.

Some commentators, such as François Delalande, believe that in the twenty-first century we should envisage reinventing the conventions of the concert. Everything needs to be reconsidered: the Italian-style hall, a seated audience, the concert starting at a set time, latecomers being turned away. In his view, since the Acousmonium in 1974, things have become entrenched. The electroacoustic concert inspired by the classical concert has passed its peak. Couldn't we return to experiments like Pierre Henry's *concert couché* with the audience lying down, or Nicolas Schöffer's spatiodynamic shows?[59] Hadn't this same Nicolas Schöffer already written in 1978, "music sank into a double excess: commercial music on one hand, asocial esotericism on the other, with their distinct hyper intellectualist chapels, and between the two the institutionalized pomposity of the political Left or Right."[60]

Christian Zanési, responsible for programming concerts at the GRM since 2002, alongside Daniel Teruggi and with François Donato,[61] and then Philippe Dao and François Bonnet, is seeking out new concert venues, new spaces, and new formulas. We have seen some breakthroughs in this direction, concerts in halls dedicated to techno, concerts in bars with free access for customers, multimedia concerts, concerts in public places (the Forum des Halles in Paris), live electronic concerts. And each time, the public has come.

5

In Search of Music Writing

Reach the mediating image I spoke of before—an image which is almost matter in that it still allows itself to be seen, and almost mind in that it no longer allows itself to be touched.
—HENRI BERGSON

—Can electroacoustic music be written down (on a score)?
—Does it possess a hidden code that must be discovered?
—If not, should we not move towards developing a code of symbolic notation specifically for electroacoustic music?
—What place should be given to graphical transcription of electroacoustic works?
—Can graphical transcription lead us to notation?

For musicologists, answering these questions would make it possible to finally assign a place to electroacoustic music with respect to the rest of the contemporary music corpus. The still often-asked question "Is it really music?" would then no longer be relevant.

In music, the French word *écriture*—"writing"—has two facets, one sonic and the other visual.* The sonic facet of writing is perceived while listening. It results from the voluntary arrangement of sounds undertaken by the composer with the help of his tools. The visual side of musical writing, which is more generally called notation, reveals the sound world to our sight, and in so doing affirms it. But the *sonic writing* and the *visual writing* of electroacoustic music, as of all music, have their own histories and may evolve separately, even if they have points of convergence. For now, the transmission of electroacoustic music has taken place in the

---

* Translator's note: Often directly rendered in this translation as "writing," even if English usage differs. The "sonic facet" of writing music is often referred to in English simply as "composition." See note on *écriture* in chapter 1.

purest oral tradition, under the guidance of accepted experts, during master classes or in workshops and practical courses, without resorting to any form of notation. At the same time, notation continues to be the subject of research throughout the electroacoustic community.

We already realize that, for the classical repertoire, notation reinforces the authority of composition. If electroacoustic music were to adopt a form of notation, its legitimacy would be all the greater. Associating images or symbols with what is heard consequently does have a real impact. It makes it possible to affirm, "I only believe what I see," because writing is linked to power, whose supreme level is divine power.

ANALYSIS AND TRANSCRIPTION AT THE GRM— INVENTORY OF THE SITUATION
Four Hundred Scores and Transcriptions

Pierre Schaeffer's idea, when at the beginning of the 1960s he began working on his *Traité des objets musicaux*, was to describe sound in order to identify the seeds of the basic laws of music. Little by little, the concept of "doing" became essential for him. In other words, it is through the very act of transforming sound matter that we arrive at the essence of musical expression. Experimentation is the key—simply obstructing the interplay between "doing" and "listening" is enough to immediately bring the creative process to a halt.

But to transform sound, it is necessary to master its parameters, namely, to describe them. In addition to verbal analyses of pieces, the description of sound has also been rendered concrete at the GRM through the existence of four hundred scores of musical works. The earliest date to 1948, listening transcriptions of Schaeffer's *Études de bruit* realized by Monique Rollin.

Of all these documents, only a few are prescriptive scores, written prior to the work, even though they are often not in traditional notation. For example, for his 1961 piece *Reflets*, Ivo Malec had first written cadences (using serial techniques) as well as a few sound morphologies, all on graph paper, set to a precise time scale. He then followed this framework to construct the piece on magnetic tape. Such a methodology was rare. Most often it was adopted by classically trained composers, neophytes in the field of music on a support. In Malec's case, it was a study executed during a workshop, his first electroacoustic work. Composers resorted less and less to preliminary scores the more they acquired practical skills in composing on a medium, the more they acquired working knowledge of the principal methods of realization and practical experience at attentive perception. In these circumstances, prescriptive scores were only warranted for the instrumental parts of mixed works, as an aid to the performers, or as a general outline of the form of the work and of the principal voices for mixing.

The four hundred GRM scores were primarily on music paper, especially those for mixed music. We also find graph paper, Bristol board with small squares, photosensitive film, and various other multimedia tools. Sometimes they were simple handwritten listening notes, or diagrams for editing or mixing. There were also curves on graphs recording sound amplitude on the bathygraphe or sonograph, especially for the early works. Use of the spectrogram and the Acousmographe was marginal until the 2000s, even if the first attempts dated back to the 1970s.

The large number of these scores, which are primarily listening transcriptions, is due to the fact that when Pierre Schaeffer founded the GRMC in 1951 he required his collaborators to write a score in advance of composing musique concrète, as he did for his own music. Unfortunately, most of the electroacoustic scores submitted by the adepts of the genre were incomplete, either presenting only the beginning of the work or written after the fact. The composers seldom even followed them, preferring to rely on their auditory sensations. But Pierre Schaeffer, son of a violinist father and a vocalist mother, had been trained at the conservatory and could not imagine inventing a new musical genre without associating it with a written form. From the beginning he had positioned his work in the scholarly musical milieu, notably by inviting the principal composers of the time to come and try their hand at this new genre: Messiaen, Boulez, Milhaud, Varèse, Jean Barraqué, and Michel Philippot. And in fact, the transcriptions also had another use, more trivial but essential; they were part of the obligatory material to be provided to SACEM when declaring each new work for copyright protection. Today, SACEM also accepts documents representing the waveform of the music instead of a graphic transcription, accompanied by a recorded version of the work.

It is necessary to differentiate between scores and transcriptions in order to refine vocabulary. Scoring and transcription function in a symmetrical relationship, with the piece of music in the middle. While the score precedes the work, the transcription follows it. The score can only be written by one person, the composer. In contrast, with pieces fixed on a medium, transcriptions can be written not only by composers themselves, but by any listener. A second point of differentiation: since electroacoustics has embraced the entire spectrum of sound, compared to other musical genres the relevant musical criteria are considerably multiplied. Given the context, the codes of Western classical notation, based essentially on representation of note pitches and durations, are insufficient for representing morphological criteria, dynamic profiles, and concepts of spatialization. A new "enlarged" code is necessary.

From Descriptive Transcription to Prescriptive Score in Electroacoustics: Why Seek Codes?

While transcription always retains an arbitrary dimension associated with the transcriber who signs it, the search for a code of notation has another ambition, to reach a kind of universality that legitimizes musical "writing."

A brief survey of the art reveals that graphic transcriptions of listening have a very long history. The best-known transcriptions in the West are those of Gregorian chant neumes, which were gradually refined from the eighth century onwards, passing through Guy d'Arezzo's theories in the eleventh century, and *measured time*, in the thirteenth century, leading to Ars Nova in the fourteenth century. The recent twentieth century was characterized by efforts to mechanize and automate graphic transcription. The most widespread tool for analysis is the sonogram, which allows visualization of the physical signal of music; presentation in terms of time-frequency is the most widely used because for the listener it is more meaningful than a sound wave form (for which the first commercial trade name was Visible Speech). The sonogram is present in most music analysis software, notably in the Acousmographe, developed at the GRM from 1978 onwards, as well as in certain sound processing software such as Audiosculpt, developed at the IRCAM.

Several recent university theses have taken stock of graphical transcription and analysis in electroacoustic music. Two of them retain our attention, those of Stéphane Roy and Pierre Couprie.[1] It is not possible here to review the contents of these theses in their entirety, but in summary, transcriptions may serve several functions:

—as a research tool for analyzing specific musical criteria or specific construction procedures, or analyzing the style of a composer or a particular piece;
—as a research tool for developing a code for graphical transcription of music from the oral tradition;
—as a guide for public performance—when there is sound projection in concert, even if it is handled personally by the composer, a transcription can be useful as a memory aid;
—as a pedagogical tool for education in listening.

And I will suggest another use:

—as a starting point for graphic and multimedia creations.

At the GRM, the question of how far to go with transcriptions has been debated from the beginning. The listening analysis and transcription software Acousmographe appears to be a promising tool thanks to its automated functions. From there to thinking transcriptions could be done completely automatically there is just a single step, which would require the development of a code. But given the multitude of musical criteria to be considered, the development of universal notation takes on the character of a real quest. In my opinion, this quest is the best bet to guarantee the richness and freedom of graphical transcription of the sound images that musicians transmit to us in their music. It is an open door to a vast realm for creativity. Ultimately interactive visual transcriptions find themselves situated between analysis, pedagogy, and creation, since the Acousmographe, as a tool, also tends to become an instrument that can be used according to the logic of creativity.

We also observe that the latest innovations in multimedia technology all point in the same direction: proliferation, diffusion, spin-offs, interactivity. They take us away from a single universal form of "writing," symbolic of the sacredness of monotheism. They bring us back to the secular, to the profane, even to animism. Multimedia plunges us into a new universe, that of the plastic and visual arts, but also of movement and interactivity. In these circumstances, the oral tradition still preserves its importance in electroacoustics.

Historical and Conceptual Background

PIERRE SCHAEFFER—MUSICOLOGY RUNNING OUT OF STEAM

In 1966, Pierre Schaeffer spoke of musicology "running out of steam."[2] In *TOM*, he evoked a "dead end in aesthetic commentary":

> Taken as a whole, the copious literature devoted to sonatas, quartets, and symphonies rings hollow. Habit alone can hide from us the poverty and the disparate nature of these analyses. When we put aside the smug comments on the composer's or the performer's state of mind that litter the work, we are left with the most tedious list, in the language of musical technology, of his methods of production or, at the very best, a study of his syntax. But there is no real critical appraisal.[3]

The concrete approach, advocated by Schaeffer, called into question not only the method of composing, but also of listening, and consequently all the sciences of music. Schaeffer added:

> In traditional music the musical was given in advance, guaranteed by the score. The *added value*—if I may use this expression—was sound. In experiential music sound is heard as a matter of course, and the ear wonders about the musical....
>
> And this is how, finally, the function of music is revealed. The work in hand is exposed to hybrid, unstable types of listening, fluctuating between simultaneous goals. The essential thing is the response, more than the purpose. Hence the expression "back-to-front music" that I suggested.* This music, which we do not know how to make but which we know how to hear, insults our will to power. We would do well, however, to learn to recognize our image all over again in this music, inverted in the mirror of sound. It is often said that immense resources of the human brain remain unknown or untapped.... Music can reveal this "of man" to us.[4]

---

* Translator's note: The original French *musique à l'envers* has been rendered by Schaeffer's official translators as "back-to-front music" (in quotes in their translation). Schaeffer is explaining that receiving this new music calls for a listening mode that is in some ways "in reverse" from the normal process for more conventional music.

As far as the search for "writing" for musique concrète is concerned, Pierre Schaeffer had made some attempts in the early 1950s. He had imagined a system with two staves, calibrated in time, and functioning in parallel. The upper staff represented the melodies and the lower staff the dynamics of the sounds. He had begun to codify the characteristics of attack, sustain, and extinction. But he did not follow up on this line of research.

PIERRE SCHAEFFER—THE CONCEPT OF SIMULACRUM

Pierre Schaeffer, armed with his extensive knowledge of media, then turned to general discussion about communication theory. He published a book in two volumes on "communication machines."[5] In his view, most situations can be explained by what he called the communication triangle—a sender, a receiver, and a third party who serves as mediator. It was the guiding working hypothesis at the Service de la Recherche. Unlike the Canadian sociologist Marshall McLuhan, who considered that "the medium is the message" (a simplification due to the fact that all messages are somewhat intermixed), Schaeffer placed himself in the position of mediator, seeking to move back to the message by starting from meaning, impressions, and even expressions that were provoked by "the causes" of the message. These *causes* were the mystery that he sought to elucidate, the mystery of music as much as the mystery of images and language. The advantage of this triangle is that it allowed for dynamic functioning and that it could be applied indifferently to all stages in research and analysis. Schaeffer even imagined applying it outside the musical sphere, to relations between institutions or to relations between individuals, etc. According to Schaeffer, everyone speaks their own language and in each case a translator is needed.

McLuhan's theories were in vogue in the late 1960s. He announced the end of the Gutenberg galaxy and the advent of mass culture. During the same period, also quite influential were Alain Baudrillart, author of works on the simulacrum, simulation, and on the consumer society in general, and Guy Debord, leader of the Situationists, philosopher critical of *The Society of the Spectacle*. Schaeffer had concentrated his reflection on the concept of simulacra. He said, "in the time it takes to understand, one is no longer the same. . . . Speaking and writing were, by definition, communicating in a certain way."[6] But with the advent of societies of mass, he deplored that the media had become "simulacra of the word and of the eye."[7]

> Speaking on the radio, appearing on the screen, addressing so many other masses of people, or even one solitary listener, this unknown spectator, is to completely change from one civilization to another without appearing to do so. It is this simulacrum of the presence of dialogue, this new conditioning of men among themselves, which is at issue, without anyone really noticing it yet.[8]

As far as music was concerned, Schaeffer stuck to his 1953 positions:

In new music, the composer is often his own performer. The score is only an outline. The work is created once and for all by a different distribution of responsibilities, reminiscent of film production teams. The contact with the public takes place under different conditions. The concert is no longer a spectacle in the way we were accustomed to.[9]

Until the end of his professional career in 1975, Schaeffer's interest in the media continued, but while always maintaining music as his operating model. For example, regarding the music of Messiaen, he declared:

> When Messiaen notated the birdsong of the Hindu mainate or the red cardinal of Virginia, from which he would create his orchestral score, he was not a listener in the way one might be a reporter, but already a musician, attentive to the real. Not to reproduce it, but to derive his work from it. Music "begins when heard" but culminates when it is made.[10]

FRANÇOIS DELALANDE—FROM PERCEPTUAL INTENTION TO LISTENING CONDUCTS

François Delalande, even before becoming head of musicology research at the GRM in 1975, had pursued the work done by Schaeffer. The concept of "listening intentions" that he had developed as early as 1972 in his first published article, "L'analyse des musiques électroacoustiques" (Analysis of electroacoustic music), extended the concept of "intentions of perception" developed by Schaeffer. Delalande, evoking electroacoustic music wrote:

> Even if their outward definition is limpid, their internal functioning, their musical meaning, and their resonance with the unconscious are a web of questions to which musicology has not provided even the glimmer of an answer.... A priori, we recognize distinctions corresponding to three listening intentions—the organization of sounds, musical values, and the meaning of the music.[11]

Delalande adopted from Schaeffer the "theory of the four listening modes"— *écouter* (to listen), *ouïr* (to perceive aurally), *entendre* (to hear), *comprendre* (to understand),* but he considered the *fait sonore* (sonorous fact) as something already established and focused only on the three other aspects of listening.[12] By not retaining the first level of listening described by Schaeffer, listening, Delalande placed himself resolutely in the field of reduced listening. This nuance is essential, because it separates the two men. Schaeffer was attached to sound phenomenon by "doing," while Delalande proceeded from "hearing." However, as early as 1970,

---

* Note from Schaeffer's translators: "Schaeffer uses four different words in French to describe the listening modes: *écouter, ouïr, entendre,* and *comprendre*. Since there are no perfect equivalents in English, the decision has been made to include the French original word in brackets." Schaeffer, *TMO*, 73–79.

Schaeffer wondered, "does music begin when one makes it or when one hears it?"[13] But he concluded, "one does not speak about music, one makes it."[14]

In the 1980s, François Delalande, influenced by his readings on Jean Molino and Jean-Jacques Nattiez's theory of tripartition, breathed new life into GRM research. He contributed to the development of *musical analysis*, where Schaeffer had concentrated on the analysis of sound phenomena and the practitioner's taming of sound. Delalande became more and more interested in the *esthesic*, while knowing that the concepts he was developing could also be applied to the *poietic*,[15] since *doing* and *listening* function dialectically. Already in the 1980s, Delalande abandoned the concept of "listening intentions" for that of "conduct of perception," and then "listening conducts." In a later article, published in 2002, he clarified the concept of "listening conducts" as taxonomic listening, empathic listening, and *figurativization* or the meaning of music. In his view:

> The *conduct* can be described at the psychological level and at the social level.... a conduct is represented by a set of features that characterize what the subject experiences at these levels—cognitive, affective, emotional, motor—these components not being independent, but coordinated by the finality that the subject attributes to his acts: it is a set of elementary acts coordinated by the search for a result that the subject expects, involving these diverse aspects.[16]

For François Delalande, "the objective of musical analysis is to study the relationship that is established between a subject and an object, within a conduct, and more precisely to explain how objects and conducts are determined in parallel."[17]

JEAN-JACQUES NATTIEZ, STÉPHANE ROY—TRIPARTITION

It should be remembered that Jean Molino's 1975 theory of *tripartition* (or *tripartite semiology*) was extended to the musical sphere by Jean-Jacques Nattiez in the late 1970s.[18] This theory was then expanded concerning electroacoustic music by Stéphane Roy in the late 1990s.[19] In the view of these authors, between *poietic* analysis, concerned with the work from the point of view of its production (compositional strategy), and *esthesic* analysis, concerned with the work from the point of view of its hearing (perceptual strategy), there is a neutral level. This neutral level poses some problems of definition and is still the subject of research in 2024. Does it have a material existence, or is it simply an operative concept to understand the music? In brief, this is the fundamental question that arises. For Nattiez and Roy, the neutral level would be a kind of taxonomic identification of sound objects, in the score, or the work, in order to explain its rhetoric.

If we refer to Schaeffer's theories, expounded in *Traité des objets musicaux*, Molino/Nattiez/Roy's relationship between *poietic* and *esthesic* is very similar to the relationship between *doing* and *listening*. In the penultimate chapter of the

Traité, titled, "In Search of Music Itself," Schaeffer evoked the constant movement back and forth

> between the conceivable and the possible, very similar in music to the confrontation I have referred to again and again: between making and hearing.... A violin, a voice, a work is a duel, a duo, judo. Knowing how to yield, outsmart the adversary, give him his own way in order to take a hold over him—this, in the great periods of history, was *playing* music.[20]

In this logic of comparison, could we not assert that *playing* music, according to Schaeffer's concepts, would be the neutral level? The neutral level would indiscriminately apply to the place and the moment of *playing* (the confrontation) between *poietics* and *esthesics*, or between doing and listening.

In a 1999 article, Jean Molino positioned his theory of tripartition in relationship with Schaeffer's ideas, playing on the term *musical object*:

> There are indeed three dimensions in the *fait musical*—the act of music.* First, the production of music. Secondly, the musical object—and I give the word here another meaning than that given by Schaeffer by making it designate the totality of the traces that musical activity produces independently of their reception by an individual, namely the physical signal as well as a recording or a score. And finally, audience reception.[21]

The neutral level, described by semiologists, raises questions because different people define it differently. According to Molino and Nattiez, it has a descriptive character. According to Stéphane Roy, we should transcend this description and try to analyze not the object but the processes that lead to the object, both from the point of view of its production (*poietic*) and its reception (*esthesic*).[22] Stéphane Roy stresses the term *induction*. He privileges the exploration of inductive *esthesics*, corresponding to the concept of conduct of reception.

JEAN-CHRISTOPHE THOMAS—THEMATIC ANALYSIS

Jean-Christophe Thomas, researcher at the GRM since 1980, gained recognition by participating in a *poietic* investigation into Bernard Parmegiani's process composing *De natura sonorum*. This examination gave rise to a 1983 book coauthored with Philippe Mion and Jean-Jacques Nattiez.[23] Jean-Christophe Thomas presented Parmegiani's work in the form of thematic analysis, a sort of synthesis between *poietic* analysis and *esthesic* analysis. As practiced by Jean-Christophe Thomas, thematic analysis is a transposition to music of the method used by Jean-Pierre Richard in the field of literature. Thomas quotes a passage from Richard's book *Littérature et sensation*:

---

* Translator's note: *Le fait musical* in poststructuralist music analysis is most often translated directly as "musical fact"; it encompasses the entire reality of music, "the thing itself."

There can exist no rift between the diverse experiences of a single man ... In the most apparently separate spheres the same patterns are revealed ... A particular slant in a sentence sheds light on the intention of a particular moral option, of a particular sentimental engagement. Obscure musings by the active or material imagination are deeply linked to the most abstractly conceptual speculation.[24]

Accordingly, Thomas collects verbal material for his research (the composer's confidences, recipes, vagaries, discoveries—a corpus of obsessive personal imagery) and treats it as he would "literary" material, musically pertinent—it is transposed into a guiding principle for "reading" (hearing) the musical work.

Since then, Jean-Christophe Thomas has continued refining this analytic model. Today his analysis has shifted to a more semiotic and even formal (rhetorical) register. In a transversal analysis of François Bayle's entire body of work, he identified fifty-three themes characteristic of the Baylian style.[25] This thematic analysis favored by Jean-Christophe Thomas also finds its roots in Bachelardian analysis, which focuses on Matter considered as an archaic and unsurpassable foundation for sensibility—Bachelard's famous "reveries" on water, fire, air, and earth. Jean-Christophe Thomas's analysis is situated between Bachelardian and Richardian themes, and between semiotics and rhetoric. In his introduction to his 2004 *Fragments pour Bayle*, Thomas wrote:

> For those who are interested in the "symbolic" ("To want to say something other than what is said, that is the symbolic function," Ricœur), a gateway exists for attempting to grasp, with words, the style of any symbolic form, including music: it is the thematic method. Barthes (in *L'ancienne rhétorique*) finds an origin for it in what he calls sensory topicality: "That which progresses by categories (...), in a word, that of Bachelard, ascending, cavernous, torrentuous, shimmering, dormant, etc.", sites which Vico already saw as "universals of the imagination".
>
> Just by using the term "sensory," it is clear that this approach is particularly pertinent to acousmatic music: nothing to see, nothing like a score to read, just let yourself keep moving through your sensations: they are your only points of reference for positioning yourself in the piece and finding a few landmarks in its structure. At least for those who still know how to savor.

For Jean-Christophe Thomas, the "archaic" side of the "music of sounds," treated "acousmatically" by the senses alone, encounters privileged echoes in the archetypal universe of Bachelard and Richard's "sensory categories."

FRANÇOIS BAYLE—SOUND IMAGES OR I-SONS—I-SOUNDS

François Bayle, also a great admirer of Bachelard, laid out his theory of sound in his 1993 book *Musique Acousmatique—Propositions ... ... positions*. Following Charles Sanders Peirce's trichotomy (icon, index, symbol), Bayle established three stages in the recognition of the form of sounds, three "images"—iconic,

TABLE 2 Charles S. Peirce and François Bayle's sound nomenclature

Charles S. Peirce	François Bayle	Image of sound (*i-son*)
Icon	Iconic sound (*im-son*)	Anecdote
Index	Diagrammatic sound (*di-son*)	Archetype
Symbol	Metaphoric sound (*mé-son*)	Abstraction

diagrammatic, and metaphorical. The iconic image corresponds to the concrete sound referred to as "anecdotal," with its recognizable and unconcealed origin. The diagrammatic image corresponds to the sound whose source is almost completely unrecognizable, but where there remains a trace in its form; it inclines towards the archetype. Finally, there is the metaphorical image, or meta-sound, leading us towards pure sensation. This becomes the now famous juncture when we attain *reduced listening*.

In line with Bayle's conceptions, researcher Marie-Noëlle Moyal described this state:

> From the moment consciousness reaches the threshold of forgetting the referent, the mind, faced with the impossibility of continuing to live in equivocation, tries to reinstate itself in a familiar world, that of a specific sensation, because all thought tends to prefer set structures.[26]

Marie-Noëlle Moyal gave this particular moment of shifting towards the senses the name of "orphan" image:

> Consciousness, as if struck by amnesia, can then experience an instant in which sound events are no longer associated with any specific experience. At this stage of the unfathomable, a veritable perceptual shift takes place, a sort of jump into the void of the unknown.[27]

FRANÇOIS DELALANDE—REVOLUTIONS IN WRITING MUSIC

Even though they were contemporaries, François Delalande's thought remained rather hermetic to François Bayle's theoretical concerns.

From the point of view of writing music, François Delalande evoked a *second revolution*.[28] The first revolution had begun at the end of the Middle Ages, around the thirteenth century, leading to a complete reversal in the application of writing music. First used as a tool for notation, it gradually evolved into a tool for invention. Seven centuries after its inception, the resulting technique of inventing "on paper" is still taught in conservatories. According to Delalande, since the end of the nineteenth century we have embarked on a second revolution, that of recording on a support, made possible by machines and electricity. This second revolution, like the first, has been characterized by the fact that tools (the tape recorder, the computer) that were historically first used for the purpose of conservation have also become tools for invention and creation, just as paper and pencil had in the past. This would explain the advent of a new approach to writing sound on new supports that conserve music—phonograph records, magnetic tape, the CD, and computer memory.

This raises numerous questions:

—How does Delalande take into account the issues of the concert and spatial projection in his analysis?

—Can we treat all media on an equal footing? There are in fact significant differences between analog recording on a vinyl record and a compressed recording in computer memory.

—Where should we position Delalande's other research—on psycho-pedagogy in young children, musical pedagogy, and the genetics of musical behavior? It would appear that an infant's early sensory-motor acquisitions are the source of future musical skills. Is it possible to verify this in electroacoustics? Delalande gives an affirmative answer, based on his research since the late 1980s conducted in Italian nurseries.[29]

MARCEL FRÉMIOT—TEMPORAL SEMIOTIC UNITS

Marcel Frémiot, a composer and member of Schaeffer's team, left the GRM in 1968 to set up the first musique concrète course in a French conservatory in Marseille. Later in 1987 he founded a nonprofit organization, the Laboratoire de Musique Informatique de Marseille (MIM), to pursue certain aspects of musical research that went beyond the framework of the conservatory. His reflection on Temporal Semiotic Units (*unités sémiotiques temporelles*, UST) is an extension of Schaeffer's research work on the typo-morphology of sounds. By approaching the subject through *esthesics*, "in the secret hope . . . of finding a technique for *explaining* contemporary music . . . or rather contemporary *musics*, to amateurs,"[30] the objective was to penetrate the meaning of music through the description of its temporal forms. From its very beginning in 1988, François Delalande collaborated closely with the project.[31] Together with Marcel Frémiot, the team brought together Jean Favory (software designer for the MIM CD-ROM about USTs),[32] Pascal Gobin, Pierre Malbosc, Marcel Formosa, and Jacques Mandelbrojt.

"What is a temporal semiotic unit? It is a musical segment which, even out of context, has a precise temporal meaning due to its morphological organization."[33] The USTs are presented in the manner of a series of stylistic patterns defining the temporal unfolding of music. Currently, the research identifies nineteen USTs. Their evocative names facilitate their use, but the researchers insist they are only labels: *Braking, Chaotic, Compressing-stretching out, Divergent, Endless trajectory,*

Fading away, Falling, Floating, Heaviness, In suspension, Moving forward, Obsessive, Propulsion, Spinning, Stationary, Stretching, Suspending-questioning, Wanting to start, Waves.

The question that the researchers at MIM are now addressing is whether there is a finite number of figures. If the answer is no, does this mean that music is an infinite domain?

One of the members of the team, Jacques Mandelbrojt, theoretical physicist in elementary particles and visual artist, has already proposed an extension to the UST, the Unités Sémiotiques Spatio-Temporelles (USST), in which both time and direction are associated.

LASSE THORESEN—A LIBRARY OF SYMBOLS

In the 1970s, the Norwegian composer and researcher Lasse Thoresen began developing a library of graphic symbols based on the criteria found in Schaeffer's typo-morphology. These very simple symbols (circles of different sizes, arrows, broken or unbroken lines, etc.), numbering about three hundred, are aimed at depicting all the configurations encountered in electroacoustic compositions. They can be used as a solfeggio code, but this requires learning them. For the moment, this initiative represents the most serious and most advanced work on the theme. An agreement has now been reached with the inventor to integrate them into the Acousmographe.

THE ACOUSMOGRAPHE—A TOOL FOR ANALYSIS AND TRANSCRIPTION

The Acousmographe was intended as an alternative situated between automatic transcription, figurative transcription, and coded transcription. The history of this software is quite involved, filled with uncertainties. Its origins can be traced back to 1975, when two young engineer trainees were given a six-month assignment at the GRM to devise an automatic graphic transcription method for electroacoustic music. François Bayle, then director of the GRM, wanted to have transcriptions available to present during his lectures and other public events, for educational purposes. The two trainees, Jean-François Allouis and Xavier Nouaille, had transformed a bathygraphe to extend the duration of the transcribed examples, beyond the few seconds that were then possible, by using a cylinder equipped with a stylus pen. In the wake of those first efforts, Jean-François Allouis also experimented with the sonagraph and began focusing on the spectral analysis of sounds, with the aim of modeling them and—why not?—synthesizing sounds based on digitized physical sound characteristics. Out of all this work emerged the 123 digital studio, and then a few years later, Syter and GRM Tools, as well as the migration of tools from analog to digital.

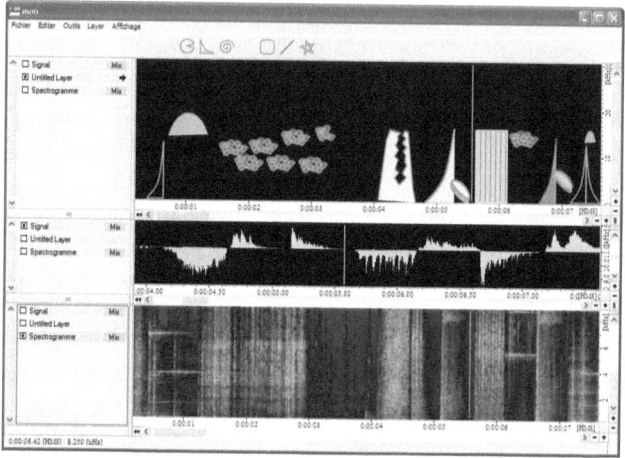

FIGURE 10. The Acousmographe version 3 software to assist with graphic transcription. Screenshots of the graphical interface. © Évelyne Gayou.

And precisely because of this redirection of research towards development of sound processing interfaces, the initial research project concerning graphic transcription was put on hold. Only a few experiments carried out by Allouis and Nouaille, remain from this initial phase, notably a montage of the spectrogram of Jimi Hendrix's version of "The Star-Spangled Banner," which was revisited when the project was taken up again a decade later by Olivier Koechlin.

In 1986, when he arrived at the GRM to replace Jean-François Allouis, the engineer Hugues Vinet, who was also an excellent jazz pianist, decided to revive the transcription project, entrusting it to another engineer, Olivier Koechlin, also a musician. Work on it began in 1988 on the Dyaxis sound platform and Macintosh II. This is when the name Acousmographe was chosen. The first operational version on the Sound Designer II board running in the Supercard environment was presented in autumn 1991 at the ICMC in Montreal. Jean Favory in Marseille continued the work until 1994. This version, somewhat buggy, survived until Didier Bultiauw took over the project in 1996.

In the beginning in 1988, Olivier Koechlin's main difficulty was dealing with computer systems that were still too slow to calculate the sonogram while also permitting the analyst to superimpose graphic symbols. Another difficulty was that the composers who frequented the GRM did not appear to embrace the project, and consequently did not contribute to enriching it with pertinent observations, as was the case with other research activities geared towards composition.

Composers ignored the Acousmographe, or even rejected it, because it went against their musical problematic, focused on sound and not on the eye. They might have found the tool useful if they could have used it to transcribe, quickly and automatically, the beginnings of their works for submitting them to SACEM for copyright. But the Acousmographe was not a robot designed with this in mind, it was a software program to assist in the analysis and transcription of listening. Except for the spectrogram, which was constructed automatically, the transcription of the symbols had to be done by hand, after listening analysis. Some composers, notably Denis Dufour, still found an interest in the tool, for pedagogical use with composition students. But the transcription work was time-consuming and tedious. Nevertheless, development of the Acousmographe led to the production of a CD-ROM featuring ten graphic transcriptions of pieces from a wide range of repertoires, from Gregorian chant to Romani music, as well as electroacoustic, rock, and jazz,[34] and even including an improved version of the first 1975 transcription of Jimi Hendrix's "Star-Spangled Banner."

This CD-ROM led to a serious misunderstanding. The transcriptions presented were quite aesthetic. There were animated sequences, in color, synchronized with the analyzed pieces of music. But contrary to a widespread assumption, they were the result of lengthy analytical and technical work. It was not automatic transcription, as was beginning to appear on the internet (generated from only one or two criteria—most often pitch and intensity). And above all, the authors of the CD-ROM, Olivier Koechlin and Dominique Besson, had not made use exclusively of the Acousmographe (too unwieldy and too buggy) to create their transcriptions. They had used commercial software such as Director, Illustrator, and Photoshop to enhance their work, while the visual transcriptions had also benefited from the talent of a graphic designer, Muriel Bonfils, who unquestionably enriched the productions.

In summary, the miscommunication, or rather the deception, arose from the facts that:

—The Acousmographe did not run automatically. As with the use of any software, it was necessary to make an effort to learn how it worked.
—The Acousmographe alone was not enough to produce a quality transcription; it was often necessary to supplement it with other software.
—Skill in musical analysis alone was not enough to produce an "interesting" transcription. It was necessary to add graphic design skills.

Last but not least, there was a copyright problem. The CD-ROM could not be marketed because of the lack of authorization from some composers. More specifically, the requested royalties were so exorbitant that it was impossible to imagine paying them with the anticipated commercial revenues of an educational CD-ROM distributed free of charge.

Olivier Koechlin left the GRM in 1995, leaving behind him the Acousmographe project, which had indeed been initiated but was still in need of much improvement.

Development of the Acousmographe resumed in 1996 with a second version, entrusted to the engineer Didier Bultiauw working on PC with the Microsoft Visual C++ compiler. Migration to Macintosh Power PC was realized by Mathieu Reinaud in 1999. For this second generation of the Acousmographe, computer systems had clearly progressed in terms of speed, making work easier. But the lack of follow-up in production got the better of the project. Several causes explain the difficulties. First, the engineers recruited as outside contractors for the GRM did not take part in the in-house synergy. They worked at home on fixed-term contracts, contracts that would not be renewed when budget restrictions forced hard decisions to be made. In addition, there was still a lack of enthusiasm among users. François Delalande, although still deeply involved as the person in charge of music analysis research, still preferred paper and pencil for his analyses (even though he had made partial use of the Acousmographe in 1997 to transcribe *Sommeil*, an excerpt from Pierre Henry's *Variations pour une porte et un soupir*). As for other composers, they still didn't see the point, since it was not a tool for composition.

The breakthrough finally came in 2001 with a third version of the Acousmographe. A combination of events, expertise, and resources made it possible to relaunch the project. The prototype Acousmographe was delivered by the engineer Adrien Lefèvre in June 2005 for testing and debugging. The official release in October 2005 was designated Acousmographe 3, version 1.0.

Among all the many factors contributing to relaunching the project and making it a success are:

—The 2001 internet launch of the multimedia publication *Portraits polychromes* brought graphic transcription back into the limelight.
—François Delalande, stimulated by a more dynamic environment, organized a research seminar in 2001–2 on the theme of interactive multimedia analysis.
—Excellent ties with the French National Education Ministry, following the success of the CD-ROM *Les musiques électroacoustiques*, encouraged the emergence of new joint projects, notably supported by financial contributions.[35]
—By completely rewriting the software in C-language, Adrien Lefèvre enabled true multiplatform architecture for both Macintosh and PC.
—A member of the GRM, Yann Geslin, took charge of monitoring the work and leading the user community.
—The circle of people interested in analysis and transcription, which had emerged over the years primarily within the music and musicology

teaching staff of the French National Education system, contributed to improvement in the software's functionalities by contributing constructive critical assessment.

—Another member of the GRM, the engineer Emmanuel Favreau, building on his expertise in the development of GRM Tools, designed all the graphic tools and contributed to defining the ergonomics of the software and its future development possibilities.

Finally, thirty years after its hesitant first steps, the Acousmographe, a tool for listening, found its rightful place next to GRM Tools, a tool for doing.

Relaunching Transcription—between Analysis and Composition

Launching the GRM website in 2001—featuring multimedia content associated with the *Polychrome Portraits* book series—generated a great deal of reflection.[36]

On the surface, the idea of using the internet to provide the widest possible audience access to a range of analyses accompanied by musical excerpts may seem trivial. The idea was to adapt existing analyses and transcriptions belonging to the GRM to new multimedia resources. But the rhythm of publication, with about two monographs every year, each devoted to a composer, quickly revealed that the stockpile of transcriptions and analyses in the GRM collection was limited. Among the four hundred inventoried scores, many were inapt for the Internet, primarily because they were "too composed," not "visual" enough, too succinct, or because of deteriorating media. It became essential to start working on new analytical transcriptions. But how to define the community's needs? Using the computer screen imposed a preference for graphic transcriptions. But in 2001, the preferred tool for transcribers, the Acousmographe, was not yet mature. The only available version, outdated and riddled with bugs, ran only on Macintosh System 9 or Windows 2000 on PC.

The musicologists solicited seemed somewhat intimidated by the request to imagine new modes of representation. They did not always feel capable of authoring analyses of electroacoustic works. Or they wanted to have lots of time, for fear of not fully mastering their subject. As to graphic transcription, they hesitated at venturing into this new field. So it was necessary to seek out new collaborators who were willing to work under tight deadlines, free from preconceptions. A solution was in part found with instructors and students from fine arts schools—écoles des beaux-arts. This put us in contact with visual artists who were at ease with computer tools and who possessed a "musician's ear," an ear that was curious and possibly even educated. Initially our request could be summarized as follows: create a graphic transcription of the work based exclusively on listening, without referring to descriptions of the piece or any other explanatory document.

Naturally, each author of a transcription conveyed his or her own sensibilities and culture. Some of them were familiar with the electroacoustic repertoire, having already explored it in their fine arts schools. A wide variety of approaches to analysis and transcription resulted. But there was still one significant stumbling block. The search for a precise graphic vocabulary for the description of sound characteristics could not be sacrificed solely for visual aesthetics.

Today the response is more nuanced. Aesthetic beauty is considered as a plus in the transcription, sometimes bringing it to the status of a work of art in itself (graphic or multimedia), without detracting from the spirit of research. "A piece of art about a piece of art," declared François Bayle in 2007 when viewing student Noémie Sprenger-Ohana's transcription of his *L'Oiseau moqueur*.

Graphic Transcription as a New Space for Creation

The 1995 CD-ROM *Les musicographies* opened the way to multimedia artistic transcriptions.[37] At that time, it had been observed that analysis of a work and its subsequent graphic transcription required as much effort as the composition of the music itself. It became common practice in 2001 at the GRM to commission transcriptions from artists for *Portraits polychromes*, just as one commissions new musical pieces, asking the transcribers to sign their work as in the tradition of classical musical transcriptions, with the exception that, in the absence of an established code, the graphic or multimedia transcription inevitably assumed its own original character. Crediting the individual analysis also reflected the observation that there are multiple possible ways of listening to a single piece of music, each unique, and each worthy of consideration.

In passing, I will evoke a contrasting approach conducted around the same time, Nicolas Donin's *écoutes signees* (literally "signed listening") at IRCAM, in which composers were asked to undertake an "inductive *esthesics*" analysis of their own music.[38] The composers were responsible for both the music and the analysis of their own music in a literary format accompanied by statistical documents and computer readouts.

At the GRM, the search for a functional code for graphic transcription engendered a new dynamic for creation with multimedia tools. Notation tools became creative tools, in a dynamic identical to that of the revolution in musical notation at the beginning of the twentieth century as described by François Delalande. Each technological breakthrough gives rise to new means that become the source for new paths for modelization and conceptualization.

Identifying Relevant Criteria for Analysis

What criteria are pertinent for analytical transcription? In 2007 it was realized that research in musical analysis carried out by the members of the GRM (essentially François Delalande, Jean-Christophe Thomas, and François Bayle) had almost never been made available through teaching, but only during lectures and seminars between specialists. It was notably absent from conservatories. The result was considerable confusion about how to approach the analysis of electroacoustic works. Most often musicologists attempted to adapt their knowledge from other repertoires, focusing on structure, or identifying the pitch of notes, or evidence of an argument or a theme. The gaps in their analysis essentially ignored the concepts of morphology, dynamics, sound color, and spatialization, the very elements that are fundamental to understanding electroacoustic music.

From a *poietic* point of view, as well as from an *esthesic* point of view, the best analysts were recruited among those who practiced electroacoustic composition. The analysis of electroacoustic works requires a type of listening and know-how that can only be acquired through prolonged experience in the studio. As for the thematic (or rhetorical) analysis as described by Jean-Christophe Thomas, it is apparently more accessible, since it indifferently combines approaches of all kinds. Its quality stems from the talent of its author as a writer.[39]

Little by little, with the proliferation of attempts to analyze and transcribe electroacoustic works, codes are being developed. Viewing the transcriptions presented on the GRM's *Portraits polychromes* site makes it possible to measure the variety of approaches and emerging trends. Each analyst/transcriber has been encouraged to develop his/her own code for describing the relevant sound criteria they have identified and to establish a personal library of graphic symbols. They can build up multiple libraries adapted to different sets of criteria. But once their work has been completed, these libraries are made available to the community, which can in turn draw on them to carry out other analyses. Gradually a corpus of "core" graphic symbols has emerged, along with methods of analysis based on identifying relevant criteria to monitor while listening.

Selecting Tools

The Bathygraphe (1948), the Sonagraphe (1975), and then the Acousmographe (1988) have all been used in turn at the GRM to automate the recording of graphic transcription.

The Acousmographe facilitates graphic transcription, based on the music's spectrogram (the term *sonogram* is also sometimes used).[40] Certain skilled spectrogram readers are able to extract considerable information from them. But one must learn to read. Analysts also use other software in complement, including Director, Graphic Converter, Photoshop, and Flash, not to mention pencil and paper.

Transcribing Dynamics, Movement, and Space

An analogy may help illustrate. Take the film techniques of freeze-framing and slow motion, which respectively let us freeze a movement, or decompose movement. Graphic transcription of music involves analogous techniques. One can freeze the forms of sounds, and also decompose their movement. In music, the sculptural dimension of space, which has been thoroughly conquered on audio media since the early 1950s, has found its visual translation, represented in graphic transcriptions. Jean-Christophe Thomas, referring to an "assault on the senses" to evoke the passage from abstract sound to the visual, asked, "Is it because our sounds today make us lose all sense of scale?"[41]

In the *Portraits polychromes*, several transcriptions aimed at conveying dynamic changes, either by means of animated drawings or color variations.[42] Other transcriptions attempted to portray the movement of sounds as they shifted from one loudspeaker to another. For example, in his graphic transcription of *La grotte*, an extract from Luc Ferrari's *Hétérozygote*, Philippe Mion's graphics run from left to right on the screen, paired with a temporal pointer; a horizontal line splits the screen in half, with the music's left space displayed at the top of the screen and the right space at the bottom. In the CD-ROM *Les musicographies*, Dominique Besson had conceived the idea of reading vertically, as in Chinese or Japanese writing. This made it possible to follow the stereo trajectories of the sounds on the left and right sides of the screen, in addition to following their temporal progression. This mode of representation was also used for the transcription of François Bayle's *Toupie dans le ciel*. Today some transcribers use the software Flash, which at a minimum permits reproducing movement in two dimensions.

But it should be mentioned that the GRM experienced a long period of latency in its theoretical research on the subject of transcription, particularly on the transcription of space. It was only following the significant period dedicated to Bayle's "Acous"—Acousmatics, Acousmonium, Acousmathèque—which occupied the GRM from 1975 to 1995, that the debate of ideas was revived in 2001, in the context of drafting specifications for developing forms of representation of musical space for the Acousmographe version 3 software.[43] The discussion was about whether it would be possible to settle for representing monophony alone, considering that the recent commissions to musicologists for graphic transcription had brought to the forefront the need to represent multidimensional space and its movement, regarded by the composers themselves as essential criteria in a majority of their works. The final conclusion was that the Acousmographe should at least offer stereophonic transcription. But the engineer in charge of development, Adrien Lefèvre, succeeded in programming for the possibility of using an unlimited number of tracks, even if, in his view, graphic transcription would become problematic beyond eight tracks, as well as requiring more powerful computer resources.

Then the question arises as to what space to represent—static space (freeze-frame) or moving, animated space? If the space is stationary, from what angle should it be presented? If it is in motion, should it be shown as a film? Or should it be multiple, allowing for interactive navigation? And how should it be dealt with when the space has been defined in 5+1, 7+1, or in eight channels?

Michel Chion, in a 1994 article, justified the value of transcribing the "spatial projection of sound works (projection in the cartographic sense) globally and synoptically, ... by virtue of the fact that it presents a total contradiction with our linear perception of time. It gives us a head start on what we hear as we go along."[44]

In 2007, graphic transcription of space in music and of movements in space is still in its infancy, certainly in the same measure as progress in their conceptual analysis.

In conclusion, I would like to again quote Henri Bergson from his book *La pensée et le mouvant*:

> But we have so much difficulty in distinguishing between succession in true duration and juxtaposition in spatial time ... This distinction cannot be clarified in too many directions at once. Let us say then, that in duration, considered as a creative evolution, there is perpetual creation of possibility and not only of reality.... so that before being real it must have been possible.[45]

IMAGED-THOUGHT

Hearing Is Seeing from Within

Using listening as the basis for transcribing puts us in contact with the subjective action of audition, in the sense described by Schaeffer in the preface to *Traité des objets musicaux*, "Hearing is seeing from within."[46] Graphic transcription of music reflects this view from within, leading us to the boundary between hearing and seeing. It should be noted that we are dealing here with sight and not with the gaze, which is situated on the side of Schaeffer's *simulacrum*.

But the debate is not new; "confusion between hearing and sight has its roots in the Bible," as Olivier Cullin reminds us. In his book *L'image musique*, he cites an excerpt from Deuteronomy 4:9, where God gives a warning to the Hebrews, frightened by the power of Yahweh's voice on Mount Sinai: "Do not forget the words your eyes have seen in any day of your life or let them fade from your heart. Teach them rather to your children and to their children after them."[47]

More recently in 1936, but still well before the current work on the graphic transcription of sound, Rudolf Arnheim commented on the subject of the art of radio: "Further research is needed to determine the extent to which the average listener develops complementary visual representations."

A little further on, Arnheim evokes "the urge of the listener to imagine with the inner eye,"[48] and in a more recent work from 1969 he borrowed Robert H. Holt's definition of the "thought image," which "includes both memory images and those of the imagination" and "may be visual, auditory, or proper to any other sensory modality, as well as purely verbal."[49]

The Visual Image as a Place of Convergence

Concordance with the visual arts has always been strongly felt by the actors of the musical process of using sounds fixed on media. *Doing*, in the dialectic of doing and listening, would consist in building mental images that could find their resolution indifferently in narrative, drawing, sculpture, choreography, film, or music. *Listening*, in this same dialectic, would consist in tuning one's perceptual system to the work one is listening to, in order to constitute imagery. The concept of imagery specific to perceptual functioning emerges as a place of convergence, situated between reality and representation, a bit like the image on film. Numerous statements by composers corroborate this hypothesis. For Francis Dhomont: "Musique concrète in particular, this sound art often compared to cinema, leads to discovering acoustic territories left fallow by instrumental composers. By never ceasing to oscillate between truth and mirage, it feeds on the strength and ambiguity of images."[50]

François Bayle, in working out the concept of the i-son (sound image), described this phenomenon extensively: "Image is on one hand opposed to reality, to concrete things, and on the other hand to the concept, to the abstract idea. Does it create a link between the two? I tend to think so."[51]

These pictures of thought are also linked to our memory. They reveal the multiple pathways and inner workings of our being. All research into cognition thus falls within the scope of our curiosity. For example, Claude Bailblé's thesis on perception and attention in cinema finds its place in this problematic.[52] In one passage of his work concerning "movements in auditory attention" while watching a film, Bailblé wrote:

> The sound score first simulates an *attentional field* . . . It also triggers *motor imagery*, to the extent that the auditory event directly reflects the action of the characters. The listener reconstitutes the (motor) cause from the (acoustic) effect. Sound is not possible without bringing energy into play—an event is linked to movement, to movement in space, to the mobilization of energy in time. The soundtrack also operates upon *mental imagery*: sound adds force and energy to the *"image in"*, in a way that silent film does not.[53]

Claude Bailblé reminds us of the importance of "convergence cells," the cortical and subcortical cells that respond to stimulation by two or more modalities (touch, sight, hearing, smell).[54] "Communication between modalities is already found at the most elementary level of coding sensory data," explains Bailblé, referring to the work of Yvette Hatwell.[55]

Another example of the fields of investigation being explored at the GRM are the theoretical advances made by the philosopher Bernard Stiegler. His concept of tertiary retention, which he adds to the primary and secondary retentions defined by Husserl, updates phenomenological thinking, which had inspired the first research in musique concrète. Primary retention (or memory) is at the level of perception, while secondary retention is at the level of imagination. In Stiegler's view, the emergence of recording techniques has made it possible to bring to light the concept of tertiary retention, or trace, which becomes a *space* because it is fixed on a support. This spatial form of memory leads us back to the concept of the image.[56]

Scores, Transcriptions, and Visual Arts

Numerous transcriptions and musical scores by contemporary artists and composers give great importance to the aesthetics of the graphic design. They are beautiful to look at and have clearly been executed, at least in part, with this in mind. According to Jean-Yves Bosseur, author of a book on the relationship between music and visual arts, the graphic scores of twentieth-century creators, with their nonlinear approach, are meeting grounds between music writing and visual writing.[57] He gives many examples of visual artists who have approached music by way of graphics—Yves Klein, Jean Dubuffet, and Tom Phillips—and conversely numerous composers who have investigated their relationship with the visual arts: John Cage, Earle Brown, Cornelius Cardew, and once again Tom Phillips, just as much a visual artist as a musician. Closer to us we find Iannis Xenakis and André Boucourechliev.[58] Jean-Yves Bosseur also explained that in the 1960s, artistic languages became so destructured that painters sometimes went so far as to write music on tape recorders, with Jean Dubuffet composing *art brut* music, without being aware of similar works by musicians.[59] All these interactions can be explained by the intrusion of the concept of time into the visual arts and, reciprocally, of the concept of space into music. Different artistic spheres encounter one another, overlapping. We are at the frontier between the arts.

Electroacoustic music, the art of support par excellence, is directly affected by this problematic, being situated directly in a privileged relationship with space and the associated arts—sculpture, choreography, theater or cinema staging, 3D animation, and architecture. This topic of reflection was the subject of a seminar at the GRM in 2000 centered on Cécile Regnault-Bousquet's research into visual representations of sound in the context of urban planning.[60]

Music from Both Ends

By examining the concept of *pensée imagée*—imaged-thought—in the light of different musical approaches, I intend to clarify the relationship between it and writing, composition, interpretation, and musical listening. This will shed light on the role of graphic transcription in music, especially in electroacoustic music. Perhaps it will also contribute to a more adequate definition of semiologists' *neutral level*.

Rather than the Holt/Arnheim term *image-thought*, I prefer employing the term *imaged-thought*, which seems to me to avoid confusion with the *material image*, represented for example by a photograph.

MUSIC WITH A CODE FOR WRITING MUSIC

If we take the example of Western classical music, it is common to assume that writing the music comes before the music itself. But this is the case only for those who have learned the code. We find an example of this in the case of the competent sight reader who manages to picture the musical piece just by looking at the score. Unfortunately, a linear representation of the musical process leads us to believe that these *poietic* and *esthesic* functions are separate and follow one another:

Composer → Score → Performer → Musical Piece → Listener

Stéphane Roy has attempted to challenge the static and purely descriptive aspect of the "immanent analysis of the work" as described by Nattiez with his Neutral Level.[61] Roy explored the dynamic aspect of the relationships between the different spheres of the musical process, production, reception, and the piece itself (at the moment it is played). It is not the stages in the musical process—score, piece—that interested Stéphane Roy, but the processes that led to them, in brief, the arrows in the little diagram above. The concept of induction, suggested by Nattiez, becomes the center of interest for Roy in the terms *inductive poietics* and *inductive esthesics*.

If we apply Stéphane Roy's approach, it makes sense to locate the activity of imaged-thought at the positions of the arrows in the diagram above. Imaged-thought would appear as a dynamic phenomenon of perception occurring among the various participants in the chain of musical communication. Images are created just as surely in the context of reception as in the context of production.

MUSICS FIXED ON SUPPORT

Analysis of music fixed on media has made it possible to go beyond the static aspect identified by Stéphane Roy in the analysis of musical phenomenon. In the absence of a score, it was necessary to probe the perceptual phenomenon involved in the experience of music, drawing on other factors. Graphic transcription of listening is one of these factors.

In his 1966 *Traité des objets musicaux*, in the chapter titled "Sound Recordist as Interpreter," Pierre Schaeffer was already examining sound recordists' dual expertise. "As long as we think only in terms of reproduction and transmission, it seems [the sound recordist] needs only to be a more or less competent technician." But Schaeffer noted that in the professional world, the distinction between "good" and "mediocre" technicians was not based solely on technical criteria, but also on "talent." Schaeffer came to the conclusion that "we must allow that the sound engineer—or the chief sound operator—must ask himself questions that are no longer purely technical but are ultimately answerable to sensitive listening and musical judgment."[62]

This sensitivity would therefore consist in the ability to use thought images, resulting from reduced listening, to serve symbolic representation, or even dreamlike fantasy. Listening "musically" to a sound without referring to its causal source is facilitated by sound recording, which establishes a physical distinction between the tangible sound source and its image, captured and memorized on media.

JAZZ AND OTHER IMPROVISED MUSIC

Jazz musicians, who use improvisation, have long been aware of this phenomenon of mental images superimposed upon compositional processes. Jacques Siron spoke of the "inner score." In his book *La Partition intérieure*, he provided a description of the phenomenon:

> How do you form your own inner score? The pathways are numerous and vary greatly from one person to another. The inner score is shaped from a blend of desire and knowledge about music, from theoretical points of reference and sound images, rhythmic precision and pleasure, solitude and sharing, finger nimbleness and deep emotion, and from rigorous, lengthy, and gradual learning and instantaneous overarching insight. Developing one's own inner score is deeply linked to personal intentions. What resonates with me? How do I relate to music and improvisation? What do I want to play?[63]

COLLECTIVE MUSICAL PRACTICES

Broadly speaking, many musicologists see the score as representing the neutral level and argue that analyzing this neutral level would make it possible to understand the music. One could conceivably admit this shortcut in the case of notated written music, due to the fact that the roles of the different actors in the process of musical communication are quite distinct. But in the case of traditional music, where everyone takes part in the musical interplay, situating a neutral level becomes more complicated. Let us imagine analytical musicologists (i.e., listening auditors) blending into the throng of singers of an African tribe. Either they fulfill their role as analysts—using their educated and cultivated memory to carry out their observations (also possibly working from recordings)—or they simply forget about their expertise. In the latter case, they would take part in the festivities to the fullest extent, just like everyone else.

Contribution to a Definition of the Neutral Level

The various preceding examples lead to identifying two forms of neutral level:

—An external material form: score, tape, or CD, exclusively possible in cases of musical practices where there is the presence of a listener who is clearly distinguishable from the performer and/or the composer.
—A virtual form, in the listening auditor's mind, in the context of listening to music from oral traditions.

These two forms of neutral level are of interest because they make it possible to reintroduce the concepts of image and imaged-thought in the context of diverse musical genres including electroacoustic music. In this way, the neutral level can be attained in two ways in musical practices where the composer and the listener are distinct: either by proceeding from the *poietic* (doing), or by proceeding from the *esthesic* (listening):

—In written music, this happens when the work, present only in its latent state in the score, is "put" into existence by the performer.
—In music on media, the neutral level is attained by diffusion of the work in physical space (with the help of an auxiliary electric energy source), and sometimes with the inclusion of interpretation of the sound projection.

In musical practices where there is no difference between composer and listener (collective practices, or even simply solitary practices—when a person plays "for their own enjoyment" or when the composer is composing—the neutral level is also accessed. The neutral level finds itself here at the point of convergence between doing and listening, between *poietics* and *esthesics*. But these two pathways are so similar, or intertwined, that we tend to confound them.

Stéphane Roy's approach here becomes even more rewarding. It leads us to examine the *dynamic* aspect of musical practice. The neutral level emerges as a point or zone of changeover. Either one moves away from it to go towards the *poietic* or, on the contrary, towards the *esthesic*. Or one attains it by proceeding from the *poietic* or from the *esthesic*. Stéphane Roy introduces a precise definition describing this instant: "Analyzing the neutral level is a point of intersection harboring the relevant information to reveal how production and reception come together, oppose each other, or "diffract" like light passing through a prism."[64]

This also brings us closer to the neutral level as defined by Gilles Deleuze in his concept of "center of indetermination," where in his view, in his reflections on cinema, "movement images" are related to a center of indetermination from which they can be divided into three possible categories—perception images, action images, and affect images.[65]

Both Ends of Writing Music

Where does electroacoustic music writing fit into in this *imaged-thought* approach? Let us examine one by one the cases of each electroacoustic music actor.

What do electroacoustic music composers do?

They do not write their music on a staff, but they "make" it directly on media (doing). Their procedures for writing music (acquired through oral transmission) consist of both technical and conceptual savoir faire, enabling them to harness the effects they select and produce while using critical listening to guide them. Critical listening is fostered by attending concerts, as well as by practicing analysis and transcription.

It is on the side of *listening* that the concept of codified music writing finds itself when using graphic transcription. But how can we notate sound flow, spatial interplay, dynamics, or sounds with complex morphologies? There is no established code. Electroacoustic notation has to be constructed based on finding a suitable balance between the analyzed sound and visual representation. It is an act of creation involving the visual design of perceived sound images. Graphic transcription of a work is then like a rough draft for a system of codification. Transcription can be executed by composers themselves, who are always their own first listeners. Or it can also be done by a listener assuming the role of a musical analyst.

> *What do listeners do when they have no intention of transcribing or commenting on their listening?*

They are simply looking for congruity between what they hear and their own individual sensibilities in order to create images for themselves. The code lies at the level of these images. It is their own "personal" code.

> *What do performers do—whether they are composers themselves or not—when they play music in concert?*

Performers must learn the piece in order to anticipate the movements that will allow them to express the images suggested by the music. These movements consist of the gestures that effect instrumental playing on the broadcasting console. In written music, the contribution of interpretation lies in the distance between the precision of the notation code and the sound effect expected by the composer—or the sound effect heard by other listeners when they are able to follow the score. In electroacoustic music, listeners do not have a score, but they can judge the quality of the sound projection by the quantity of thought images they perceive, and consequently by the pleasure they derive from it.

But in every instance, the kinesthetic actions associated with sound projection in physical space, both dynamic and movement, are necessary to "give life" to the music. This explains the importance of research into the transcription of space, where one of the avenues of exploration is drawing a parallel between the initial sensory-motor gestures acquired in early childhood and the spatial sound effects created by composers. Wouldn't spatial interplay be a projection of bringing together these primitive gestures?

CONCLUSION

The ultimate objective is to understand the *fait musical*—the act of music—as a whole, viewing it as an activity of thought taking part in the process of communicating. One observes that all the systems of analysis advocated by the different authors are three-dimensional, and all suggest a dynamic approach.

One of the initial questions was whether it would be possible to codify thought images that are formed in the minds of composers and listeners. The typo-morphology formulated by Schaeffer, the Temporal Semiotic Units of the MIM, Lasse Thoresen's symbols, and Delalande's research into early sensory-motor acquisitions as archetypes of sound morphologies are the first answers brought to the search for a form of notation in electroacoustics. Establishing a fixed form of notation seems, if not impossible, at least of no great interest. On the other hand, the search for a form of notation can serve pedagogically as a means for mastering the forces of creation. I share Jean-Yves Bosseur's opinion expressed in his book on the history of musical notation:

> Can today's sound vocabulary, which has become limitless, be satisfied by a music writing system shared by all, applying standardized music theory? Given the variety of implications of the sound phenomenon that we are capable of actively experiencing—if we resolve to steer clear of restrictive dogmatism, schoolbook prohibitions or prejudices—can we respond in any other way than with a plurality of modes of transmission for these musical undertakings that are known for their polyvalence? Notation is no longer necessarily just about indicating a pitch or a rhythm—notation is also about inventing a new form of writing.[66]

But to go from imaged-thought, produced in the brain, to its graphic transcription, it is necessary to make use of one's hands. Along with the auditory and visual senses, a third also comes into play, touch. The convergence between the audible, the visual, and touch passes through kinesthesia and finds its fulfilment in the field of the mental image. In fact, in the liner notes of their 1995 CD-ROM *Les musicographies*, Dominique Besson and Stéphan Dunkelmann opted for the subtitle "hearing images, seeing sound, touching music." They had already reconstituted the dynamic triangle, sight-sound-touch.

One of our hypotheses would be that sound spatialization, carried out by composers during the sound projection of their work in concert, is an inverted and enlarged image of the imaged-thought developed in their own minds in the studio at the moment of composing the music. It is:

—An inverted image, because in the studio, it is *doing* that stimulates *listening*, while in concert, it is *listening* that stimulates *doing*.

—An enlarged image, because the spatial projection of the music amplifies the psychic phenomena that triggered the composition.

There is still much to be done in terms of research into graphic transcription of sound, in particular in terms of transcribing spatial criteria and movement.

PART TWO

Memorializing the Facts

A Chronological Approach

THIS SECTION PROVIDES A HISTORICAL overview of musique concrète, focusing on references and dates, offering a chronological account of the GRM's history, decade by decade and often year by year. This encompasses easily identifiable elements such as works, publications, and technological advancements, as well as more nuanced aspects including the formulation of new concepts, research efforts, and the evolution of the institution.

The narrative begins with the official inception of musique concrète on the radio in the late 1940s and concludes with developments in the early 21st century.

6

1948–1958

The Avant-Garde of Musique Concrète

It is the artist who is truly rich. He rides in a motorcar. The public follows in a bus. How can we be surprised if they follow at a distance?
—JEAN COCTEAU

It was 1948. Pierre Schaeffer had just "invented" musique concrète. This musical avant-garde first appeared on the radio. If we examine the works and their premieres, what do we see? First of all, historical but low-profile premieres—*Études de bruits*, the first musique concrète concert. Very rapidly there was widespread curiosity on the part of the foremost composers of the moment: Olivier Messiaen, Pierre Boulez, Michel Philippot, Jean Barraqué, André Hodeir, and Karlheinz Stockhausen. Then suddenly a succession of scandals arose: *Orphée 53* and *Déserts* in 1954, and even derogatory remarks such as Pierre Boulez's commentary in the *Fasquelle* encyclopedia in 1959.

What explains this phenomenon of attraction/repulsion? Luc Ferrari vividly recalled the atmosphere of the postwar musical milieu: "You were either serial or you were reactionary!"[1] This position, more political than aesthetic, came straight from Germany, where the trauma of Nazism had contributed to consecrating the serialism of "the three Viennese"—Schoenberg, Berg, and Webern—as the only possible path to progress.

In a 2001 article, the musicologist Danièle Cohen-Levinas described the link between serial aesthetics and the psychological situation of the listener at the time: "Listeners' memories were suddenly suspended in a here and now of serial combination, atomization of sound, and its disarticulation in time and space. An acoustic urgency beating at the door, searing into our consciousness by freeing our references from a politically determined past/future axis."[2]

Under these conditions, didn't the new "concrete" avant-garde simply come into competition with an already established avant-garde, the serialists? Beginning in the mid-1950s, electroacoustic music pursued a vein radically different from serialism, dodecaphonism, and other forms of atonality. But in its early days, the situation was not exactly the same. Numerous bridges existed and composers cheerfully crossed them, with the idea of visiting new musical territories. Practically all the serial composers of the 1950s spent time at the GRM at one point or another to dabble in musique concrète. One of the points of convergence between serial music and musique concrète lay in the desire shared by all these composers to achieve abstraction.

Iannis Xenakis, in a brief article published in the special issue 244 of *La Revue musicale* in 1958, reminded us that the return to abstraction observed at the beginning of the twentieth century in the different arts, painting and music (the beginnings of serialism), was contemporaneous with the birth of modern algebra around 1910.

As for the public, it felt a bit baffled confronted with all these innovations, both serial and concrete. As Cocteau had so well put it, "the public follows in a bus."

Historic Milestones: A Succession of Premieres

On October 5, 1948, in its "absolute public premiere," Pierre Schaeffer played *Concert de bruits* on the radio waves of the station Paris-Inter at 9 p.m., during a broadcast of Club d'Essai.

On the program:

Étude no. 1 *Déconcertante*, or *Étude aux tourniquets*, 3:00;

Étude no. 2 *Imposée*, or *Étude aux chemins de fer*, 3:25;

Étude no. 3 *Concertante*, or *Étude pour orchestre*, 6:00;

Étude no. 4 *Composée*, or *Étude au piano*, 3:30 (later redone as *Étude noire*, 4:00);

Étude no. 5 *Pathétique*, or *Étude aux casseroles*, 4:10.

In his introductory remarks, Pierre Schaeffer used the words *musique concrète* for the first time: "If the term did not risk appearing pretentious, we would entitle our pieces *attempts at Musique Concrète*, to emphasize their general character and that we are not strictly concerned with noises but rather with a method of musical composition."

These five études had already been presented in private two days before over loudspeakers, using turntables, on October 3, 1948, at the François Devèze studio in Paris. The broadcast recording had been completed at the Club d'Essai on June 20, 1948.

In his journal, partially published in 1952, Pierre Schaeffer related how he composed these different études in the studio, on "acetate discs."[3] He employed recordings created specifically for the occasion—passages at the piano, railway sounds, diverse noises produced at the microphone. He also "recycled" previously recorded excerpts, such as a fragment of Sacha Guitry's voice saying "sur tes lèvres" ("on your lips") from a radio drama—this recording, interrupted by the coughing of the script supervisor, and consequently deemed a "dud," had been discarded in the studio. Schaeffer mixed these elements, superimposing them by repeated recopying while also using all the known techniques of electroacoustic composition—fragmentation, repetition, transposition, looping, and sound played back in reverse.

In the radio audience of the October 5 concert was Pierre Henry, a young orchestral percussionist who was also a student of Messiaen and Félix Passerone. He was particularly impressed by *Étude Pathétique*. What happened next is well known: In October 1949, Schaeffer was looking for someone who could play drum rolls. Messiaen suggested Pierre Henry. And so in 1950 Henry became Schaeffer's principal collaborator. They composed their first piece in common, *Bidule en ut*.

THE FIRST MUSICAL WORKS

1948. *Concertino Diapason* (9:30) by Pierre Schaeffer. The original version of this work, a sort of mixed piece ahead of its time, consisted of an improvisation at the piano by Jean-Jacques Grünenwald, and a "concrete orchestra." The pianist had to respond to "sequences themselves extracted from orchestral material." With these initial experiments, Schaeffer already observed that "concrete manipulations create forms which contrast with normal musical style."[4]

This piece was removed from the *Études de bruits* to follow its own path (initially it was Étude no. 3, *Concertante*), and was subsequently reworked several times. The final revision is dated 1970. The new version on 2-inch 38cm/s tape is now only 4:15, in four movements: Allegro, Andante and Intermezzo, Andantino, Final.

1949. *Variations sur une flûte mexicaine* (7:20) by Pierre Schaeffer. The 1971 revision reduced the duration of the piece to 2:56.

1949. *Suite for 14 instruments* (25:20) by Pierre Schaeffer. This work premiered on the radio between November 6 and December 3, with a single movement presented each broadcast: Prélude, Courante/Roucante, Rigodon, Vagotte/Gavotte, Sphoradie. A version reduced to 9:10 was created in 1970 by François Bayle and Pierre Schaeffer.

1950. *Bidule en ut* (1:50) by Pierre Schaeffer and Pierre Henry. This first piece in common between the two composers was created using prepared piano sounds. The original work was produced by direct engraving on a 78 rpm acetate disc. It is worth mentioning that the French were not aware of John Cage's use of the prepared piano, which dated back to 1938, and even before him, to Henry Cowell in the early 1930s.

1950. *Symphonie pour un homme seul* (first version, 46:00) by Pierre Schaeffer and Pierre Henry. The original was produced by direct engraving on an acetate disc, consisting of twenty-two sections. The best-known version, in twelve movements, was revised by Pierre Henry for the 33 rpm record released in 1966, lasting 21:30. In 2010, Pierre Henry composed a special-edition *Symphonie collector* based on the basic sound elements from the original symphony, as a tribute for the one hundredth anniversary of Schaeffer's birth. This new work of 38:49 premiered in Paris on January 9, 2010. Pierre Henry had just celebrated his eighty-second birthday.

The composer Jean-Etienne Marie related in his book *Musique vivante* the debates that had followed hearing the first version of *Symphonie pour un homme seul* by Pierre Schaeffer and Pierre Henry, in 1950 at the Club d'Essai.[5] Someone had declared at the end, "Finally, here is a piece of surrealist music!" The discussion did not settle the question, but the strong analogies between musique concrète and surrealism had been clearly identified—in particular the use of the technique of collage, as well as the quest to notate dreams. *Symphonie pour un homme seul* was immediately recognized as a musical achievement, and still contributes to international recognition of the genre. The 1955 ballet version, staged by Maurice Béjart, established the piece's reputation as a masterpiece.

FIRST MUSIQUE CONCRÈTE CONCERT

On March 18, 1950, the first concert of musique concrète took place at the École Normale de Musique in Paris.[6] It was presented by the Tryptique, a philosophical organization based at the Sorbonne that organized chamber music concerts.[7]

In his presentation of the concert, Serge Moreux provided the audience with a definition that would prove to be seminal: "Musique concrète is the use of sound in its native state, as provided by nature, recorded by machines, and transformed by machine manipulations."

On the program:

Three of the *Études de bruits*, by Pierre Schaeffer:
  Étude aux chemins de fer (2:50);
  Étude violette (3:18);
  Étude pathétique (4:10);

Two movements from *Suite pour 14 instruments*, by Pierre Schaeffer, a work from 1949:
  mvt. 2 *Courante/Roucante* (6:40);
  mvt. 3 *Rigodon* (5:31, three loops of different lengths played simultaneously);

Symphonie pour un homme seul (21:30), by Pierre Schaeffer and Pierre Henry;

Bidule en ut by Pierre Henry and Pierre Schaeffer, which had its premiere two days before this first concert, during a conference at the Sorbonne, in the Richelieu amphitheater.

SOUND RELIEF: THE FIRST STEREOPHONIC RADIO BROADCAST

On June 18, 1950, the Club d'Essai attempted the first experiment in stereophonic broadcasting on Radio Paris IV, under the direction of the filmmaker René Clair. The promoters of this experiment, Jean-Wilfrid Garrett and José Bernhart, authored a radio drama together with René Clair, *Une larme du diable* (A tear from the devil), which was awarded the Prix Italia the following year in 1951.[8]

In order to perceive the stereo effect, listeners were asked to gather together two radios in the same room, tuning the first to the channel France 1, where one of the two tracks of the program was broadcast, and the other to France 2, where the second track could be received. Then the listener needed to adjust the levels and tone settings of the two radio receivers to be able to experience the spatial effects—essentially shifting between the left and right receivers. These effects were quite novel for the time.[9] Unfortunately, these experiments encountered their limitations in differences in quality of the transmitters, which did not allow for precise adjustments.

EARLY DIFFERENTIATION OF WORKS CREATED FOR RADIO

The early 1950s saw the beginning of distinguishing between musique concrète intended for concert and that intended for radio. But work for radio continued to flourish and innovate. For example, Bernard Blin's radio works at the Club d'Essai in 1951, based on a technique reminiscent of the "waking dream" cherished by the surrealists: *Terre de poussière et de nuit* (Land of dust and night, on life in the mines) and *Ciel d'orage et d'azur* (Stormy and azure sky, on impressions of a plane trip).[10] When recording his commentary, the author withdrew into a studio without any light and, focusing on one specific memory, related it spontaneously into the microphone, in all simplicity, allowing images to flow from one to another, recreating the once-experienced emotion. These reveries were enhanced by astute editing and sequences of musique concrète.

In parallel with the radio works, the Club d'Essai programs broadcast musical pieces composed in its studios. *Le microphone bien tempéré* was the title of a series of ten programs broadcast on Paris IV in 1951. These broadcasts, presented by Monique Rollin and Pierre Schaeffer, with Jean Wiener at the piano, featured early works. Later Pierre Henry would again use the title *Microphone bien tempéré* (which he himself had devised) to group together his own early works.

1951. *Le microphone bien tempéré* (first version, twelve sections, 33:40) by Pierre Henry. The best-known version is from 1978, with sixteen sections:

1. *Batterie fugace*, 2:12;
2. *Bidule en ut*, 2:00 (with Pierre Schaeffer);

3. *Fantasia*, 4:39;
4. *Dimanche noir I*, 2:35;
5. *Tam Tam II*, 2:25;
6. *Tam Tam concret*, 2:34;
7. *Tam Tam III*, 3:30;
8. *Micro rouge I*, 2:30;
9. *Tabou clairon*, 1:25;
10. *Tam Tam IV*, 3:30;
11. *Vocalises*, 2:41 (serial study from 1952);
12. *Sonatine*, 4:20;
13. *Antiphonie*, 3:00 (serial study from 1951);
14. *Micro rouge II*, 3:15;
15. *Tam Tam I*, 3:50;
16. *Dimanche noir II*, 2:43.

FIRST SPATIAL RELIEF PROJECTION IN CONCERT

On July 6, 1951, the first concert of sound projection in spatial relief by potentiometric control panel took place at the Théâtre de l'Empire in Paris. Pierre Henry "staged in relief" the works that he had composed with Pierre Schaeffer, *Orphée 51* and *La Symphonie pour un homme seul*.[11]

Between *Symphonie pour un homme seul* and *Toute la Lyre*, little time passed, barely a year. Yet a technological revolution separated these two works. While the source sequences for *Symphonie* were produced by the closed groove technique, *Toute la Lyre* benefited for the first time from transformations on phonogènes, using tape.[12]

Orphée 51 ou Toute la lyre (first version, 45:00), by Pierre Schaeffer and
  Pierre Henry, is a lyrical pantomime. A mixed work for voice, instruments, and tape. With Maria Ferès, contralto; Denise Benoît, mezzo; Geneviève Touraine, soprano; Jean-Christophe Benoît, tenor; Habib Benglia, narrator and mime; Maurice Le Roux, preacher. This piece constitutes the kernel of the concrete opera *Orphée 53*. The theme of Orpheus was dear to Pierre Schaeffer; he evoked it frequently in his writings and conferences.[13]

FIRST MUSIQUE CONCRÈTE FOR COMMERCIAL CINEMA

1952. *Astrologie*, by filmmaker Jean Grémillon, music by Pierre Henry.[14] In this short commercial film, Grémillon used four musical sequences composed by Pierre Henry: *Les étoiles*, *Les cataclysmes*, *Les machines*, *La guerre*. Later, this music would serve as the theme for Maurice Béjart's ballet *Arcane*.

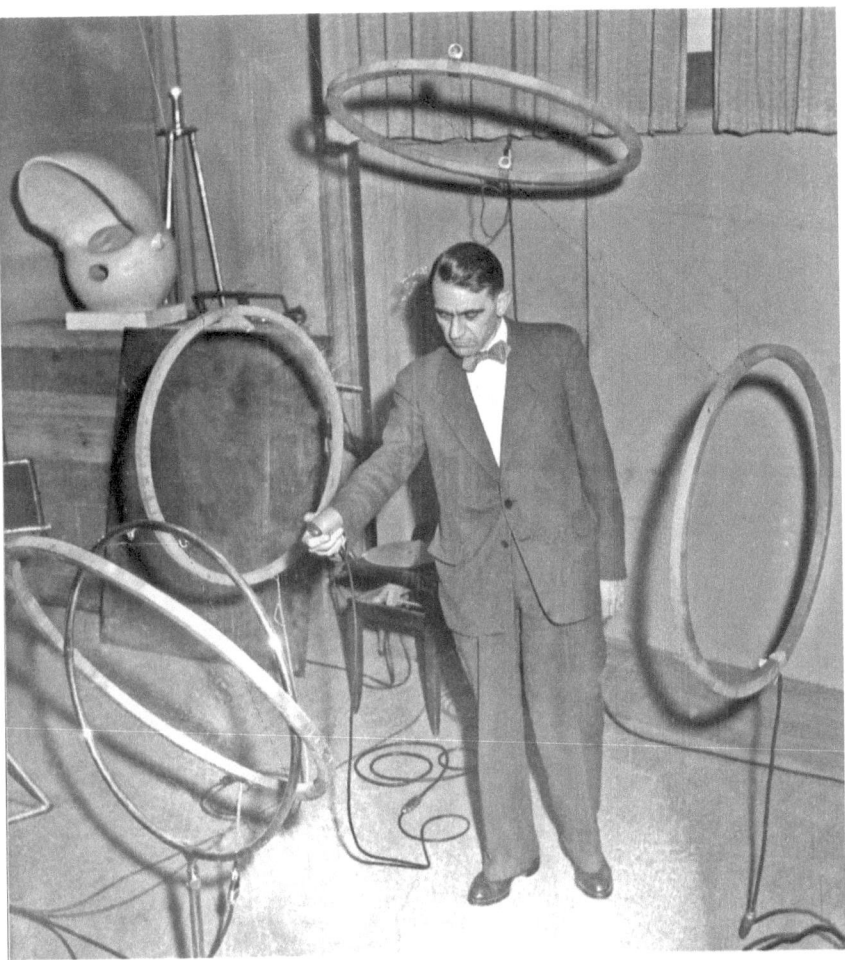

FIGURE 11. Pierre Schaeffer at the controls of the spatial device. © INA/Maurice Lecardent (1951).

An Influx of Curious Composers

In 1951 began an influx of composers who would later achieve considerable renown: Olivier Messiaen, Michel Philippot, André Hodeir, Pierre Boulez, and Karlheinz Stockhausen. They came to try their hand at electroacoustic techniques with the aim of going even further in the refinement of their compositional techniques, particularly serial. The first workshop of the Groupe de Recherches de Musique Concrète (GRMC) in 1951 exploring these early undertakings was attended by Pierre Boulez, Michel Philippot, Jean Barraqué, and André Hodeir.

FIGURE 12. Pierre Schaeffer at the keyboard phonogène; on his right, the slide phonogène. © INA (1951).

Messiaen and Stockhausen appeared as "guest" composers. The second workshop took place in 1954, with Darius Milhaud and Hermann Scherchen.

In the beginning of 1952 Olivier Messiaen composed his first studio work, *Timbres-Durées*, 15:05, realized by Pierre Henry. In this piece based on serial concepts, each duration was linked to a timbre, a nuance, or an attack. For the May 21, 1952, premiere at the Salle de l'Ancien Conservatoire, the sound system consisted of three channels played from the studio on the six-reel tape recorder, with three monophonic tapes synchronized by sharing a single capstan, and then retransmitted via a telephone line, along with a fourth channel (which was the aggregate of the other three) that was "in motion." It was the first attempt at spatial organization of a work. Pierre Henry described the technique:

FIGURE 13. Francis Coupigny at the three-track tape recorder that was used for the "spatial relief" of Olivier Messiaen's *Timbres-Durées* during its premiere May 21, 1952. © INA (1950).

Each sound was edited on different voices and we had a set of responses and dialogues on 4 voices: 3 static voices and a "kinematic" voice that was mobile—something exceptional for the time. So this monophonic piece was heard from all directions, in a very fragmented way, and yet with strict precision, since the tape was played back by a synchronous multiband device.[15]

Jean-Étienne Marie, president of the Cercle Culturel of the Conservatoire, commented on the May 21, 1952, premiere:

Messiaen had determined a certain number of "timbre-durations" to be applied serially and monodically. On hearing the work, it became clear that the rhythm of the spatial splitting took on a very strong predominance over the timbres-durées themselves. Despite all the care Pierre Henry took with the spatial division, it is regrettable that Messiaen did not do it himself. The essential eluded him.[16]

Étude 1 (2:25) by Pierre Boulez, serial study on a sound, composed in 1951.
Étude 2 (2:41) by Pierre Boulez, serial study on seven sounds, composed in 1951.

For these études, Pierre Boulez, like Pierre Henry in *Antiphonie* (1951), could have used a tape recorder to assemble systematically. Commenting on this final work on the program, Pierre Boulez's *Étude 1 & 2* on May 21, 1952, at the Salle de l'ancien Conservatoire, Jean-Étienne Marie captured the impressions of the time:

> In contrast, P. Boulez had organized both his rhythmic cells and his spatial distribution: the listener felt trapped in an iron vise ... but, although elaborated from concrete material (dense sound) that had been developed sufficiently to appear highly abstract, the work we heard stemmed from an abstract approach: the score had preceded the realization and Boulez had introduced into it this thirst for simplicity, this discipline, which so characterizes him.[17]

On the same date, Michel Philippot's first studio work, *Le joueur de bruits*,[18] premiered in the form of a ballet-pantomime with Yves Baudrier.

Also noteworthy, in the program of these concerts of May 21 and 25, 1952, is *Étude vocale* (1:08), based on an anonymous thirteenth-century motet and composed by Monique Rollin, the first female composer to appear in the GRM's catalog of electroacoustic works.

And finally, let us mention the pieces by Pierre Henry also on the program of these two long concerts (each with two intermissions). The same works were presented in both concerts, but in a different order:

—*Vocalise*, a serial study by Pierre Henry, which lasted 2:41;[19]
—*Dimanche noir*;
—*Batterie fugace*, in kinematic relief directed by the author;
—*Tam Tam IV*, in kinematic relief directed by the author;
—*Symphonie pour un homme seul*, by Pierre Schaeffer and Pierre Henry;
—*Bidule en ut*, by Pierre Schaeffer and Pierre Henry.

FIRST MIXED WORK(S)

1951. *Jazz et jazz*, André Hodeir (3:02) for piano and tape. This work is considered to be the first mixed work in the GRM repertoire, competing for the position of first historical work of mixed music with *Musica su due dimensioni*, for flute and tape from 1951 (revised in 1958) by the Italian Bruno Maderna—where flute and tape alternated without mixing.

Just to put the notion of "first" into perspective, let's not forget the existence of another first mixed work at the GRM, *Orphée 51*, for voice, instruments, and tape.

STOCKHAUSEN, FIRST—AND FINAL—VISIT (FOR COMPOSING)

The year 1952 also saw the first (and final) visit, for the purpose of composition, of the young German Karlheinz Stockhausen. He composed *Étude 1*, also known as *Étude aux mille collants*, lasting 1:13. It is an experiment of millimetric microediting on magnetic tape, applying serial laws. The term *mille collants*—a thousand splices— evokes the impressive number of editing splices performed by Stockhausen.

FIRST FILM MUSIC COPRODUCED WITH TELEVISION

1953. *Léonard de Vinci*, by Enrico Fulchignoni, music by Pierre Henry and Pierre Schaeffer, premiered in June at the First Decade of Experimental Music.

1953. *L'art précolombien*, short film by Enrico Fulchignoni, premiered at the Venice Biennale, music by Pierre Henry.

1953. *Les fils de l'eau*, feature film by Jean Rouch, music by Pierre Henry, presented at the Cannes Festival in 1954, during a conference on new sound techniques and the notion of object.

Scandals

But several scandals came to blemish the emergence of early musique concrète. In order to understand them, we must try to picture the challenges of the time, when the audacity of young composers of musique concrète was equaled only by the certainties of serialist avant-garde.

The first scandal took place in Germany, with *Orphée 53*.

Orphée 53, the first concrete opera by Pierre Schaeffer and Pierre Henry, which premiered in Donaueschingen, Germany, on October 10, 1953, was a thorough reworking of *Orphée 51 ou Toute la lyre*. The version given in Donaueschingen lasted an hour and twenty-six minutes. Original libretto by Pierre Schaeffer, music for quarter-inch 76cm/sec. tape. Performing live were Jeannine Collard (Orphée), Andrée Lescot (Eurydice), Heinz Fischer-Karwin (the voice of love), Philippe Arthuys (harpsichord), and Janine Grutter-Rauch (violin).

Philippe Arthuys, who had just joined the Schaeffer team, recalled, "For the first time, I saw Germans laughing."[20] But this laughter was at the expense of the Schaeffer-Henry team, who embarrassed themselves and were mocked at the festival. In all honesty, one must admit that Schaeffer's project had been a bit too ambitious and maladroit. Producing a first opera, mounted with such limited means and so few rehearsals, while integrating concrete sequences into passages sung in traditional operatic style, meant taking ill-considered risks with a connoisseur audience well-versed in the grand repertoire, and what's more, under the watchful eyes of the media, which was on the lookout for the slightest misstep—forty-four journalists were in the audience.

Philippe Arthuys, who performed the harpsichord part, recounted his terror when he came on stage—candlestick in hand, dressed in sixteenth-century baggy breeches, with buckled shoes on his feet—only to discover that in the place of the concert harpsichord on which he had rehearsed was a bentside spinet.*

Following this concert, the press did not refrain from disparaging comments. Schaeffer reacted in an open letter to the German press in which "Waterloo of musique concrète" was the term he himself used to describe the episode. "This is how we lost the battle of Donaueschingen and how we were relegated to international reprobation for years."[21]

The following year, on April 8, 1954, Pierre Henry premiered *Voile d'Orphée* at the Théâtre des Champs-Elysées. This reprise of *Orphée 53* would be reworked several times, finally serving in 1958 as the conclusion to Maurice Béjart and Pierre Henry's ballet *Orphée*. The first version to be recorded on vinyl in 1969 (Philips 836 887) ran for 27:15; the final remix of the short version (Philips 472204–2), lasting 15:35, dates to March 2004.

In his presentation on the evening of the premiere, Pierre Schaeffer declared:

> Independently of any aesthetic pretension, *Voile d'Orphée*, which we are about to hear, is an important step for Musique Concrète. Perhaps for the first time, following his 5 years of research, Pierre Henry, both its composer and interpreter, has been able to assemble the equivalent of an orchestra—which in terms of Musique Concrète consists of a limited number of sound materials, recorded in loops, and whose chromatic phonogène modifies the rhythm and tessitura while electroacoustic filters adjust their timbres towards the low or high registers. Essentially each of these loops is the equivalent of an instrument or a group of orchestral instruments adding its own independent part.
>
> These imaginary instruments are then brought together in an orchestra made up of several tapes, and the interpreter's or conductor's baton is nothing other than a 3-track tape recorder ensuring that all of these parts are rigorously synchronized.[22]

In his presentation, Schaeffer also employed infinite oral precautions aimed at preparing the audience's reaction (the scandal that had occurred in Donaueschingen rendered him particularly prudent):

> Make no mistake about it, we are already aware that the reaction of the audience cannot be perfectly adapted to a work based on techniques they are unfamiliar with,

---

* Translator's note: The full-sized harpsichord had been replaced with an *épinette*, a reduced amateur or home version of the harpsichord with a different mechanism, inadequate for the concert context. The destabilizing act was perceived as an intentional affront, even sabotage.

where the audience's prior musical experience does not necessarily prepare them for careful listening and dispassionate curiosity.

Whatever the reaction to this genre of pieces, the first impression is not necessarily the right one. It is important to underline that certain technical difficulties have been resolved, perhaps for the first time, and that the material is less elusive, and even if it often remains brutal, it is beginning to soften, to take shape, to sketch out vast forms, and to take on an architectural structure, of which it must be said that it is the violence itself that pleases or displeases, that provokes adhesion or indignation, applause as well as outrage.[23]

The scandal of Edgar(d) Varèse's *Déserts* occurred on December 2, 1954, at the Théâtre des Champs-Élysées in Paris. Varèse's *Déserts*, a twenty-four-minute work, was premiered by the Orchestre National under the direction of Hermann Scherchen. Philippe Arthuys, under the technical direction of Pierre Henry, was placed in the middle of the orchestra with the tape recorders to play the interludes—three electroacoustic sections of "organized sounds" intercalated between the sections played by the orchestra.[24] This encounter between orchestra and organized sounds was a first. The electroacoustic parts, realized over a month and a half just before the concert with the participation of Pierre Henry in the studios of the Club d'Essai, revealed sounds of friction, percussion, whistling, breath, and grinding,[25] all filtered and transformed with the intention of bringing them closer to the colors of the orchestral material. These sounds had been recorded in foundries, sawmills, and assorted factories near Philadelphia, USA.[26] The second interlude consisted of percussion sounds recorded by a small ensemble. The diffusion of this part sought to create a feeling of stereophony. Finally, the third interlude mixed the first two, with instrumental and concrete sounds rubbing shoulders. What's more, for the first time in France this artistic creation would be broadcast live in stereophony on the radio. The host of this radio broadcast was none other than Pierre Boulez, and for this grand premiere, Stravinsky was in the audience. Unfortunately, this concert gave rise to a memorable scandal, of which we still have a trace since it was recorded for the radio and conserved in the archives.[27] At each electroacoustic interpolation occurrence, the audience shouted and whistled. Philippe Arthuys related that during the concert, each time that the hubbub in the hall increased, he employed a technique well known to professionals: he lowered the playback level, forcing the audience to quiet down to be able to hear. It is necessary to note that the programming of this work could not have escaped being problematic, because it was aesthetically so far removed from the other music played that evening—Mozart's *Grand Overture in B-Flat Major* and Tchaikovsky's 6th, the *Pathétique*.

This sad adventure marked for Varèse a halt to his project of returning to France, after an absence of twenty-one years. For his part, Hermann Scherchen noted that this scandal meant that he had to wait four years before being again entrusted with the direction of a work for broadcasting. In 1958, Varèse again attempted to persuade the French minister of culture, André Malraux, to honor his promise to entrust Varèse with important means at the helm of a laboratory at French radio, with equipment, studios, and technicians (perhaps at the Club d'Essai), but without success.

In the end, it would be at the 1958 Brussels World's Fair that Varèse would give his last major performance in Europe. The Philips Pavilion, designed by the Swiss architect Le Corbusier (who had delegated much of the work to his collaborator Iannis Xenakis), was the setting for the premiere of *Poème électronique*. The piece had been prepared in the Philips laboratories in Eindhoven. Varèse assembled untransformed sounds of machines, bells, piano, and percussion, along with pure electronic sounds associated with the same sounds, filtered or transposed, all on tape. He envisioned having the sounds rotate in the space of the pavilion, played back on three channels and 450 loudspeakers cleverly placed over the 360°. Between each performance of *Poème électronique* and a series of poetic texts associated with a montage of photos of Le Corbusier, visitors could hear a short electroacoustic work by Iannis Xenakis (2:45). This brief piece, *Concret PH* (*Paraboloïdes hyperboliques* / Hyperbolic paraboloids) had been composed by Xenakis from recordings made in stereo, for the very first time, at the GRMC studio, 37 rue de l'Université in Paris.

During the World's Fair in Brussels, Iannis Xenakis met the young François Bayle, who was on leave from the military, defying the ban on French conscripts leaving the country during military service at the time of the Algerian War. Bayle wanted to visit the Pavilion and especially to try to meet Schaeffer. After the exhibition, the Philips Pavilion was demolished. Seven years later, in 1965, Varèse died in New York, his adopted city, without ever having returned to his native Paris.

Other Premieres

1953. *Étude* (5:02) by Jean Barraqué. Used for the film *Objeu*, directed by Monique Lepeuve, in 1961, text by Francis Ponge.

1954. *La rivière endormie* (7:35) by Darius Milhaud. A piece of a dramatic nature, due to the presence of its somewhat declamatory text, realized with the assistance of Jean-Étienne Marie.

1955. *Le crabe qui jouait avec la mer* (22:05) by Philippe Arthuys. Musique concrète composed for the tale *Les Histoires comme ça* (Just so stories) by Rudyard Kipling. This was Arthuys's first important work. The concert version titled *Pau Amma* lasted twelve minutes. This work was released on vinyl in 1955 by La Boîte à Musique.

1955. *Six pièces en un acte*, stage music by Henri Sauguet, text by Jean Tardieu. First work composed by Sauguet at the GRMC. Since this piece has little in common with musique concrète, it is perhaps preferable to mention *Aspect sentimental* (7:08), which he composed in 1957 with the assistance of Philippe Arthuys.

1955. *Symphonie mécanique* (14:21), musique concrète by Pierre Boulez (among his most successful works of musique concrète) for the film by Jean Mitry.

1956. *Mavena* (10:43) by Ivo Malec. This first electroacoustic work by Ivo Malec took on a dramatic quality due to the presence of Radovan Ivsic's text, read by Anne Pérez.

1957. *L'Amen de verre* (4:46), *Passacaille* (9:10), and *Exergue pour une symphonie* (6:35) were the only three pieces composed by Roman Haubenstock-Ramati at the GRMC. He then resumed his solitary path as a composer.

1958. *Diamorphoses* (6:50) by Iannis Xenakis. For his first work composed at the GRM, Xenakis (with the precious assistance of Alain de Chambure) used the slider Phonogène to realize very gradual glissandi, which he called *trames orientées* (guided grids). This piece premiered in the French pavilion of the Brussels World's Fair on October 5, 1958. In the notes for his work, Xenakis also stated that his piece was "an investigation aimed at making people forget the primary character of sounds of an anecdotal nature."[28]

FIRST MULTIMEDIA WORK

1955. *Spatiodynamisme*, by Nicolas Schöffer, music by Pierre Henry.

Considered the first interactive multimedia work in the world, this spatiodynamic and cybernetic tower was created by the artist Nicolas Schöffer (1912–92). Realized with the assistance of Pierre Henry and Jacques Bureau, an engineer at Philips, the tower was a fifty-meter-high installation that reacted in real time to its environment through a system of sensors.[29] Its lightweight, elongated metal structure, with its airy architecture, was coupled with colored rectangular plates installed at different levels of the sculpture. It was built by the Mills company in June 1955 for a public works exposition, and was situated outdoors in the Parc de Saint-Cloud on the banks of the Seine.

For the sound, Pierre Henry first recorded percussive sounds on the tower's metallic plates, then produced twelve sound loops using his habitual techniques, using one of the first multitrack tape recorders (six stereo tracks) made by the Philips company.

Thanks to the recording devices placed within the tower, any change in the environment was communicated to the homeostat (interactive electronic system using photoelectric sensors), including changes in temperature, humidity, wind, color, light, sound, and movement in the surrounding area. The sensors provided continuous information to an "electronic brain" designed by Jacques Bureau. This brain piloted the tape recorder in real time, as well as the associated amplifiers, by influencing sound diffusion within the tower. The interactive design then executed real-time mixing, sound processing, and spatialization on more than a dozen speakers. These changes constantly triggered new combinations of sound, alternating with silences.

Thousands of people gathered to admire and listen to this tower that composed its own music with its own sound material.

FIRST MUSIQUE CONCRÈTE FOR DANCE

On June 14, 1952, at Brandeis University in the United States, Merce Cunningham was the first to choreograph to musique concrète during the first performance of this kind of music on American soil. He danced to excerpts from *Symphonie pour un homme seul* by Pierre Schaeffer and Pierre Henry, well before Maurice Béjart, who would also choreograph to the same composition, but not until 1955, at the Théâtre de l'Étoile in Paris with the Ballets de l'Étoile. This "first" French choreography to musique concrète was a resounding success, helping to "launch" Maurice Béjart.[30]

After this successful adaptation for dance, Pierre Henry continued his collaboration with the choreographer. In 1955, Béjart adapted Pierre Henry's *Concerto des ambiguïtés* (1950) under the title *Voyage au cœur d'un enfant*. Henry would go on to compose musique concrète specifically for ballet: *Arcane* (1955), *Haut voltage* (1956), the first work to use electronic sounds, and finally *Le cercle* (1956).

For his part, Philippe Arthuys also continued to collaborate with Béjart on *Voilà l'homme* (1956) and *Ils récolteront la tempête* (1956).

In a short text from 1979 regarding Maurice Béjart, Pierre Henry revealed his discovery of the world of dance and its influence on his future music:

> After *Symphonie pour un homme seul*, the percussions of *Cercle* and the flocculent music of *Arcane*, I understood that musical phrasing and its articulation could be integrated with gestural rhythmicity. As for example in *Haut-Voltage*. In fact, while composing I was affected by what was happening when a choreography was being created. So when I adapted *The Tibetan Book of the Dead*, I first created a breakdown of the scenes, which was very choreographic. Right away I sensed *Le Voyage* as ballet music. It became second nature. There was a kinship between my music and dance—the rhythm, the accents, the falls, the pauses. Before meeting Béjart, I did not like dance, in fact I knew nothing about it. I considered ballets to be conventional—or even ridiculous. Then I started to look at the great solos and *pas de deux* from the repertoire. As a result, my music became more dynamic.[31]

MUSICAL CONTEXT: OTHER AVANT-GARDES

While for the French, until the Second World War (1939–45), the avant-garde scene was essentially limited to Europe (Russia, Italy, Germany, Belgium, France), after 1945 the horizon widened to the American continent. This opening was all the easier since Dada and surrealism had been preparing the relay between America and Europe since the First World War (1914–18) through the presence of numerous European artists exiled in the United States.

Since his first study trip to Germany in 1936, Pierre Schaeffer had been monitoring the experiments carried out in that country. In 1948 he was informed of Werner Meyer-Eppler's work at the University of Bonn producing synthetic sounds with the help of the Bode Melochord and the Trautonium.[32] Schaeffer also wanted to invent an instrument, "the most general musical instrument possible, ... to enter into direct contact with sound matter" and especially "without electrons intervening."[33] But contrary to the Germans, who were looking in the direction of synthesis, Schaeffer conceived his research from direct recording of the material, using microphones.[34]

The French Musical Avant-Garde, Its 1953 Turning Point

Everything was going smoothly until 1953. The Club and then the Studio d'Essai were quite inventive. From the early days of musique concrète, key works had emerged: *Symphonie pour un homme seul*, the *Etudes de bruit* ... Jean-Étienne Marie had launched the first course in experimental music at the Schola Cantorum in Paris in 1953. Other artistic practices linked to recording on a support saw the light of day. As early as 1948, Jean Thévenot organized the first amateur sound recording competition. This competition still existed in 2005 under the name "Chasseurs de son," a term inaugurated by Pierre Schaeffer in his 1942 article *Esthétique des arts relais*. Concrete musicians were part of the modernity of their time, in the same way as serial musicians. Links were gradually formed between different genres.

But with the scandal of *Orphée* at the Donaueschingen festival in 1953, Boulez's affection for Schaeffer suddenly cooled. Relations became more and more strained. In correspondence with Cage in July 1954, Boulez wrote:

> Stockhausen is more and more interesting! He is the best of them all in Europe! ... At the same time it must be admitted that he is able to work tranquilly in Cologne. Eimert, the director of this studio, is very liberal and lets him do what he wants, when he wants and how he wants! (This is not the case with the beloved Schaeffer, with whom I am now totally on the outs! In fact, I even refused to work with him, although he has asked me more than once. The concrète studio is now vegetating more or less badly and less rather than more! We don't talk about it much.)[35]

In contrast, it should be noted that the conductor Hermann Scherchen, who visited Pierre Schaeffer and Pierre Henry for the first time in 1953, would always remain attentive (until his death in 1966), a faithful supporter of the research at GRM and of Iannis Xenakis.

Another remarkable person also made her discreet entry into the musical avant-garde, where she would long reside. The young Éliane Radigue, twenty-three years old in 1955, sporadically assisted Pierre Schaeffer, sometimes representing him at conferences exploring the newly created musique concrète. Their collaboration lasted about two years, during which Éliane Radigue was granted the use of the GRMC studios in exchange for her services as an assistant. Later in 1967, after returning from a ten-year stay in the United States with several personal works in her repertoire, she became Pierre Henry's assistant in his Apsome Studio, notably for *Apocalypse de Jean*. And there again she was able, in her spare time, to use the equipment for her own composition.[36]

The Domaine Musical

On May 5, 1953, Pierre Boulez inaugurated the Concerts du Théâtre du Petit Marigny, which would become the Domaine Musical and which he would direct until 1967, with Suzanne Tezenas as president (benefactor). Schaeffer attended the concerts of the Domaine Musical, as did many other personalities, artists, and composers of the time. He was even a member of the organization. But Boulez would not allow any composer of musique concrète to be included in his program. We can use the example of Michel Philippot—his serialist works were performed at the Domaine, but he was immediately excluded after his rapprochement with Schaeffer. Xenakis, for his part, did not have his work performed until leaving the GRM in 1963.[37]

It proved impossible to be close to both Boulez's Domaine Musical and Schaeffer's GRMC. So when John Cage returned to France in 1954 for the Paris premiere of his *Concerto for Prepared Piano* organized by the GRMC at the Salle Gaveau on October 25, he did not even see Boulez.[38] The apotheosis of the divergences between "serial musicians" and "concrete musicians" materialized in 1959 in Pierre Boulez's article for the *Fasquelle Encyclopedia of Music*. Alongside an article signed by Schaeffer defining musique concrète, Pierre Boulez availed himself of his privileges as a member of the editorial team. His comments occupied as much space as Schaeffer's text, and above all they adopted a defamatory tone, entirely unexpected in an encyclopedia.

> Our "concrete musicians" have compiled a weighty brief against them. They are accused of producing sounds which, technically speaking, from the sole point of view of quality, are ugly.... The work of an amateur on a pilgrimage, musique concrète cannot compete, even in the field of "gadgets," with "sound effects" fabricators working in the American film industry. Interesting neither from the point of view of sound nor from the point of view of composition, one is justified in questioning the aims and utility of musique concrète.[39]

The New York School

Cage and Schaeffer first met in 1949. John Cage (1912–92) had received a grant from the Guggenheim Foundation to spend three months in Paris. He was received by Boulez. The Paris premiere of Cage's *Sonatas* and *Interludes* took place at the home of Suzanne Tezenas, whose salon was renowned at the time. Schaeffer discovered Cage's *prepared piano*, which made it possible to create "melodies" out of timbres.[40]

In 1951, back in the United States, Cage, along with his pianist friend David Tudor, joined Louis and Bebe Barron in their private studio in New York, approaching them for funding to launch *Project of Music for Magnetic Tape* in New York. After considerable legwork, they finally obtained help from their friend Paul Williams. The Barrons served as sound engineers. The group was expanded with the addition of jazzman Earle Brown (1926–2002), composer Morton Feldman, and the seventeen-year-old Christian Wolff. The most beautiful moment of this adventure, which lasted until 1954, was the Cage-Brown-Wolff group composition, *Williams Mix*, an electroacoustic piece consisting of eight mono tapes to be played simultaneously on eight loudspeakers.

According to Earle Brown's account, the 1952 composition of *Williams Mix* lasted for months, with Brown arriving at Cage's home around 10 a.m. and working with him until around 5 p.m. On Fridays, they went together to attend Suzuki classes at Columbia University. Christian Wolff, whose father was a publisher, had given Cage the first English translation of the *I Ching—The Book of Changes*.[41] These episodes evoke the origins of the attraction of Cage and the other members of the New York school for Eastern philosophies and the concept of "chance."

This interest is reflected in the preamble to the score of Earle Brown's so-called "open form" piece *December 52*, for chamber ensemble.

> The score consists of a single sheet of unruled paper, on which lines and rectangles of various lengths and thicknesses are arranged horizontally and vertically at varying intervals. This sheet is both score and part. It can be placed on the music stand in any direction, and the performer is free to move from any point on the score to any other point.

The performer took on an important role in Brown's work in the making of the piece. Needless to say, Earle Brown came from the world of jazz and improvisation, where the performer is often also the composer.

Earle Brown, in a private interview in 1998, reminded us of the fact that Americans composed open-form works earlier than Europeans. Europeans, like Boulez with his *Third Sonata* (1955–57) or Stockhausen with his piano piece *Klavierstück XI* (1956), came after. Brown remembered having hosted Boulez in 1953 when he came to New York with the Renaud-Barrault theater company. At the time, Brown was writing his piece *Folio* and showed it to his French friend, who would later be influenced by this writing model. A few years later, Brown visited Boulez in Paris in 1956, a year before the latter wrote his article *Aléa* on the concept of the aleatory in music.[42] Behind the friendly exchanges, competition was always a bit evident.

This theme of chance and randomness in music was the subject of polemics between Americans and Europeans, notably Boulez, but also Schaeffer, who never took the work of the American friends very seriously. Was it due to the ambiguity of the word *chance*? Was it a chance one is subjected to, or calculated chance? Even the Americans did not use the same concepts among themselves. While Cage spoke of *indeterminacy*, Brown preferred the term *open form*. Boulez, on the other hand, adopted the term *guided improvisation*.

In spite of everything, relations remained courteous. In 1953, Cage and the Barrons were scheduled by Schaeffer for the concerts at *Première décade de la musique expérimentale*. And later in 1963 Earle Brown was commissioned by the GRM to compose *Times Five* in Paris. For his part, Boulez would turn a little later during the 1970s to other Americans, in the field of computer sound synthesis.

The Cologne Studio, Birth of Electronic Music

The history of the Cologne Studio in Germany interests us because it has many similarities with that of the Paris Studio—their lifespans were contemporaneous and relatively comparable. During an interview in 2001, Karlheinz Stockhausen, evoking the closing of the Cologne Studio, concluded, "Let's hope the same thing doesn't happen to the Paris Studio!"[43]

The electronic music studio in Cologne took form between October 18, 1951, and May 26, 1953. Herbert Eimert, Werner Meyer-Eppler, and Robert Beyer broadcast a program titled *Die Klangwelt der Elektronische Musik* (The sound world of electronic music) on German radio, where one could hear examples of experiments in creating sounds by synthesis. On May 26, 1953, the official founding document of the studio was issued at Nordwestdeutscher Rundfunk (NWDR, which a few years later, after the reorganization of the German radio networks, became Westdeutscher Rundfunk, WDR). This program was based on the conference "Music and Technology," given in the summer of 1951 in Darmstadt by Meyer-Eppler and Beyer. The first director of the studio, Herbert Eimert (1897–1972), was a music critic, radio producer, and composer. He had composed his first electronic pieces in collaboration with Robert Beyer (1901–89), whom he met in 1950 at the Darmstadt conferences.[44] That same year Varèse was invited to give a seminar. During these three days, August 21, 22, and 23, 1950, Eimert, like Edgar Varèse, Karlheinz Stockhausen, and many others, listened to lectures by the physicist and phonetician Werner Meyer-Eppler (1913–55) of the University of Bonn and the acoustician Robert Beyer, on the subject of "The Sound World of Electronic Music."[45] This summer course continued the following year in 1951 under the title "Music and Technology." Robert Beyer evoked the importance of seeking new sounds at the limits of music and raised the issue of space in electronic music. Werner Meyer-Eppler presented the first examples of sound synthesis realized with a Melochord and an AEG tape recorder.[46]

Until 1999, when it was mothballed by its parent institution, the Cologne electronic music studio remained one of the most important sites for avant-garde musical creation in the second half of the twentieth century. Numerous guest composers including Karel Goeyvaerts, Franco Evangelisti, György Ligeti, Mauricio Kagel, Bruno Maderna, Henri Pousseur, Ernst Krenek, Johannes Fritsch, Peter Eötvös, Jean-Claude Eloy, Iannis Xenakis, York Höller (who would also be the last director of the studio), and Marco Stroppa came to work alongside the small core of engineers and young assistants, who would themselves be promoted to careers as composers: Heinz Schütz, Gottfried Michaël Koenig, Konrad Bœhmer (1958), Leopold von Knobelsdorff, Eugeniusz Rudnik, Volker Müller, and Paulo Chagas. But the figure who forever dominated this venture in the eyes and ears of everyone was none other than the composer Karlheinz Stockhausen, who occupied the premises almost permanently.

THE MOVE TOWARDS ELECTRONIC MUSIC AT THE COLOGNE STUDIO

In the wake of the Second World War, young composers in Germany, and in the world in general, were firmly resolved to forget the past, or more precisely to seek new musical paths. This is why new theories such as serialism, which Arnold Schoenberg and his followers had pursued since the beginning of the century, found a favorable echo. The same can be said about the emergence of electronic music. The personalities of its pioneering members—Herbert Eimert, Werner Meyer-Eppler, and Robert Beyer—also demonstrated that the field of musical inquiry had been extended to the realm of sound, and in particular to language in the case of Germany. Meyer-Eppler and Beyer, at the Institute for Information Science of the University of Bonn, were engaged in research on language recognition and automatic translation. The expected implementation did not concern music but, more tangibly in the context of the Cold War, involved robots translating conversations intercepted during radio or telephone transmissions. Secondly, it was also aimed at helping the young United Nations Organization to automate simultaneous translation. Numerous other civil applications were also envisioned, such as assistance to hearing-impaired and mute individuals wounded in the war. The first step in synthesizing the voice is to know how to analyze it—not only the phonemes but also the rhythm of the sentence and even the melody of the entire conversation. By analogy, music was regarded as a rudimentary form of language; the musical skills acquired in their youth by both Meyer-Eppler and Beyer therefore proved useful.

This research effort was also coordinated with the work of some important American organizations, especially the Massachusetts Institute of Technology (MIT) in Boston and the University of Illinois, where Lejaren A. Hiller and Leonard Isaacson were among the first to produce computer compositions, notably the *Illiac Suite for String Quartet*. The Columbia Princeton Electronic Music Center, officially founded in 1958 by Vladimir Ussachevsky, Otto Luening, and Milton

Babbitt, was also in contact with German researchers. Ussachevky had been teaching at Columbia since 1947. The earliest experiments were in 1952. Luening's *Fantasy in Space* and Ussachevsky's *Sonic Contours* were "music for tape" pieces that included electronic and concrete sounds. These compositions were presented in a concert at the Museum of Modern Art in New York. The network of expertise also encompassed figures such as Max Mathews, a specialist in information theory; the linguist Noam Chomsky; and Louis and Bebe Barron, experimental composers.[47] The Barrons' private studio in New York had been equipped since 1948 with numerous devices for synthesis, wired together based on the model of the animal brain as an artificial intelligence project, the rudiments of which they had first encountered at the university in the cybernetics course of the German Norbert Wiener. They too were trying to recreate the voice.

In 1951, twenty-three-year-old Stockhausen, who was studying with the phonetician Meyer-Eppler in Bonn, went to the United States to present the research carried out on voice synthesis in Germany, during a series of thirty-two lectures at various universities. This emphasis on voice and phonetics explains why Eimert always tried to link the activities of the electronic music studio with the drama department at Cologne Radio. The guiding cornerstone was work on the voice. The same applied to the Studio di fonologia musicale at Radiotelevisione italiana (RAI) in Milan, founded in 1955 by Luciano Berio and Bruno Maderna, who approached voice primarily from a dramatic point of view rather than from the point of view of synthesis. However, Stockhausen would later prove to be a fervent defender of a purely musical approach at the WDR studio in Cologne, independent of the production of radio dramas.

THE BEGINNINGS OF THE COLOGNE STUDIO

At the beginning of 1954, Cologne Radio broadcast works by Eimert and Beyer, as well as Stockhausen's first two electronic études, *Studie I* and *Studie II*.

Later that same year, the composer Gottfried Michaël Koenig joined the staff of the studio. In addition to his decisive assistance in the realization of Stockhausen's famous *Gesang der Jünglinge* and *Kontakte*, he himself composed *Klangfiguren No. 2*, the first serial work in which the parameter of timbre was entirely structured. Koenig was pursuing aesthetic innovation but also new methods.

In May 1956, a second large concert was organized in Germany, where Karlheinz Stockhausen's *Gesang der Jünglinge* (The song of youth) was performed, one of the founding works of electroacoustic music but also a prime example of the quest for spatialization in sound. *Gesang der Jünglinge* was the first work to illustrate the fusion of electronic music and musique concrète.[48] Exclusive use of sinusoidal sound had been abandoned in the Cologne Studio in 1955, and efforts were made to add other sound materials, in particular, concrète sounds. This successful rapprochement between the two "hostile" genres rendered the aesthetic polemic with the Paris Studio null and void—without, however, eliminating the related competition between the two studios.

The Milan Studio

Luciano Berio (1925–2003) had met Bruno Maderna, as well as Cage, Stockhausen, Boulez, and Pousseur, in Darmstadt in 1953. He married the singer Cathy Berberian in 1950. She sang in many of his works, notably in *Thema (Omaggio à Joyce)* (1958), a work given in concert at the GRM in 1960. The RAI studio in Milan had only loose contact with the Paris studio; it was much closer aesthetically and politically to the studio directed by Stockhausen at the WDR in Cologne. However, the people in charge of the Milan Studio were careful to avoid taking a position in the competition between the two pioneering studios in Paris and Cologne. Berio accepted the invitation from Paris to come to the GRM studios to mix *Laborintus* in 1965, and he returned in 1975 to compose *Chants parallèles*.

Bruno Maderna (1920–73), the other leading figure at the Milan Studio, had first met Hermann Scherchen in 1948, and subsequently turned to dodecaphony. André Boucourechliev, in his monograph on Maderna, presents him as "one of the great pioneers of electronic music, which he immediately steered towards a quest for the physical properties of sound, thus joining the preoccupations of musique concrète." Maderna founded the phonology studio at RAI in Milan with his friend Luciano Berio in 1955, after having succeeded in persuading Italian Radio of the necessity of a musical and sound research unit.[49] The studio was also expected to produce functional music and other elements for radio. As in all music studios, behind the figure of the master stood a highly skilled technician. In Milan, this was Marino Zuccheri, who said that the very first words he heard upon arriving at the studio in 1955 were "Abbiamo nove oscillatori" ("We have nine oscillators"), which in itself demonstrated the modesty of their initial resources.[50] Nevertheless the studio developed, hosting numerous composers in residence including Luigi Nono, Niccolo Castiglioni, Aldo Clementi, John Cage, and André Boucourechliev.

Berio directed the Milan Studio from 1955 to 1961. Much later, after his stint as director of the electroacoustic department with the initial IRCAM lineup from 1975 to 1980, Berio returned to Italy to direct the organization Tempo Reale in Florence. RAI permanently shuttered the musical phonology studio in Milan in 1983.

The Experimental Studio in Gravesano

Conductor Hermann Scherchen's private studio opened in Gravesano, Switzerland, in 1954. The site became a summer rallying point for composers curious about technical innovations and eager for encounters within the professional milieu. The attraction of this experimental studio, equipped with the best material, was strong among often penniless composers. Scherchen financed the operation of the studio from his own personal fortune, and he also initiated publishing activities in the form of a biannual journal, *Gravesaner Blätter*. Practically all the composers of modern music of the 1960s were invited to Gravesano at one time or another, including Xenakis, Ferrari, and Mâche, among others.

EVENTS, MUSICAL RELEASES, AND PUBLICATIONS
Conference at the Sorbonne

The first concert of musique concrète, organized by Tryptique at the École Normale de Musique on March 18, 1950, was associated with a preparatory lecture given two days earlier by Pierre Schaeffer at the Richelieu amphitheater of the Sorbonne. At the time, and until late in the 1960s, concerts were either accompanied by commentary or coupled with lectures, with the intention of enlightening listeners. Two days before, the March 16 lecture at 8:45 p.m. at the Sorbonne was titled *Initiation to Musique Concrète*. A list of the topics addressed by Pierre Schaeffer attested to the immensity of the research process he had launched:

For a New Definition of Music
Is There Unexplored Musical Material?
Unforeseen Instrumental Possibilities?
Unexpected Areas for Composition?
Is Music a Plastic Art?
Man and Machines
Fission of the Note
Classical Music Seen as a Special Case
Towards Generalized Music[51]

In Search of Musique Concrète, the Book

After the radio presentation of the first *Concert de bruits* on October 5, 1948 (where the term *musique concrète* was first evoked orally), and in particular following a number of articles in the sixth issue of the review *Polyphonie* published by Richard-Masse (1950), Pierre Schaeffer embarked on writing an entire book on his music research. *À la recherche d'une musique concrète* was published by Éditions Seuil in 1952. The first half of the book featured excerpts from Pierre Schaeffer's first and second journals, written from 1948 to 1949 and from 1950 to 1951. The second part penetrated more deeply into the description of musical phenomena through the experimental method dear to the author. The "concrete solfeggio" that appeared at the end of the work was sketched with the collaboration of

André Moles, who had just defended his thesis on "the physical structure of the musical signal."[52]

"First Decade of International Experimental Music"

June 8–18, 1953, Première décade internationale de musique expérimentale was organized by the GRMC of French Radio at UNESCO and at the Centre d'Études Radiophoniques located at 37 rue de l'Université in Paris.

The numerous lectures and public gatherings for this event aimed at informing as many people as possible about the first milestones in the evolution of this new musical genre.[53]

Below is the list of events, demonstrating the eclecticism of research in music at the time:

—André Moles and Jean-Wilfrid Garrett, "Music Machines" (from phonogène to Vocoder);
—Robert Barras, director of programs at Radio-France Asia, "Exotic Music";
—Vladimir Ussachesvsky, from Columbia University, "Music for Tape" (in English);
—Raoul Husson, Doctor of Science, "Psycho-physiological Conditioning of Musical Aesthetics";
—Pierre Boulez, "Trends in Recent Music";
—Dr. Herbert Eimert, musical director of the Nordwestdeutscher Rundfunk, Cologne, "Trends in Electronic Music";
—Antoine Goléa, "Tendencies in Musique Concrète"; and
—Debate: "Thought and the Instrument" with Maurice Martenot, Olivier Messiaen, and Boris Schlœzer.

Three concerts were presented alongside the lectures:

—At Salle Erard, "Going beyond the Orchestra," with the symphony orchestra of French Radio-Television under the direction of Hermann Scherchen (works by Bartok, Stravinsky, Chavez, Klebe, Riegger, Varèse, Webern).
—At Salle des arts et métiers, "Sound and Image," presented by Enrico Fulchignoni, from Unesco. Projection of sequences of previously unreleased films made in collaboration with avant-garde composers: *Léger* (Varèse), *Calder* (Cage), *Bells of Atlantic* (Louis and Bebe Barron), *Le Tempestaire* (Baudrier), *Mac Laren*; *Masquerage* (Schaeffer), *Les désastres de la guerre* (Grémillon), *Léonard de Vinci* (Pierre Henry).
—At the Centre d'Études Radiophoniques, "Sound and Space," presented by Jacques Poullin. Concert of spatial music with concrete, exotic, and poetic works. Executing the spatialization were Pierre Schaeffer and Pierre Henry.

Involvement with the Cannes Film Festival and Other Venues

On April 2, 1954, Pierre Schaeffer gave a conference at the 7th Cannes Film Festival, as part of the international encounter of film music authors and composers organized by UNESCO.

Pierre Schaeffer, sometimes accompanied by Jacques Poullin, regularly participated in international conferences (Venice, London, Barcelona, Aldeburgh, Gravesano) to present the current state of their research.

First Workshops for Discovering the Studio's Resources

Given that the workshops for composers were destined to expand, Philippe Arthuys was appointed to take charge of them. After the first workshop in 1951 (Philippot, Barraqué, Hodeir, Boulez), a second workshop was held for Milhaud and Scherchen. The purpose of these workshops was to put composers in direct contact with the new tools of the studio and allow them to work on sound material from a concrete perspective.

First Record Releases

In 1955, two vinyl records were released by Ducretet-Thomson in a series titled *Premier panorama de la musique concrète*. It consisted primarily of all the early works of Pierre Schaeffer and Pierre Henry, from *Bidule in C* to *Voile d'Orphée*, including *La symphonie pour un homme seul* and the *Études de bruits*. Another record, presenting the work *Le crabe qui jouait avec la mer* by Philippe Arthuys, was released the same year by La Boîte à musique.

Radio Broadcasts

One of the production activities of the Groupe consisted in realizing radio broadcasts, whether in the form of didactic series on experimental music to be broadcast on national networks, or by using musique concrète techniques and approaches in radio dramas.

During one of these broadcasts in 1955, the poet Jérôme Peignot pronounced the word *acousmatic* for the first time.[54] He suggested using it to designate a "sound that one hears without detecting its causes," as suggested by the dictionary. The word would then be taken up again by Pierre Schaeffer in *Traité des objets musicaux* (*TOM*), and then by François Bayle in the 1970s.

CONCLUSION: THE END OF THE MUSIQUE CONCRÈTE AVANT-GARDE, FROM GRMC TO GRM

In 1951, Pierre Schaeffer created the Groupe de Recherches de Musique Concrète (GRMC) within French Broadcasting. The Articles of Association were "read and approved" November 21, 1951, by Wladimir Porché, director general of French

Radio and Television. The GRMC was given the means to carry out its research in the following six directions:

—invention and realization of devices;
—processes of analysis, transformation, and sound synthesis;
—methodology;
—composition and performance of works;
—concert technique (spatial interpretation—sound projection); and
—practical implementation.

But in 1954, Pierre Schaeffer, "creator of impossible and necessary institutions," as he himself put it, was called upon to found and direct the overseas component of France Radio, which would in 1956 become known as Sorafom (Société de Radiodiffusion de la France d'Outre-mer) and then Ocora in 1962 (Office de coopération radiophonique). For five years, from 1953 to 1958, he was in charge of the overseas radio network and was therefore forced to abandon his research at the GRMC. During his absence, Philippe Arthuys was appointed as delegated director of the GRMC and Pierre Henry as project director. Subsequently, Pierre Schaeffer retained this practice of delegating the directorship (but not the title of director) of the group to collaborators: François-Bernard Mâche (1959–60), Michel Philippot (1960–61), Luc Ferrari (1962–63), Bernard Baschet and François Vercken (1964–66). From 1966 on, François Bayle took over the direction for thirty-one years until his retirement in 1997. He was then replaced by Daniel Teruggi.*

Following his numerous absences from Paris, away from the GRMC for professional missions abroad on behalf of Radio Télévision Française, Pierre Schaeffer regained a more stable situation in 1958. In reality he had been dismissed from his functions, and after the Sorafom he was without an assignment for nearly a year. For lack of anything better, he decided to assume effective direction of the GRMC, choosing to orient it less towards composition and more towards research. This became the new vocation of the Groupe de Recherches Musicales (GRM).

Pierre Schaeffer's reinstatement at the GRMC was not without turmoil. Philippe Arthuys, delegate director of the GRMC, and Pierre Henry, director of projects, were severely criticized by Schaeffer. He reproached them for preferring to devote themselves to individual activities rather than to new group research projects. On March 26, 1958, in the middle of a department-wide meeting, Pierre Henry handed Pierre Schaeffer a letter containing his unrenewed contract—or, more precisely, threw it at him. Pierre Henry then decided to set up his first small studio at home—limited to his means. On the strength of his numerous musical successes, notably in his collaborations with Maurice Béjart's ballet, in 1960 Henry eventually and

---

* Translator's note: Daniel Teruggi would be replaced at the helm of the GRM by François Bonnet in 2018.

FIGURE 14. Pierre Henry at the GRMC studio. © INA (1951).

successfully founded the Apsome studio (Applications de Procédés Sonores de Musique Électroacoustique) in association with Jean Baronnet.[55] In 1982 Apsome would become SON/RÉ.*

Other members of the team also resigned, including Philippe Arthuys, who would pursue his career as a film composer and sound engineer for Rossellini, Becker, Rivette, Mitry, etc.

This rupture represented a major loss for the Groupe. In my opinion, it marked the end of the musique concrète avant-garde. But new collaborators—and not the least—would arrive to take their places. Among others, there were Iannis Xenakis, Luc Ferrari, and Bernard Parmegiani.

Today we can see that the sparks of genius of Pierre Schaeffer and Pierre Henry were complementary. One was more of a theorist, the other more of a practitioner. And we must not forget the third man, the engineer Jacques Poullin. In just the ten years from 1948 to 1958, every key research topic was tackled by the entire Schaeffer team: the concepts of the sound object and reduced listening, sound relief, the development of machines for transforming sound, the electroacoustic concert,

---

* Translator's note: Pierre Henry passed away July 5, 2017.

mixed music, music applied to the image and to ballet, musical pedagogy, analysis, and transcription.

Pierre Schaeffer summed up this period of the musique concrète avant-garde: "The two antagonistic forms of music in 1950–55, concrete and electronic, had finished in a draw. Both had been ambitious, one dreaming of conquering sound in one fell swoop, the other aiming at producing all music through synthesis."[56]

In the end, one can ask if the avant-garde of musique concrète did not owe its demise more to Pierre Henry's departure than to the waning competition with electronic music.

7

1958–1968

Birth of the GRM

The decade began with the birth of the GRM, followed by years of intense activity essentially oriented towards fundamental research and collective works, and less towards individual artistic creation. The results were impressive, even citing only the publication of the book *Traité des objets musicaux* in 1966 and the creation in 1960 at the ORTF of the Service de la Recherche, which would continue its work until 1975.[1]

The sixties saw also the birth of many artistic movements: new realism, kinetic art, arte povera, art/technology, happenings, Fluxus, pop art, nouvelle figuration, support-surface, minimalism, new wave. Creative and often contentious effervescence was everywhere, rebelling against the dominant abstraction and "petty-bourgeois" subjectivity. All these new tendencies were the fruit of regroupings of artists, cooperatives, collectives, workshops, and studios.

Numerous ramifications connect these movements to experimental music at the GRM, and in a general fashion to all of the ORTF's Service de la Recherche, because artists circulate. The Research Department became an unavoidable site for encounter.

HISTORICAL CONTEXT

The GRM was founded in 1958 by Pierre Schaeffer, in the context of Club d'Essai directed by Jean Tardieu, and under the institutional patronage of Radio Télévision Française (RTF).[2] Its founding did not give rise to the drafting of an official document; it consisted simply of internal recognition for its research, legitimized by the allocation of personnel and technical resources. The birth of the GRM was

LE GROUPE DE RECHERCHES MUSICALES
de la Radiodiffusion-Télévision Française

présente

CINQ MANIFESTATIONS
de
MUSIQUE EXPÉRIMENTALE
★

1ᵉʳ et 30 Juin - *Concerts* - Salle Gaveau, 45, Rue La Boétie, à 21 heures
MUSIQUE CONCRÈTE - ELECTRONIQUE - EXOTIQUE
Bério, Boucourechliev, Ferrari, Ligeti, Mâche, Maderna, Mayuzumi, Philippot, Pousseur, Sauguet
Schaeffer, Varèse, Xenakis

16 et 26 Juin - *Expériences Musicales* - Salle Guimet, Place d'Iéna, à 21 heures
"LES IDÉES ET L'OBJET"
"L'IMAGINATION ET L'INSTRUMENT"
projections sonores commentées par Pierre Schaeffer

23 Juin - *Image et Mouvement* - Salle Guimet, Place d'Iéna, à 21 heures
Films de : Alexeieff, Bourguignon, Brissot, Fulchignoni, Grémillon, Hirsch, Lalou, Le Corbusier
Max de Haas, Patris, Sarrut, Schwab et Gout, etc..
avec les musiques de : Ferrari, Henry, Mâche, Philippot, Schaeffer, Varèse, Xenakis
et la compagnie de Mime Jacques Lecoq

avec la participation des studios :
Fonologia di Milano - *Apelac de Bruxelles* - *Westdeutscher Rundfunk de Cologne*

★

LOCATION : R.T.F. 18, rue François-1ᵉʳ, - ELY. 57-50
SALLE GAVEAU, 45, rue de la Boétie - BAL. 19-14
Chez DURAND, 4, place de la Madeleine, dans les Agences et par téléphone à S.V.P.
Prix des places : 400 - 500 - 600 - 800 - 1.000 - Tarifs réduits : J.M.F. - A.M.J. - Etudiants
organisation de concert : M. DANDELOT

FIGURE 15. First public appearance of the GRM logo on a poster. © Évelyne Gayou (1959).

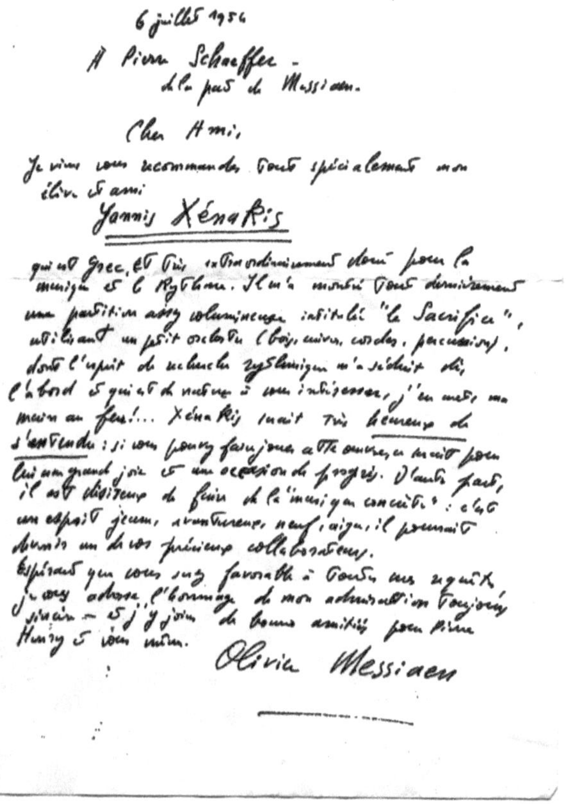

6 juillet 1954
A Pierre Schaeffer –
de la part de Messiaen.

Cher Ami,
Je viens vous recommander tout spécialement mon élève et ami
Yannis Xénakis
qui est Grec, et très extraordinairement doué pour la musique et le Rythme. Il m'a montré tout dernièrement une partition assez volumineuse intitulée "le Sacrifice", utilisant un petit orchestre (bois, cuivres, cordes, percussion), dont l'esprit de recherche rythmique m'a séduit dès l'abord et qui est de nature à vous intéresser, j'en mets ma main au feu!... Xénakis serait très heureux de l'entendre: si vous pouviez faire jouer cette oeuvre, ce serait pour lui une grande joie et une occasion de progrès. D'autre part, il est désireux de faire de la "musique concrète": c'est un esprit jeune, aventureux, neuf, aigu, il pourrait devenir un de vos précieux collaborateurs.
Espérant que vous serez favorable à toutes mes requêtes, je vous adresse l'hommage de mon admiration toujours sincère – et j'y joins de bonnes amitiés pour Pierre Henry et vous même.
Olivier Messiaen

FIGURE 16. Letter of recommendation concerning Iannis Xenakis written by Olivier Messiaen and addressed to Pierre Schaeffer. © Évelyne Gayou (1954).

6 July 1954
To Pierre Schaeffer
From Messiaen

Dear Friend,

I am writing to recommend, in particular, my student and friend Yannis Xénakis who is Greek and very extraordinarily gifted for music and rhythm. He recently showed me a rather voluminous score entitled "Le Sacrifice," using a small orchestra (woodwinds, brass, strings, percussion), with a sense of rhythmic exploration that seduced me from the outset, and which is likely to be of interest to you. I'm putting my hand into the fire!

Xénakis would be delighted to hear if you could have this work performed: it would be a great joy for him and an opportunity for development. Besides, he is eager to compose "musique concrète." He is a youthful spirit, adventurous, fresh, and acute. He could become one of your precious collaborators.

Hoping that you will be supportive of all my requests, I extend to you the testimony of my always-sincere admiration, and I extend warm regards to Pierre Henry and to yourself.

Olivier Messiaen

FIGURE 17. Iannis Xenakis at the GRMC. © INA/Laszlo Ruszka (1957).

the culmination of a lengthy process. It was the successor to Studio d'Essai (1943–46), Club d'Essai (1946–60), and the Groupe de Recherches de Musique Concrète (GRMC, 1951–58), that last one also being a group within Club d'Essai.

According to Pierre Schaeffer, following the public events of *Décade internationale de la musique expérimentale* in 1953, no event of real importance took place at the GRMC until 1957. In his letter to Albert Richard, dated May 18, 1957, he wrote:

> If Musique concrète has not, in fact, made great progress, if it has not revealed any masterpieces, nor yet established a doctrine, the electronicists have had the time to equip themselves, to express themselves. Confronting these results, even if they seem convincing to me, I could therefore declare forfeit and persuade myself of a failure. But this is not the case.[3]

From 1953 until the end of 1957 Pierre Schaeffer was absent from the GRMC, occupied with setting up Radiodiffusion d'Outre-Mer, the vast French overseas broadcasting network. During this time his collaborators—Pierre Henry and Philippe Arthuys in particular—did not remain inactive. On the contrary, they composed numerous works for ballet (Béjart, Dirk Sanders), for cinema (Mitry, Fulchignoni), and for the stage (texts by Tardieu, Mossy, Kenan, Peignot, Vian, Desnos). But on his return, at the end of 1957, Pierre Schaeffer declared his disagreement with the directions taken:

> I had dreamt of collectively questioning processes and systems, whether they were those of the Conservatory, of atonality, or of cybernetics. Instead of concerts and festivals where snobbery rules, instead of these spectacular displays where the tangible success of *Son et Lumière* and TV only blur the picture,* I had dreamt of an honest approach to the phenomenon of hearing, of experimentation with diverse audiences and of an ethics of the listener in which the musician would find, in the end, his own rules and morals. None of this happened.[4]

Pierre Schaeffer took over the reins of the GRMC and decided to dedicate it to fundamental research and experimentation. He wanted to refine working methods. He thought that it was necessary to insist on further efforts to systematize all the discoveries made since the 1950s. He wanted to "rethink music, all music,"[5] which would soon lead to founding the Service de la Recherche de l'ORTF (1960) and to his writing *Traité des objets musicaux* (*TOM*) and then *Solfège de l'objet sonore* (*SOS*).[6]

Here is how Pierre Schaeffer, with a touch of false naïveté, presented in the third person the founding of the GRM:

> Abruptly dismissed from his position in 1958, he willingly returned to Musique Concrète. He noticed that use of the resources he had created ten years earlier was almost exclusively dedicated to secondary purposes, and was in danger of being left behind by foreign techniques. He then decided, with the approval of the management of Radiodiffusion Télévision Française, to completely renew the spirit, methodology and personnel of the Groupe, to return to fundamental research and to provide the necessary conditions for the increasing numbers of young composers who were curious about these new possibilities for musical creation.[7]

The GRMC turned out to function like a centrifuge. The shattering departure of Pierre Henry and his entire team at the beginning of 1958 demagnetized the avant-garde of musique concrète. In seizing the destiny of the GRM in 1958 with an iron grip, Schaeffer would lead research on the sound object, and more broadly seek to pierce the secret of music. He obliged his new collaborators to curb their inclinations for composing, to prioritize experimental research. This harsh period would be highly instructive for those who lived through it. The result of this abnegation is difficult to evaluate as it would have long-term repercussions. A new avant-garde was being born, that of experimental and then electroacoustic music.

Other Sixties Avant-Gardes Related to the GRM

VISUAL ARTS: NEW REALISM, NEW FIGURATION

New realism, defined by the critic Pierre Restany as a set of "new perceptive approaches to reality," reached full bloom in 1958–63. It affected all the arts.

---

* Translator's note: *Son et Lumière*, literally "Sound and Light," is an outdoor nocturnal form of popular entertainment melding site-specific lighting effects, narration, and music, developed in France beginning in the 1950s and most often presented at venues of historical significance.

Certain of its figures would collaborate with the Service de la Recherche: Pierre Restany, Martial Raysse, Christo, Gérard Deschamps, César, Jean Tinguely, Niki de Saint-Phalle, Spoeri, Arman, Yves Klein, Jacques Dufrêne, Rotella, Goldsmith, Raymond Hainz, and Jacques de la Villeglé. The last two, friends since their studies together at Beaux-Arts, the school of fine arts of Rennes, made a film in 1960 to Pierre Schaeffer's *Étude aux allures*.

Narrative figuration, or new figuration, appeared in France during the 1960s, embodied by the visual artists Adami, Erro, Fromanger, Klasen, Monory, Rancillac, and Télémaque. As its name indicates, the new narrative figuration sought to devise narrative approaches, in contrast with advocates of abstraction such as Hartung, Manessier, Poliakoff, and Bazaine. But narrative figuration found it difficult to impose itself in the face of the onslaught of American pop art by Andy Warhol, Lichenstein, John Chamberlain, Jim Dine, Jasper Johns, Frank Stella, and Robert Rauschenberg, which highlighted comic books, portraits of stars, news items, suspended time, and captured moments. The composer Luc Ferrari acknowledged he had been fascinated with pop art when he composed *Hétérozygote* in 1963.[8] Ferrari, also influenced by his experiences in radio with Hörspiel, often incorporated recordings of sound sequences from everyday life into his compositions, or long silences inhabited by only a few rustles, or snatches of conversations in different languages—in a nutshell, anecdotal sounds.

THE SITUATIONIST MOVEMENT

The Internationale Situationniste, founded by Guy Debord and a few friends in 1957, following the Letterist International (1952), set itself the revolutionary goal of creating "constructed situations." Debord advocated nomadic thinking. Asger Jorn suggested engaging in discussions with engineers and moving towards an "imaginist Bauhaus" to escape from automation.[9] In certain aspects this philosophy corresponded with Schaeffer's concept of experimental music. One must, of course, not imagine there was any revolutionary political agenda in Schaeffer's work, but in contrast his approach to musical investigation corresponded to the spirit of the times—breaking out of the established order, engaging in collaborative artistic work, making use of modern instruments, and testing their limits in order to go beyond mechanical reproduction, if necessary through *détournement*.

Détournement has always mesmerized concrete musicians—détournement of studio tools as much as of musical ideas and sounds: détournement of tools such as the tape recorder, by making it run in slow motion or by braking the motors, or by recording the same material repeatedly in superimposition, or by playback of the tape in reverse; détournement, as well, of the measuring devices and generators of all kinds used in research laboratories and even sometimes in industry.* This approach of détournement, or at least of recycling, can be found in the design of the console/synthesizer of Studio 54, transferred in 1975 to Studio 116C of the Maison de la Radio, and exhibited today in the musical instrument museum at the Cité de la Musique in Paris (La Villette).[10] It was completely constructed by Enrico Chiarucci, Francis Coupigny, and Guy Reibel from electronic components originally designed for medical equipment.

Détournement is a state of mind found in the contrarian use of identifiable sound material. Ordinary listening—oriented towards recognition of sources—can be "turned" towards abstraction. This was the case in Luc Ferrari's 1963 piece *Hétérozygote*, where snippets of conversation in different languages, animal cries, recordings of waves, airplanes, and slamming doors were offered up for détournement, solely on the basis of their incongruous presence in the music. The composer proposed détournement as a means of questioning established musical truths, of raising questions in listeners, and of provoking the musical connoisseur. Luc Ferrari drew an analogy, for example, between the bleating of sheep heard in his piece and the "Moutons de Panurge" attitude—"following like sheep"—that he reproached in his fellow human beings. In the same fashion, in 1940 the American John Cage had already applied détournement to the piano, by "preparing" it, as had Pierre Henry in 1948 (as well as the futurist Russolo even earlier in 1927). All these musicians operated with the same intention, attempting to produce sounds that could claim musical value, thus transcending the established musical order of tonal music and its instrumentarium.

CINEMA: FRENCH NEW WAVE AND CINÉMA VÉRITÉ

In cinema, the French *Nouvelle Vague* movement gained international resonance. After starting their careers as critics for the magazine *Les cahiers du cinéma*, the young filmmakers Claude Chabrol, Pierre Kast, Jacques Rivette, François Truffaut, Jean-Luc Godard, Eric Rohmer, and Jacques Doniol-Valcroze stepped behind the camera to make their first feature films between 1958 and 1960. Among these newcomers, there were also young and talented directors of short films such as Jacques Demy, Alain Resnais, Georges Franju, and Chris Marker. In the beginning, they all shared in common working with small budgets, emphasizing improvisation, and rejecting cumbersome technique—all in the name of realism, passion, and simplicity. They were also trying to make a place for themselves in the sleepy French cinema of the time, which had not yet taken note of the existence of new

---

* Translator's note: *Détournement*, a French word translated variously as "rerouting," "diverting," "hijacking," "misappropriating," or "perverting," was adopted first by the situationists to describe the use of existing imagery, concepts, or techniques against themselves, or at least deviating from their original usage.

tools such as Kodak's 16mm camera (which could be handheld, even in the street), "daylight" film, and the Nagra portable tape recorder. These filmmakers were eventually able to develop professional methods comparable to those of the nascent television industry. It is not surprising to find them in the studios of Pierre Schaeffer's Service de la Recherche de la Radio Télévision, which in a few months became the meeting place for avant-garde art in general and audiovisual art in particular.

Cinéma direct, also known as *cinéma vérité*, primarily represented by Jean Rouch in France and Richard Leacock in the United States, also made its appearance in the 1959–60 period. It corresponded to the desire of ethnologists, sociologists, and reporters to investigate directly in the field, thanks to recent technical improvements, including portable, silent, and robust equipment, good sound quality, and easy synchronization between sound and image. In France, this documentary style in cinema found its counterpart in sound in "films for the ear" produced, from 1969 onwards, at the Atelier de Création Radiophonique de France Culture (ACR), directed by Alain Trutat, himself a product of the Studio d'Essai.

All this synergy between music, radio, cinema, and television, partly under leadership by the Research Department, helps explain the intrusion into Pierre Schaeffer's office one Saturday afternoon, during the events of May 1968, by a team from *Cahiers du cinéma* led by Jean-Luc Godard and Eric Rohmer. They wanted to hold Schaeffer captive to extort film equipment from him. However, a few hours later, around two o'clock in the morning, the situation calmed down. In the end, it was agreed that the equipment would be "lent" to them and then returned.

THE MUSICAL CONTEXT

Apsome—Pierre Henry's Studio

Pierre Henry composed several masterpieces in his Apsome studio, including *Voyage* in 1962, *Variations pour une porte et un soupir* (Variations for a door and a sigh), 1963, *Messe pour le temps présent*, 1967, and *Apocalypse de Jean* in 1968. His career transpired independently from the GRM, while aesthetically remaining quite close. His reconciliation with Pierre Schaeffer did not occur until 1966, and then only thanks to intercession by François Bayle, newly appointed director of the GRM. According to Bayle, "it was not difficult, because they appreciated each other very much." During the artistic effervescence of the 1960s, Pierre Henry collaborated with many artists. Besides his well-known connection with the dance world, in particular with Maurice Béjart, one should not overlook his numerous contacts with visual artists: Nicolas Schöffer, for *Tour spatiodynamique*, and Arman, to whom *Variations pour une porte et un soupir* was dedicated; with writers such as Weyergans; and with poets like the Letterist François Dufrêne, who as early as 1953 had used a "pen recorder" to record his poems. Henry used the voice of Dufrêne to compose *Granulométrie* in 1967, a work that Michel Chion considered "one of the summits of electroacoustic expressionism," a "masculine counterpart" to Luciano Berio's *Visage*, where the voice of the singer Cathy Berberian was also as "extraordinarily present" as that of Dufrêne.[11]

Domaine Musical

It is no secret that the Domaine Musical never opened its doors to musique concrète. During the 1960s, serialists and classical contemporary composers continued to meet among themselves "at the Domaine." But under the growing pressure of the new avant-gardes, Pierre Boulez chose to broaden his palette of guests. Jesus Aguila described this evolution:

> Between 1960 and 1963, as the Darmstadt community showed signs of unraveling under the shock of John Cage's ideas, as the electroacoustic tool gained in refinement, and as the alternative approaches of Györgi Ligeti and Iannis Xenakis received recognition, Pierre Boulez's position changed significantly. He himself was no longer the figurehead of modernity, but the symbol of retreat into post-Webernian values.[12]

Boulez would finally withdraw from the Domaine Musical in 1967; Gilbert Amy assumed the difficult role of his successor as director of the association.

The Cologne Studio

From 1956 to 1972 the Cologne Studio experienced its golden period, hosting a multitude of composers eager to experiment with this new electronic tool. Key works in the history of electronic music emerged: Stockhausen's *Kontakte* (1960), Ligeti's *Artikulation* (1958), Eimert's *Epitaph für Aikichi Kuboyama* (1962), and more. With *Mikrophonie I* (1964) for tam tam, microphones, and live transformation of concrete sounds by electronic means (essentially filtering), Stockhausen introduced the concept of live electronics to Europe. He renewed the same experience of live transformation of sounds, this time with voices, in *Mikophonie II* (1965), with four ring modulators. This concept of live electronics—already used in pop music—had originated in the United States, where Stockhausen had spent time. John Cage had composed *Cartridge Music* based on this principle in 1960. Stockhausen's innovation lay in the fact that he integrated it into an expanded serial conception. During this period, intense creative activity engulfed the Cologne Studio, and its reputation rivaled that of the Paris and Milan studios.

In 1963, Karlheinz Stockhausen officially took over direction of the Cologne Studio from Herbert Eimert, who was retiring, while Koenig went to Utrecht in the Netherlands to assume artistic direction of a new electronic music studio (the future Institute of Sonology). A page had been turned. Stockhausen, left alone, obtained a new studio from the management of WDR. Radio, in addition to the original smaller Studio 11. The Cologne Studio became a place of encounter and exchange, all the more so since it was equipped with two new 4-track tape recorders, a new console, and a new "rotation table."*

During the period 1966–68, a new circle of artists from the Fluxus movement came into closer contact with the Cologne Studio: Nam June Paik, John Cage, and David Tudor. Fluxus, the international movement created in 1961 by George Maciunas in the United States, advocated a philosophy that can be summarized as follows: art and life are one, each moment of life is an instant of art. As an example of the euphoria that reigned in Cologne at the time, here is an anecdote shared by Leopold von Knobelsdorff, sound engineer at the studio and an excellent jazz pianist:

> We named our Studio "eleven," the mythical number in the Chinese Tao, 11. It is the *unus mundus*, the One (of dozens). We tried to discover this number everywhere, for example at the Chinese restaurant where we went for lunch, we amused ourselves calculating it from the dish numbers on the menu, figuring it out using serial techniques. We forbade ourselves from ordering dishes where the sum of the numbers or the difference did not equal 11.[13]

In a 1992 interview with Jean-Yves Bosseur, Nam June Paik revealed that as early as 1958, being interested in collage, he had wanted to come to France to make musique concrète at the GRM:

> So I sent a letter to Pierre Schaeffer, but he never answered me. I met him later and told him: "Say, Pierre, you didn't answer my letter." He replied, "Oh, I'm so sorry." He regretted it—meaning he remembered it. I had done everything I could to get into the Groupe de Recherches that he directed. Failing that, I left for Cologne.[14]

But the two studios, both engaged in intensive production, didn't waste energy on politeness. For example, Pierre Schaeffer evoked the work done at the Cologne Studio in his lecture *Retour aux sources* during the inauguration of the Festival de la Recherche on May 26, 1960: "Take a small wooden bow. Pinch the string in front of your mouth, which serves as a resonator ... If you possess a bit of talent and a lot of patience, you will get music that is strangely reminiscent of some of the products of electronic music studios, those of Cologne or Milan."[15]

A little later during the same lecture, he concluded his introduction:

> The excesses of experimental modernism reopen our eyes to the fundamental gestures of prehistoric musicians, lending us the ear of these primitive *homo faber*.

---

* Translator's note: Stockhausen's invention placed a loudspeaker on a rotating table, creating a sensation of movement in the listener by recording with four separately-placed microphones. This technique was notably used for the tape for *Kontakte*.

We would like to offer you the opportunity to retrace, in a single evening, this journey of ten years and of a hundred millennia. We cannot guarantee its precision, but a hypothesis is better than indifferent courtesy.[16]

Such was the tone of interaction between the Paris and Cologne studios during the 1960s, bittersweet collusion and serious probing into the nature of music. But unlike Schaeffer, who was able to create an institutional environment favorable to the continuation of research, Stockhausen remained focused on musical creation. Between his frequent travels, he returned to Cologne to compose the electronic parts of his works. In his absence, the studio hosted composers preparing new works. However, from 1966 onwards, the director of the WDR, himself an opera specialist with little interest in electronic music, imposed an administrator on the studio to "second" Stockhausen. This was the first sign of the gradual disengagement of German radio WDR from the adventure of electronic music.

Sound Synthesis in the United States

In the field of sound synthesis and computer music, America was ahead of Europe. In 1957 at Princeton University, analog synthesis was already available to students in the music department. Milton Babbitt had just acquired an analog synthesizer, the Mark II, from RCA. In an interview with Joel Chadabe in 1997, Milton Babbitt recalled: "The Mark II was a floor-to-ceiling wall-to-wall synthesizer built by Harry Olson and Herbert Belar at RCA's Sarnoff Laboratories in Princeton."[17]

In 1957, again in the United States, in the state of Illinois, Lejaren Hiller and Leonard Isaacson composed *Illiac Suite for String Quartet*. The score of this piece was written by computer from a set of rules; this "listing" output was then scrupulously transcribed into musical notes. That same year, at the Bell Telephone Laboratories in New Jersey near New York, in a completely different approach, the musical engineer Max Mathews was working with John Pierce, Henk McDonald, and Newman Guttman on digital voice synthesis, generated by the computer, and various other processes for telephony. With the support of his director John Pierce, Mathews decided to extrapolate his work to music. He began writing the Music I computer program, the first in a long series. Varèse paid close attention to this undertaking, in the spring of 1959 bringing to the attention of his audience one of the first pieces of music using computer-synthesized sounds—Newman Guttman's *Pitch Variations* (1:00)—during his carte blanche performance at the Village Gate in New York. Guttman's first real work of digital sound synthesis was the nineteen-second 1957 piece titled *The Silver Scale*.

These events constitute the "official" beginning of the long epic of computer music.

In 1959, Mathews wrote Music III, a synthesis program that prefigured Music IV, V, X, C Music, C Sound, modular synthesizers, and object programming.

Beginning in 1964 the Frenchman Jean-Claude Risset participated in the pioneering period of American computer music, spending three years with Max Mathews at Bell Labs as part of his doctoral research into analysis and digital synthesis, notably succeeding in using synthesis to imitate the timbre of the trumpet.

In 1969 Risset composed the sounds for *Mutations* using the Music V software at Bell Labs, in response to a music commission from the GRM. It was the very first work realized on a computer to enter the GRM's repertoire.

EVENTS, PERFORMANCES, PUBLICATIONS, PEOPLE

Expo 58—Brussels World Fair

Stimulated by the deadline of the Brussels World Fair in 1958, Schaeffer returned to the studio to prepare new things to present. He composed *Étude aux allures* (3:28) and *Étude aux sons animés* (4:12). These two études premiered October 5, 1958, at the French Pavilion of the Brussels International Exhibition, along with Luc Ferrari's *Étude floue* (2:40), *Étude aux accidents* (2:12), and *Étude aux sons tendus* (2:44), as well as Schaeffer and Ferrari's *Continuo* (2:38) and Iannis Xenakis's *Diamorphoses* (6:50). Simultaneously, but this time in the Philips Pavilion, Iannis Xenakis presented another composition, *Concret PH* (2:39) as a prelude to Edgar Varèse's *Poème électronique*.

In 1959, Pierre Schaeffer would present in Paris his study of musical objects *Étude aux objets* (16:40, 17:10 in its 1971 revision). He had not had the time to finish it for the Brussels exhibition. This etude consisted of five parts:

Objets exposés (3:34), the objects introduced;
Objets étendus (2:54), objects expanded;
Objets multipliés (3:02), objects accumulated;
Objets liés (3:07), objects linked; and
Objets rassemblés (4:19), objects assembled.

Pierre Schaeffer wrote about his work in *La Revue musicale* in 1959:

> My three new études [*Étude aux Allures, Étude aux sons animés,* and *Étude aux objets*] are based on this threefold asceticism: a deliberately limited number of sound bodies, manipulations essentially oriented towards "montage" and no longer towards deformation, and finally, a compositional bias consisting in submission to the object rather than torturing its modulation based on preconceived considerations. It is the *object*, I think, which has something to tell us, if we know how to make it speak and assemble it accordingly based on its family relationships and concordance in its characteristics.[18]

After these three études, Schaeffer the composer fell silent again until 1975, when he would compose *Trièdre fertile*.

Following Pierre Henry's departure, a new group of composers gradually formed: Michel Philippot, Ivo Malec, François-Bernard Mâche, Luc Ferrari, Iannis Xenakis, Mireille Chamass-Kyrou, Bernard Parmegiani, and François Bayle.

At the beginning of 1958, Luc Ferrari joined the GRM almost the same day that Pierre Henry had left. He immediately began to compose extensively. While his first pieces were still exclusively instrumental—such as *Visage III*, his first new work registered in the GRM repertoire—the following works in the wake of *Visage IV* immediately reflected his interest in the problematic of sound objects. And finally, his first works on a support, studies on typo-morphological criteria, arrived at the end of 1958: *Étude aux accidents* (Étude of accidents) and *Étude aux sons tendus* (Étude of sustained sounds).

1958, *Visage III* (Ferrari), *La prose du transsibérien* by Blaise Cendrars (32 minutes) for violin, cello, clarinet, percussions, and narrator. Poem by Blaise Cendrars, read by Jean Topart. Recorded by the Ensemble Instrumental de Musique Contemporaine de Paris (EIMCP), conducted by Konstantin Simonovic, in 1962.

1958, *Visage IV* (Ferrari), or *Profiles* (10:05) for ten instruments: two flutes, trumpet, trombone, bass trombone, string bass, piano, three percussions. This piece premiered at the Musik der Zeit in Cologne in 1959 and was awarded the prize of the Paris Biennale in 1962.

1959, *Visage V* (10:32) was Luc Ferrari's first work of musique concrète. According to Schaeffer, writing in 1961, "*Visage V* follows four other compositions for orchestra. Luc Ferrari reconnects naturally, through new means, with his personal preoccupations."[19]

Mireille Kyrou, a student of Olivier Messiaen and an excellent conservatory-trained musician, pursued experimental sound research activities (listening, classification) when she joined the GRM.

Bernard Parmegiani was a sound engineer for television. Schaeffer invited him to join the GRM in 1959. His first activities at the Groupe consisted in assisting Luc Ferrari and Iannis Xenakis. At the beginning, Parmegiani composed mainly music dedicated to the image: *Jours de ma vie*, a film by Max de Haas, and *Steinberg*, a film by Peter Kassowitz, based on drawings by Steinberg.

François Bayle came into contact with Schaeffer in 1958, but Schaeffer did not introduce him to the GRM until 1960, in the context of the Service de la recherche, as secretary general. "Yet I bravely showed him my first attempts at musique concrète, which he listened to attentively with Luc Ferrari (to tell me that it was not worth much, obviously!)."[20] Like many composers/researchers affiliated with the

GRM, Bayle first composed instrumental music. It was not until 1962–63 that he truly engaged in serious studio composition techniques. He later commented, "The first achieved piece in my repertoire was *Espaces inhabitables*, in 1966, but my musical beginnings were earlier, in 1962, with *Trois portraits d'un Oiseau-Qui-N'Existe-Pas*, a score for a film by Robert Lapoujade.²¹

During the first two years of the GRM's existence, Schaeffer focused on implementing his musical research project. In June 1959 the GRM organized a series of six concerts at Salle Gaveau in Paris. The work of the Paris studio was presented alongside work from comparable foreign centers: Tokyo, Brussels, Cologne, Milan. Two remarkable musical premieres were presented to Parisians: Gyorgy Ligeti's *Artikulation*, an electronic composition realized at the WDR studio in Cologne in 1958, and Luciano Berio's *Omaggio à Joyce*, realized at the RAI studio in Milan in 1958, based on vocal elements extracted from a reading of Joyce's *Ulysses*. We also note the presence of the composer André Boucourechliev with his first work realized at the GRM, *Texte 2*.²² This brief aleatory piece, lasting from four and a half to five minutes, was composed on three tape tracks that could each be launched at will.²³

The success of these public events created favorable circumstances for the management of RTF to promote a more ambitious project. Schaeffer proposed creating the Service de la Recherche by regrouping, in a single service, the activities of the Club d'Essai, the Centre d'Études de Radio-Télévision (CERT), and the young Groupe de Recherches Musicales. In addition, he advocated extending the activity of fundamental research to the visual image, "the common language of cinema and television," "in an approach combining artistic objectives with technical means."²⁴ The project came at the right time: the RTF needed to modernize, and at a lower cost. The Service de la Recherche responded to two imperatives. It created a pole of high technology without disturbing the rest of the institution, and there was another advantage: it kept the "turbulent" Schaeffer within a circumscribed perimeter.

Birth of the Service de la Recherche at the ORTF

The RTF Research Department, or Service de la Recherche, was officially created on January 1, 1960, and operated until January 1, 1975. In 1964, it survived the change of status of RTF, which became the ORTF by putting an end to its monopoly on production.²⁵ Schaeffer was appointed director of the Service de la Recherche, in a few short months finding himself at the head, not of the modest GRM within the Club d'Essai, with its fifteen people, but of a large research department of more than a hundred people, most of whom he himself had hired. For the first time in its existence, the GRM had bylaws. They provided the basic institutional model for the entire Research Department, of which it was a part: with the exception of the statutory technical and administrative staff, all the other contributors received contracts, either temporary for the duration of a production, or renewable annually. In

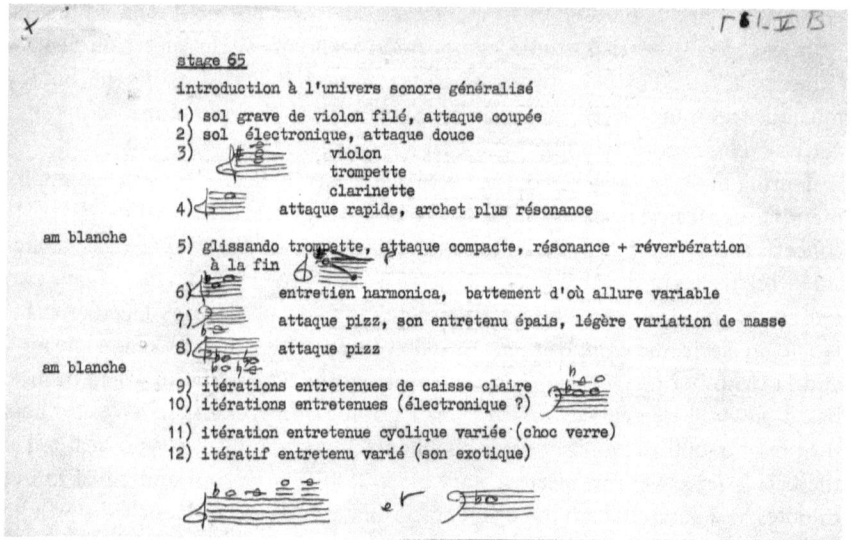

FIGURE 18. File card of sound examples, one of thousands at the Research Department. © Évelyne Gayou (1961).

consideration for their "functional" activities, collaborators were entitled to also undertake creative activities. Research, with no limits, was the order of the day, in synergy with the great cultural effervescence preceding the events of May 1968.

With this in mind, researchers were brought together with creators and all the categories of audiovisual professionals to collaboratively advance and develop production and innovation. Based on the example of the Groupe de Recherches Musicales, Pierre Schaeffer created the Groupe de Recherches Image (GRI, Group for Visual Research), the Groupe de Recherches Technologiques (GRT, Technological Research), and the Groupe Recherches Langage (Language Research), which became the Groupe d'Études Critiques (GEC, Critical Studies Group).[26] One of the main objectives of this system was to foster mutual training of the different actors in the new audiovisual sector, which was developing very quickly and was cruelly lacking in technical, artistic, and managerial skills. This institutional structure devised by Schaeffer proved to be particularly productive.

The GRM's First "Grand" Workshop

Ivo Malec, Iannis Xenakis, Luc Ferrari, Michel Philippot, and Mireille Chamass-Kyrou all joined the GRM in 1957–58. They were trained on the job and during training sessions lasting a few weeks under the patronage of Pierre Henry and Philippe Arthuys. In 1961, the first GRM "grand" training course, led by Pierre

Schaeffer, Luc Ferrari, and Mireille Chamass-Kyrou, adopted the same pedagogy as the previous courses: polyvalence and listening and composition exercises. But this time it was to last two years. Among the first trainees were François Bayle (who would become director of the GRM in 1966), Bernard Parmegiani, Edgardo Canton, Gilbert Amy, Claude Ballif, Philippe Carson, Juan Hidalgo, Ivo Malec, and N'Guyen Van Tuong. Then would follow Guy Reibel, Enrico Chiarucci, Romuald Vandelle, Beatriz Ferreyra, and Francis Régnier.

Festival de la Recherche 1960

The 1960 "Research Festival" was the first public event held by the GRM after becoming part of the newly created Research Department. This high-level and multidisciplinary festival would leave a strong impression on people's memories. Pierre Schaeffer inaugurated the festival on May 26 at the Salle Gaveau in Paris with a concert accompanied by commentary, under the title *Retour aux sources* (Return to the roots). During this conference, recorded to be broadcast later on radio, he presented his research hypotheses in the form of a comparative study of a large number of musical excerpts from the classical and traditional repertoires from around the world—Asia, Africa, and Europe. In his introduction, Schaeffer asserted:

> If you want grace, Pascal told musicians, make yourself stupid. But they were all so intelligent . . .
> But it isn't stupid, once you have reached the end of the tape reel, to return to roots! It isn't stupid, once the musical scale has been cut into twelve and all the permutations exhausted, to see how the others, the naive, the primitive, the great Ancients, use it, or used to use it.

It was during this well-known conference that Schaeffer articulated for the first time his famous words: "One can ask oneself which, man or nature, is the musician. Or more to the point, since there is music only through our ears and our understanding, if music begins when we make it or when we hear it."[27]

For a month, a dozen conferences, six concerts and technical demonstrations, and screenings of experimental films followed one another. All these events were organized jointly by the different research groups—musical, image, and technical. Among the themes addressed in the conferences:

—Music and Acoustics;
—Music and Machines;
—Networks and Message;
—Music, Physiology, and Psychology;
—Memory and Processing; and
—Anamorphosis between Music and Acoustics.

The lecturers were selected from among the most renowned specialists in their fields: Professor Fritz Winckel (from the Berlin Technical University), Abraham Moles, Émile Leipp, Maurice Martenot, Louis Busnel, Arno-Charles Brun, Jean Yanowski, Pierre Schaeffer, Michel Philippot, Jacques Poullin, among others. And the filmmakers included René Kassovitz, René Laloux, Jacques Brissot, Piotr Kamler, and Raymond Hainz.

The Festival de la Recherche reflected the magnitude of the exploratory activities underway. But it is important to understand that this was merely the visible part, shown to the public, of the multitude of research subjects that had been launched simultaneously. All these undertakings would continue for several years.

Following the success of the Festival of Research in 1960, Pierre Schaeffer considered using this model to hold a research biennial. In the end, he instead decided to orient the Research Department towards producing television programming on research.

Concert Collectif, 1959–1963 Group Concerts

This original compositional experiment of the *Concert Collectif* lasted from April 1959 to March 1963. The participants were numerous and varied. Not all participated in the experiment from beginning to end. In chronological order of arrival were Mireille Chamass-Kyrou, Luc Ferrari, François-Bernard Mâche, Iannis Xenakis, Claude Ballif, François Bayle, Edgardo Canton, Ivo Malec, Bernard Parmegiani, Michel Philippot, Romuald Vandelle, N'Guyen Van Tuong, Philippe Carson, and Jean-Étienne Marie.

Using nine basic sequences that were used in common, each author was required to compose a work of approximately ten minutes in length. Most of the basic sequences were taken from recordings made by the Ensemble Instrumental de Musique Contemporaine de Paris (EIMCP), directed by Konstantin Simonovic.[28] This ensemble collaborated with the GRM, both in creating sounds and in performing finished compositions.

Noteworthy in this group venture was the participation of Iannis Xenakis, which lasted only a few months from November 1961 to May 1962. His dynamism, combined with that of Luc Ferrari, made him a major architect of the project's success. His contributions stimulated the musical debate.

At the beginning of 1959, Pierre Schaeffer evoked a "concerto grosso," which he would direct with the help of the principal collaborators in the group. On April 9, 1959, he launched the *Concert Collectif* project with the participation of three members of the GRM: Mireille Chamass-Kyrou, Luc Ferrari, and François-Bernard Mâche. After a dormant period from April 1959 to November 1961, the project was revived by Iannis Xenakis and Luc Ferrari, following discussions initiated by Pierre Schaeffer on the idea that one could not devise a piece from "musical objects," but rather only from musical structures. Xenakis and Ferrari decided to develop a plan for the realization of this concerto grosso. The idea was to organize a one-hour concert that would constitute a "group piece." The proposal detailed the available resources: classical orchestra, new lutherie, recorded orchestra, and sounds transformed by electroacoustics. A temporal breakdown was provided, as well as a profile of the group piece and its spatialization.

On January 3, 1962, Pierre Schaeffer entrusted responsibility for this concert to Iannis Xenakis, who further refined the project—each composer was expected to write three sections of the piece, ensuring that they could be interlinked with the sections written by the other composers. The first team of composers brought together Ballif, Bayle, Canton, Ferrari, Mâche, Malec, Parmegiani, Philippot, and Xenakis. They met several times a week to listen, criticize, and compare their sequences. Pierre Schaeffer declared that he did not believe in the idea of group work, but did believe in the *Concert Collectif*. He distinguished between the search for musical structures resulting from an objective study and the search for form or musical architecture that comes from an instinctive methodology. A contradiction emerged. While Pierre Schaeffer saw it as a research experiment into the concept of structure, the composers above all envisioned a creative experience.

To break the deadlock, Xenakis then proposed using probability tables to determine the passage between composed sections, leaving the composer with complete creative liberty within each section. This provoked a great deal of excitement and enthusiasm for the project. Each composer would now have to provide nine sequences, each following an exacting table of criteria imposing three levels of intensity, three levels of density, three levels of variation, and three levels of originality.

In May 1962, the second team, composed of Xenakis, Bayle, Canton, Ferrari, Malec, Parmegiani, and Philippot, began listening to and testing the fifty-four sequences. It was a setback since the sequences did not sufficiently respect the criteria and they retained the stamp of each individual's style to an excessive degree. It was therefore not possible to apply the stochastic method to their sequencing. Pierre Schaeffer then proposed to return to a more empirical way of working, by separating the theoretical from the realization. "The criteria turned out to be erroneous, but served as an effective stimulus for creation," observed Ferrari.[29]

On May 30, 1962, Iannis Xenakis, the project leader, decided to withdraw.

The *Concert Collectif* finally took place on July 20, 1962, at the Salle du Ranelagh in Paris. The third team consisted of Bayle, Canton, Ferrari, Malec, Parmegiani, Carson, Marie, Philippot, N'Guyen Van Tuong, and Vandelle. Each author composed a work of about ten minutes, using sequences shared in common and considered as raw material. A second and more finished version of the *Concert Collectif* was performed on March 18, 1963, at the Salle des Conservatoires in Paris.

Today the traces of this experiment still remain. In the sound archives of the GRM are all the carefully classified musical sequences, together with their preparatory drafts. And in the participants' memories remain the trace of an incalculable number of meetings and exchanges which, in the final analysis, constituted the most significant aspect of this unique musical enterprise.[30] But it is necessary to admit that the musical quality of this group production was not up to the level of the richness of the exchanges generated between the composers.

The following works were performed on March 18, 1963, for the *Concert Collectif* at the Salle des Conservatoires, with Charles Bruck conducting the instrumental ensemble:

—Bernard Parmegiani, *Alternances* (7:47), 4-track 1-inch version. This composition was entirely electroacoustic, as was Tuong's. All the other pieces featured instrumental parts performed live.
—Philippe Carson, *Collages* (Tape, 3:36; mix with sixteen recorded instruments, 5:57).
—Edgardo Canton, *D'un bout à l'autre* (tape, 5:59; mix with sixteen recorded instruments, 6:45).
—Jean-Étienne Marie, *L'Expérience ambiguë* (tape, 6:22; mix with twelve prerecorded instruments, 6:20).
—N'Guyen Van Tuong, *Éventail* (4-track version, 8:30; 2-track version, 7:47).
—François Bayle, *Pluriel* (tape, 3:21; mix with seventeen recorded instruments, 8:23).
—Ivo Malec, *Tutti* (Tape, 7:16; mix with nineteen recorded instruments, 7:58).
—François-Bernard Mâche, *Synergies* (Tape, 7:26; mix with twenty recorded instruments, 7:44).
—Luc Ferrari, *Composé* (Composite) (tape, 8:26; mixed version with twenty precorded instruments, 8:26).

TOM and SOS

The *Traité des objets musicaux* (*TOM*) was published in 1966 by Éditions Seuil.[31] Its companion recording, *Solfège de l'objet sonore* (*SOS*), was released the following year in the form of a boxed set of three vinyl records, accompanied by a booklet in three languages: French, English, and German.[32]

In the *TOM*, sound objects were analyzed in terms of perceptual criteria, and classified by type, morphology, and so forth. But for Schaeffer, the essential question was to pierce the mystery of the passage into music. The *SOS* enumerated, in the form of sound examples commented orally by Schaeffer himself, the entire range of the parameters of sound perception described in the *TOM*. But as Schaeffer himself lamented in the accompanying booklet, what was ultimately missing, in addition to a few selected musical excerpts, were more numerous and irrefutable examples of this famous passage into the musical. In his view, the work was unfinished. But in spite of everything, it represented an immense and almost inconceivable body of work. The *TOM* and the *SOS* were the fruit of the preceding twenty years of research and experimentation. They were the outcome of collaborative work that had begun in 1964, directed by Pierre Schaeffer and conducted by a small team composed of Bernard Baschet, Henri Chiarucci, Beatriz Ferreyra, Pierre Janin, Guy Reibel, David Rissin, and Simone Rist.

Beatriz Ferreyra, speaking in 2001, recounted:

> Starting in 1964, I was part of a group that we jokingly called "the Solfeggio Group," which included Bernard Baschet, at that time director of the GRM, Enrico Chiarucci, a brilliant physicist, Guy Reibel, an engineer and composer, and the singer Simone Rist. Our task was to analyze sounds for the development of the typo-morphology, confirming or invalidating Pierre Schaeffer's intuitive, theoretical and descriptive suggestions for drafting *Traité des Objets musicaux* and later for producing *Solfège de l'Objet Sonore*.[33]

Guy Reibel adds another perspective:

> That was indeed the spirit in which we experienced our activities. Listening, greedy listening to the most incredible sounds, eagerly created and appreciated, and an attempt at theorizing a posteriori. Theorizing focused only on perception and not on the methods of production. Any attempt related to the modes of production would have been doomed to failure, so great was the distance between cause and effect.
>
> Our empiricism was vindicated as a method, a way of being. On one side there was "doing," a set of gestures and artisanal procedures linked to machines, on the other side "listening," a set of codes defined gradually by Schaeffer, inscribed in his famous *Traité des Objets Musicaux*, where typology and morphology constituted the most concrete chapters.[34]

Traité des objets musicaux has become a work of reference for composers, pedagogues, and musicologists interested in music on support and in the concrete approach focused on perception. But partly because it would not be translated into other languages until much later, nor associated with teaching, and partly because it was not described as an essay by its author himself, it has been the subject of multifarious exegeses and interpretations that have sometimes contributed to clouding understanding of it.

Another publication should be evoked, much more modest, but highly symbolic of the growing notoriety of musique concrète. In 1967, the book *La musique concrète* was published in *Que sais-je?*, a series featuring books for the general public written by specialists in each field. It was authored by Pierre Schaeffer with the collaboration of Pierre Janin and David Rissin, and then of Michel Chion for the 1973 reedition.

Technical Exploration

Modernization of the studios was extensive between 1958 and 1968. Analog sound transformation tools continued to be refined. The *phonogène universel*, developed in 1961 by the engineer Jacques Poullin, allowed for independent variations in duration and pitch. It is comparable to the Springer machine, which uses the same patented "rotating head" as the one used in the Cologne Studio. Today we might speak of a *harmonizer* to describe this device. The GRM, like the rest of the Service de la Recherche, also acquired new equipment when it moved to the Bourdan Center in 1963. Francis Coupigny and Enrico Chiarucci, assisted by Alain de Chambure, completely rethought the two GRM studios. They decided to dedicate Studio 52 to research into music spatialization and to equip Studio 54 with analog synthesis equipment dedicated to music composition, with the construction of a modular synthesizer prototype, coupled in 1965 to a mixing console. This latter equipment, fully operational beginning in 1971, would strongly mark the aesthetics of the works of the following decade.

In the meantime, another major mutation was already beginning to take shape: the transition to digital. The composer Yannis Xenakis, fond of mathematics and a friend of Pierre Barbaud and the now defunct Groupe de Musique Algorithmique de Paris (GMAP), was the first to attempt, as early as 1962–63, to convince Pierre Schaeffer to take an interest in the nascent field of computers.[35] In vain. Xenakis even founded a "club," somewhat "dissident" from Pierre Schaeffer: MYAM, an acronym formed by the initials of the first names of its members—Mireille Chamass-Kyrou, Yannis Xenakis, Alain de Chambure, and Michel Philippot. MYAM focused on the application of information theory and cybernetics to music. Schaeffer, who at first tolerated this initiative, ultimately rejected it by pronouncing the dissolution of the "club." The divergence of viewpoints between Pierre Schaeffer and Iannis Xenakis concerning information technology was one of the causes for Xenakis's departure from the GRM in 1963.[36] And yet, Schaeffer placed high hopes on "cybernetics," in particular in terms of hybrid synthesis. He thought of computers as large instruments for the transformation of sounds.

GRM Musical Creation in the '60s

Within the Research Department, GRM members were called upon to compose a multitude of works for use in film, theater, dance, and television as part of their experimental activities. A great diversification of genres developed inside electroacoustics itself. During the period of the Service de la Recherche, the output of works for the visual and performing arts exceeded that of pure music, more so than at any other time in the GRM's past or future. The overall number of works produced between 1960 and 1974, in other words during the entire lifetime of the Service de la Recherche, reached an annual average of thirty-five to forty. The peak year of 1962 reached seventy works, half of which were of applied music. After 1968, works of pure music would predominate again.[37]

1960, *Tête et queue du dragon* (9:08), musique concrète by Luc Ferrari.

1960, *Orient-Occident* (11:09) by Iannis Xenakis, music for Enrico Fulchignoni's film by the same name, which premiered at the Cannes Film Festival the same year.

1960, *Volumes* (12:28) by François-Bernard Mâche, for instrumental ensemble and 12-track monophonic tape.

1961, *Reflets* (2:33), a workshop étude. Ivo Malec's second piece of musique concrète, following his *Mavena* in 1956. This étude was subsequently accompanied by images in an abstract style by Piotr Kamler (reflections of water on the Seine).

1961, *Dahovi* (6:30, and 7:15 for the second version from 1962). Ivo Malec composed this music for the film *Structures* by Piotr Kamler. The Serbo-Croatian word *dahovi* signifies breath, alluding to the white noise that served as the initial sound material for the composition.

1961, *Tautologos 1* (4:20) by Luc Ferrari. This work, commissioned by Hermann Scherchen, was realized in the Gravesano studio and mixed at the GRM. Ferrari was the first to have access to a 4-track tape recorder. He chose to present his piece using a circular installation.

1961, *Turmac* (9:43) Philippe Carson composed this piece as sound for the Peter Stuyvesant exhibition at the Museum of Modern Art in Amsterdam.

1961, Stockhausen Concert, November 27, at Salle Gaveau:
François Bayle still takes great pride in having succeeded in persuading Schaeffer to hold the first full concert dedicated to Stockhausen in Paris, presenting *Kontakte*, *Refrain*, and *Zyklus*. According to François Bayle, "Stockhausen had begun having problems with Domaine Musical. Boulez was quite friendly with him, but at the same time wary of him as a burdensome competitor. So Boulez would program one work by Stockhausen, but not more."[38]

The GRM also gave the Parisian premieres of Stockhausen's *Mikrophonie 1* at the Salle de l'Ancien Conservatoire on December 9, 1965, and *Stimmung* on December 9, 1968, at the Maison de la Radio.

1962, *Bohor* (22 minutes) by Iannis Xenakis. First work conceived for eight tracks (quadruple stereo). This work, dedicated to Pierre Schaeffer, evoked the Knights of the Round Table. Insiders realized that King Arthur represented Schaeffer, and Bohor was Xenakis. Schaeffer declared that he hated this piece, that it "ripped eardrums."[39] It is true that *Bohor* was often performed by its composer at a very high amplification level.

1963, *Sigma* by Ivo Malec. This work for large orchestra—published by Breitkopf und Härtel and which premiered on May 16, 1963, with the Zagreb Philharmonic under the direction of Igor Gjadrov at the Zagreb Biennale— is not in the GRM catalog, but I am citing it because of its great impact on the international music scene at the time. In this work, Ivo Malec, then a member of the GRM, succeeded for the first time in transferring the composition techniques specific to studio work to composing for orchestra. The inspiration for this process came to him fortuitously during the experience of the *Concert Collectif*. One day while walking outside under the open windows of the GRM studios, where everyone was working on the sequences of the *Concert Collectif*, he heard his own instrumental sequence played simultaneously in several studios, giving the effect of a radio mix. The idea then came to him to apply this mixing process to orchestral writing—and by extension to transfer other studio effects to written music as well—editing, manipulation of potentiometers, attack substitutions, loops, etc.

1963–64, *Hétérozygote* (26:20) by Luc Ferrari, the first so-called "anecdotal" work, because it uses sounds with recognizable sources—bleating animals, snatches of conversations. But these sounds are arranged in such a way that the listener is constantly caught in the ambiguity between musical listening and being tempted by documentary listening. *Hétérozygote* is also remarkable because it uses realistic recordings made outdoors, in the spirit of the Nouvelle Vague, or pop art, or even situationism, as we shall see below. These anecdotal sequences alternate with instrumental passages written in a pure classical spirit. Considered as the founding work of the anecdotal "genre," *Hétérozygote* was to contribute to reviving the debate concerning the abstract/concrete relationship in electroacoustic music. But Schaeffer welcomed it coldly, inciting Ferrari to distance himself from the GRM. This separation would lead to his brilliant and original career.

1963, *Times Five* (15:13) for 4-track 1-inch tape and instrumental ensemble, by Earle Brown (1927–2002), the only work realized at the GRM by this American composer from the New York school.[40] Frontal sound projection—two stereo systems at two different heights. Performed on February 15, 1963, at the Salle des Conservatoires, under the direction of Gilbert Amy.

1963, *Trois portraits d'un Oiseau-Qui-N'Existe-Pas* (8 minutes) by François Bayle. This music, composed for an animated film by Robert Lapoujade, blended concrète sound with instrumental sound played by the Ensemble Instrumental de Musique Contemporaine de Paris (EIMCP). Later the third section of this film music, *L'Oiseau chanteur*, took on its independence from the film, becoming the first section of *Trois rêves d'oiseau* in 1971.

1964, *Violostries* (16:45), mixed work for tape and violin by Bernard Parmegiani, with the violin part composed and performed by Devy Erlih, and the tape part composed from nine basic sounds from the violin.

1965, *Opérabus*, by François Bayle, Edgardo Canton, Luc Ferrari, Ivo Malec, and Bernard Parmegiani. Musical theater, for tape and instrumental ensemble, to a text by Pierre Schaeffer. Premiered at the 1965 Zagreb Biennial.

1965, *Laborintus* (31 minutes) by Luciano Berio, in homage to Dante. Radio version of the concert version *Laborintus II*, for tape, orchestra, three voices, choir, and announcer, under the direction of the composer. With the Swingle Singers, and Jean-Pierre Drouet on percussions. Texts by Eduardo Sanguinetti, Dante, Ezra Pound, and T.S. Eliot, read by the composer. The preparatory work for this piece was done in part at the RAI in Milan, with the final mix being done at the GRM.

1965, *Voix inouïes* (9:45) by Edgardo Canton. Premiered the following year at the Exposition des Musiques Éxpérimentales.

EXPERIMENTAL MUSIC EXPOSITIONS

Beginning in 1966, these monthly events programmed by François Bayle and Ivo Malec (five concerts each season) presented premieres and other works from the electroacoustic "repertory" at Studio 105 of the Maison de la Radio.[41] From 1966 to 1969 there were four such exposition seasons.

1966, *l'Instant mobile* (9:30) by Bernard Parmegiani. First entirely electronic work in the GRM repertoire,[42] notably conceived in quadriphony with the aid of the shape modulator, with which he extracted fragments of the musical continuum, projecting them to several points in space. On the occasion of the premiere of this major piece in his repertoire, Parmegiani quoted a phrase by Gaston Bachelard that would long possess him: "Time is not noticed except for instants, duration is not sensed except for instants."

1966, *Jassex* (12 minutes) by Bernard Parmegiani. Premiered at the Festival de Royan. First experience of free-jazz with interplay between the instrumentalists on stage and tape, with Jean-Louis Chautemps, sax; Bernard Vitet, trumpet; Charles Saudrais, drums; and Gilbert Rovere, double bass.

1966, *Cantate pour elle* (14:35) by Ivo Malec, for soprano, harp, tape, and electroacoustic installation. Pierre-Albert Castanet described the piece:

> As in the case of *Reflets* (1960), the composer imposed three distinct procedures on himself, three "obligatory paths":
>
> a) "concrète" recording of the harp (using close-up mic, and most importantly, unconventional performance techniques) provided the pool of sounds necessary for constructing the tape;

b) the harp would also be performed "live," in the same unconventional manner, enriched and developed, as well as traditionally;

c) the shapes thus obtained, the "sound objects," would also serve as material for the voice, in turn integrated into the ensemble's motion, which was sometimes strict, sometimes improvised, but always clearly defined by the extremely rigorous score.[43]

1966, *Und so weiter* by Luc Ferrari, first mixed work for tape and instruments utilizing contact microphones.

1966, *Variations en étoile* (15:30) by Guy Reibel. First electroacoustic work composed using sounds from a single sound body. Marcel Frémiot, in his research into Temporal Semiotic Units (UST), suggested that listeners should

> not analyze [music] in terms of melodies, or even motifs, but in terms of sound figures. In this respect, I cannot recommend too strongly that you listen carefully to the original version of Guy Reibel's *Variations en étoile*. It is often studied for its extensive catalog of "concrète" manipulations. It is much more than that. One discovers incarnated within it a major focus: the construction of sound figures, the methods for such construction, and their diversity, but always in reference to the solfeggio of musical objects. It is, for me, a foundational work . . . which paradoxically founded nothing. This research path has been abandoned, even by Guy Reibel.[44]

Reibel has reworked his piece several times over the years. But like Frémiot, I also recommend the original version.

1967, *Espaces inhabitables* (18 minutes) by François Bayle. This work was also performed in a version with choreography by Vittorio Biagi in 1972 at the Lyon Opera. A description of the piece by Régis Renouard Larivière:

> This is, in his own words, the composer's "first real piece." The five movements that make it up were composed consecutively and form a whole. *Espaces inhabitables* raises, among other things, the problem of "anecdotal" or "figurative" sounds and their relationship to "abstract" sounds. One of the starting points for the piece was the use of outdoor stereophonic sound recordings (which was a technical innovation at the time).[45]

1967, *Capture éphémère* (11:53) by Bernard Parmegiani. 1-inch 4-track tape. This piece was also performed in a version with choreography by John Andrews at the Museum of Modern Art in Vienna. According to Parmegiani, "Originally this sound capture was that of the flapping of the wings of a passing bird, crossing the virgin sky in the Dead Sea desert."[46]

1967, *Nuit Blanche* (12:05) by François-Bernard Mâche. The original version was in Polish. It was commissioned by Warsaw Radio and realized at Studio Eksperimental. Text by Antonin Artaud spoken by Alain Cuny.

1967, *Étude de rythme* (3:50) by Edgardo Canton. Concrète song sung by Mouloudji. Text by Jean Tardieu.

1967, *Demeures aquatiques* (7:30) by Beatriz Ferreyra. 1-inch 4-track tape.

1967, *Oral* (34 minutes) by Ivo Malec. For actor and large orchestra, published by Breitkopf und Härtel. First performed at the Zagreb Biennial in 1967 by the Orchestre National de France, conducted by Ernest Bour. Although the title and the texts are borrowed from André Breton's *Nadja*, it is not an adaptation of the surrealist work.

1968, *Luminétudes* (12:17) by Ivo Malec. Work for tape dedicated to Pierre Schaeffer.

1968, *Chants des nuits désertes* (24:30) by Catherine Bir. First performance at the Maison de la Culture de Bourges, February 14, 1968.

1968, *Chemins d'avant la mort* (4 minutes) by Francis Régnier. Premiered at the Maison de la Culture de Bourges, February 14, 1968.

1968, *Lumina* (13 minutes) by Ivo Malec, for tape and twelve string instruments, published by Salabert. This work symbolizes the lights that cover the cities at night. The concrete sounds are very discreet; they melt into the instrumental ensemble.

1968, *Médisances* (7 minutes) by Beatriz Ferreyra, premiered in Studio 105 of the Maison de la Radio in Paris, during the 1969 Exposition des Musiques Expérimentales. A version with choreography by Pierre Oca would be performed in Arras in 1970.

Four Former GRM Members

LUC FERRARI, 1929–2005

Luc Ferrari began his musical studies by learning to play piano, but he turned to composition at the age of twenty. His classical training at the Conservatoire de Versailles, then with Arthur Honegger and in the classes of Olivier Messiaen (1953–54), and finally with Varèse led him to classical instrumental composition. Long before joining the GRM in 1958, Ferrari had been spotted by Schaeffer in 1950 during a concert given by the young pianist. At the time Ferrari was a serialist, like all the young people of his time who wanted to be part of the contemporary avant-garde. "One was necessarily serial if one was avant-garde—and necessarily avant-garde if one was not a reactionary."[47] His encounter with Varèse in 1954 was decisive for his career. Curious about everything, he attended the Darmstadt summer courses in Germany in 1952 and 1956. He was also interested in Cage and the New York school, as well as pop art and the American avant-garde of the 1960s. When he arrived at the GRM at the beginning of 1958, Ferrari discovered a musical universe that corresponded to his aesthetic aspirations. He composed his first electroacoustic work, *Visage V*, in 1959, while serving as the GRM's delegated director.

But Ferrari left the GRM in 1966 to continue his career independently. He taught in California, composed extensively for German radio, notably Hörspiele, and became musical director of the Maison de la Culture in Amiens as well as of an experimental music workshop in Stockholm. He was also artistic director of the Ensemble Instrumental de Musique Contemporaine de Paris (EIMCP), directed by Konstantin Simonovic. In 1972, he founded his own studio, Billig, and in 1982 the nonprofit organization La Muse en Circuit, whose direction was taken over by David Jisse in 2000.

There is an obvious desire to charm in Ferrari's music, tinged with humor and virtuosity. Ferrari was never afraid to use recognizable concrète sound elements—texts, snatches of conversations, exterior ambiences recorded on the spot. He is even considered the inventor of "anecdotal" music with his composition *Hétérozygote* in 1963–64. But he should not be reduced to a single genre, as he always made it a point of honor to try everything. Throughout his career Ferrari produced works for instruments, for voice, pieces for radio, musical theater, and installations. From 2000 onwards, many young people of the techno generation, who were looking for aesthetic roots, named Ferrari as one of their forefathers, along with several other pioneers of musique concrète—Henry, Parmegiani, and Schaeffer. During the final years of his life, Luc Ferrari gave numerous concerts in France and around the world, accompanied by famous DJs, notably the American DJ Olive. His catalog includes about two hundred works.

MICHEL PHILIPPOT, 1925–1996

A member of the French Resistance during the Second World War, imprisoned by the Nazis at the age of seventeen, Michel Philippot entered the Paris Conservatory after the Liberation in 1945 as a student in the composition class of René Leibowitz, after experiencing the revelation of serial music while listening to Schoenberg's *Ode to Napoleon*. While continuing to compose, he became a *musicien metteur en ondes* (MMO)—a musician sound engineer for radio—at Radiodiffusion française, and then at RTF, from 1949 to 1959.* From 1953, the year of the composition of his serial *Étude 1* (5:18), through 1962, he collaborated with the Club d'Essai and then the GRM. He was even secretary general of the GRM, assistant to Pierre Schaeffer, for a few months in 1960. He left the GRM in 1963, at the same time as Iannis Xenakis, in the midst of developing the *Concert Collectif*. In the meantime, he was involved in numerous pedagogical activities such as training sound engineers for

---

* Translator's note: On *musicien metteur en ondes*, see the translator's note in chapter 1.

Office de Coopération Radiophonique (OCORA) productions from 1958 to 1964, but also teaching musicology at the Sorbonne University from 1969 to 1976.[48] In 1970, he was appointed professor of composition at the Paris Conservatory. In 1976, he founded and directed the Music Department at the State University of São Paulo in Brazil. A personal friend of Iannis Xenakis, he was president of the UPIC Ateliers for many years, while being himself one of the pioneers of computer-assisted composition in France. A friend of Pierre Henry, he was the first president of Son/Ré at its creation in 1982. Very attached to teaching, to which he devoted a large part of his life, he also always dedicated himself to musical composition as well as to painting. Michel Philippot's style was strongly influenced by his serial origins and his immoderate taste for rigor and method. He composed about fifty works, published by Salabert (in Paris), Boelke-Bomart (in Hillsdale, NY), Billaudot (Paris), and Novas-Mestas (Brazil).

FRANÇOIS-BERNARD MÂCHE, 1935–

François-Bernard Mâche began associating with the GRM in 1958, immediately after completing his *agrégation* in literature at the École Normale Supérieure and his musical training in the class of Olivier Messiaen. At the GRM he was introduced to electroacoustic techniques and actively participated in the *Concert Collectif* between 1959 and 1963. In a sense, he was its "secretary," in charge of keeping its journal. In 1963 he left the Groupe to continue teaching classical literature and Greek poetry, without ever ceasing to compose. He developed personal musical concepts based on concepts of natural and linguistic sound models. After *Prélude* (1959) for three tape tracks, most of his works would be mixed works: *Volumes* (1960) for tape and orchestra, *Nuit blanche* (1966) for narrator and tape, *Rituel d'oubli* (1969), and *Korwar* (1972) for harpsichord and tape, among others. Mâche then composed at least one work per year. He also published several books and numerous articles on music, notably for the *Nouvelle Revue française*. His musical works are published by Durand.

A close friend of Xenakis, Mâche became president of the nonprofit organization Association des Amis de Xenakis.

IANNIS XENAKIS, 1922–2001

Even though between 1954 and 1962 the name of Iannis Xenakis was intimately linked to the history of the GRMC and then the GRM, he never became a full member, but simply an "associate" member. This status granted him access to the studio facilities and enabled him to participate freely in activities and debates, but did not entitle him to regular remuneration. He could only be paid in the form of honoraria or commissions. At the GRMC, he associated with Abraham Moles, Michel Philippot, François-Bernard Mâche, Jean-Étienne Marie, and Alain de

Chambure. He certainly attempted, but without success, to obtain a leadership position in the group focused on a project for stochastic music. But this was without reckoning on the strength and determination of Schaeffer to carry out his gigantic project of group research into sound perception. The two men did not have the same conceptions of how to approach music. Through contact with Schaeffer, Xenakis acquired the ability to use his perception to the utmost. He would later exploit this ability to ultimately rectify the formal and conceptual rigidity that one senses underlying many of his compositions.

At the GRM, Xenakis composed *Diamorphoses* (1957), *Concret PH* (1958), *Analogique B* (1959), *Orient-Occident* (1960), and *Bohor* (1962). After 1962, he returned to the GRM sporadically for specific productions, in 1967 for the tape of the *Polytope de Montréal*, in 1982 to make the 8-track mix of *Pour la paix*, and in 1993 for the preparation of a CD of his works on tape.[49]

In 1997 François Delalande, a researcher at the GRM, published a book of interviews with Iannis Xenakis under the title *Il faut être constamment un immigré*.[50]

8

1968–1978

End of the Schaeffer Era

Without music, life would be an error.
—FRIEDRICH NIETZSCHE

The decade from 1968 to 1978 was crucial in the history of GRM because it marked the beginning of a new phase in the life of the group that would last more than thirty years. But the decade was not seamless, because the dissolution of the ORTF (replaced by seven new audiovisual entities on January 1, 1975) represented a radical rupture. Before 1975, within the ORTF, the GRM had been part of the Research Department, for which it had been the original model; after 1975, the GRM was integrated into the Institut National de l'Audiovisuel (INA), becoming a department among others. It no longer served as a model, particularly because Schaeffer, who had reached retirement age, had not had his mandate renewed.

The avant-garde period was far behind. How would François Bayle, the new director of the GRM, manage to "hitch" the Groupe to its new life? There would be no shortage of pitfalls, but unlike many other research centers, notably in Cologne (but also Milan and Tokyo), which would gradually and inevitably begin to decline, the GRM would find new life. But geographically closer to GRM, a new entity appeared. In 1975, the creation of IRCAM completely changed the French musical landscape.

HISTORICAL CONTEXT

After the widespread contestation of the 1960s, which culminated in 1968, the arts mirrored the return to order that followed. We witnessed the assimilation of former protest movements, a return to individualism, a withdrawal into the private sphere, and the quest for simple values such as nature or hedonism.

Pieces composed at the GRM during this period were no exception to the rule. No one wanted to compose "studies" or engage in musical experimentation anymore, as they had "back in the days of La Recherche" with Schaeffer. Each composer wanted to assert his or her own personality.

From the point of view of both the quantity and quality of musical production, the 1970s constituted an extremely fertile period, the apogee of electroacoustic music at the GRM. After the ordeal of imposed fundamental research at the inception of the Service de la Recherche in 1960 and its consecration by *TOM* and *SOS* in 1966 and 1967, Pierre Schaeffer had shifted his attention towards television, leaving practical responsibility for the GRM in the hands of François Bayle from 1966 on. Bayle restored the focus on musical creation.[1] From the beginning of the 1970s, there was a frenzy of production and a proliferation of works in all sorts of new hybrid styles ranging from mixed pieces to live electronic to acousmatic. The length of pieces was also increasingly linked to the existing commercial format, the vinyl record, with its two twenty-minute sides.

Research took on a new face, placing itself at the service of musical creation. The enthusiastic atmosphere of the Research Department in the 1960s endured for a few more years after the events of May 1968, but based more on momentum than on new projects. Even the opening of the class at the Conservatoire National Supérieur de Musique de Paris (CNSMP) under the aegis of the GRM, which was at first called *Classe de musique fondamentale et appliquée à l'audiovisuel* (Class in fundamental music and music applied to the audiovisual), quickly lost its appeal. After two or three years of existence, it became normalized, and eventually merged into the general instrumental composition teaching. The breach opened in the conservative edifice of musical education closed, in part. It was time to usher in new directions: the issue of reexamining music teaching for children emerged. François Delalande, since 1972 its main advocate, along with Guy Reibel and a few teachers on loan from the French Ministry of Education, inaugurated a series of radio broadcasts that were to have a profound influence on French music teaching in schools.

The GRM was increasingly well integrated into the musical milieu, but institutionally it remained linked to the audiovisual sector. The end of the Research Department in 1974, coinciding with the retirement of Pierre Schaeffer, marked a turning point oriented more towards musical creation, pedagogy, and development of new tools for composing, rather than the focus on experimentation and general theories of language and communication so dear to Schaeffer. But this tendency was somewhat mitigated—from 1969 to 1975—by the reorganization of the ORTF that took the revolt of May 1968 into account.

To gauge this rupture between the two periods, let's examine a few documents. First of all, two tape recordings of work sessions at the end of the "Recherche" period.

FIRST DOCUMENT: THE DECEMBER 3, 1969, SEMINAR

The opening of the electroacoustic class at the Conservatoire in 1968 brought in a new generation of composers/researchers: Jean Schwarz, Robert Cohen-Solal, François Delalande, Robert Cahen, Jacques Lejeune, Michel Chion, Roger Cochini, and Jack Vidal.

The seminar of the composition class at the Conservatoire on December 3, 1969, recorded on ¼-inch tape, is archived in the "general" sound library of the GRM like all the other conferences and seminars, as well as the radio broadcasts.[2] Because at the GRM, everything is saved, sometimes in multiple copies. Lectures were even deciphered and transcribed before being archived.

The interest of this December 3 seminar lies in the fact that it took place at the beginning of the second year of the conservatory class. An assessment of the first year was consequently possible. The first- and second-year students were present, as well as auditors. Pierre Schaeffer was present but silent for the first hour of the session, which was conducted by Beatriz Ferreyra and an assistant. The students were invited to listen to and comment on a series of sound examples taken from different musical genres: a sound excerpt of Japanese Noh played alone, and then the same excerpt filmed, an excerpt of music from Togo, followed by an excerpt of an Indian raga and then the same one filmed, an excerpt of Stockhausen's *Stimmung* without film, and then with, and so forth. Following this listening session, student comments referred back to the previous class where the concept of opposition between nature and culture had been discussed. Some students (among whom we recognize the voices of Michel Chion and Jack Vidal) differentiated between music/performance and the piece itself. The instructor tried to orient the discussion by emphasizing the difference between the rational character of Western works and the functional character of primitive music. She also noted that the acousmatic listening context of the session introduced an additional dimension of questioning. Someone contrasted hearing the sound alone with the emotion created by the images; another pointed out that primitive music is socialized while contemporary music is individual.

And then suddenly in the middle of the session, Pierre Schaeffer, probably sitting in the back of the room, took the floor, commenting on the observations. He deplored the fact that these sessions were "muddy," when in his opinion the presence of the previous year's students should have contributed to elevating the debate. He seemed disappointed by how little the returning students had learned.

And yet we can observe that in the course of this simple session, all the preoccupations of Schaeffer's "Research" were addressed. Here they are, just as they were listed on the archive entry for this December 3 seminar, stored inside the box along with the magnetic tape recording:

—determining factors in the current state of music;
—accelerated evolution;
—the advent of electroacoustics;
—returning to sources;
—the concept of the musical piece in West and non-Western civilizations;
—the importance of the instrument;
—conditioning;
—the introduction of the image;
—comparisons between films and recorded examples;
—links between gesture and music;
—intervention of the camera;
—Pierre Schaeffer's critique of the level of comprehension of the last year's students;
—difficulty of talking about music;
—different modes of looking and listening;
—positioning music as phenomenon;
—conception of installations; and
—we are overlooking the subject.

We notice that the subject of multidisciplinarity as suggested in the subtitle, "interdisciplinary essay," of his *Traité des objets musicaux: Essai interdiscipline* was omnipresent in Schaeffer's thought, as it was in the activities of the Groupe at that time. Music was only a pretext for reflection on the much broader theme of audiovisual communication. This seminar session clearly illustrates how Pierre Schaeffer used the GRM as a laboratory to test his ideas—before sharing them with the other "groups" under his supervision, concerning theoretical concepts just as much as "management" techniques.[3]

SECOND DOCUMENT: THE 1970 FRAMEWORK PLAN

This second document is a recording of the general meeting of the Groupe d'Enseignement Recherche (GER, Teaching Research Group), on July 3, 1970, a day devoted to the development of the 1970–71 Master Plan.[4]

Pierre Schaeffer was at the blackboard, organizing the repartition of tasks between different GER members for implementing the Framework Plan. The objective was to set up procedures for archiving, indexing, and conservation and for access to the archives. The aim of this archiving was to favor availability of a rational selection of television productions to universities and other sites likely to promote training in the audiovisual and mass media fields. Pierre Schaeffer's idea was holistic, as usual. He wanted to link production, conservation, criticism, and training. For that he challenged all his collaborators to look for "valuable" partners to widen their role of mediation—between audiovisual production and the public—to a wider sphere than just radio and television. He wanted to integrate new mediators or organizers, for example a few motivated professors, to help university students, and even the general public, to enter the world of audiovisual communication, to understand it and become familiar with it.

Once again Schaeffer made use of the GRM as an example, with its "bridge" class at the conservatory. But he admitted in passing that the class did not work as well as he would have liked, finally concluding that "an entire new pedagogical approach needs to be invented."

Beginning in 1970, Schaeffer reorganized "research" at the GRM. He created four new research groups, with new research topics under the responsibility of three new researchers: Solfège expérimental and analyse musicale (Experimental solfege and musical analysis) with François Delalande, Musiques ethniques (Ethnic music) with Jean Schwarz, and Informatique musicale (computer music) with Francis Régnier.

François Delalande focused on identifying new research subjects. With a degree in engineering, coupled with his belated but solid studies in music, François Delalande showed up at the GRM after having read—or rather devoured—Schaeffer's *TOM*. Initially hired as a half-time administrative assistant, after a couple of years Delalande was awarded his first research contract. He published his first article, "L'analyse des musiques électroacoustiques," in the journal *Musique en jeu* in 1972.[5] Extending Schaeffer's ideas as formulated in *TOM*, Delalande established three categories of listening intentions: the organization of sounds, musical values, and the meaning of the music. Drawing on Jean-Jacques Nattiez's terminology,[6] Delalande concluded his article by hypothesizing that electroacoustic music is more often a music of signification than of meaning. Faithful to the tradition of musique concrète, Delalande also situated himself in relationship with phenomenology and linguistics (Merleau-Ponty, Husserl, Saussure, Jakobson), which had so influenced Schaeffer's thought. But Delalande also took an interest in psychology (Jean Piaget) and perception (Robert Francès). These different problematics nourished his research into childhood music pedagogy and musical analysis based on listening.

The ethnic music atelier led by Jean Schwarz primarily concentrated on closely following the work of the ethnomusicologists at the Paris national ethnology museum, Musée de l'Homme. The computer music atelier would see considerable continuing development.

End of the Service de la Recherche, Creation of the INA

On December 31, 1974, simultaneously with the closing of the ORTF, the Research Department ceased to exist. Some of the groups founded by Schaeffer survived the change of status: the Groupe de Recherches Techniques (GRT), still directed by Francis Coupigny, the Groupe de Recherches Images (GRI), and the GRM.

The dissolution of the ORTF, by a law voted in 1974, was followed by the creation on January 1, 1975, of seven new audiovisual companies: Radio France (National Radio and regional network), TF1 (National Television, first network channel), Antenne 2 (National Television, second network channel), FR 3 (Regional Television, third network channel), TDF (Télé Diffusion de France, in charge of the network of transmitters), SFP (Société française de production—cinema and television), and INA—with the poet and academician Pierre Emmanuel as its first president. Let us note in passing that the INA had not been planned in the initial project of reform of the ORTF, but that thanks to the vigilance of Pierre Schaeffer, who had noticed the shortcoming in time, at the last moment this last organization was put to the vote of the deputies. The INA encompasses activities of conservation of radio and television archives, professional training of audiovisual personnel, and research.

The GRM became part of the INA as a research and creation department. It moved to the Maison de Radio France, because on the one hand there was not enough room for it in the new premises in Bry-sur-Marne (located to the east of Paris), and on the other hand its activities naturally linked it to the activity of Radio France.

The INA has the legal structure of an EPIC: an *établissement public à caractère industriel et commercial* (a public institution of an industrial and commercial nature). This means that it receives subsidies from its supervisory ministry, Communication, but that it may, and indeed must, carry out commercial activity that generates revenue. This revenue is derived primarily from the sale and transfer of rights from the audiovisual radio and television archives and from professional training for audiovisual personnel in the form of paid workshops. The GRM itself contributes revenue through the sale of its discs, software, and books and by obtaining subsidies for its research and development work.

Birth of the IRCAM, as Seen from the GRM

"It's simple. From the moment the Ircam appeared, no one believed in the GRM anymore." This brief comment by François Bayle sums up in itself the context of IRCAM's emergence in the music scene in 1975.[7]

The Institut de Recherche et Coordination Acoustique Musique had first been conceived by Pierre Boulez in 1969, taking shape during the summer of 1973 at the Abbaye de Sénanque with the first official work session by the team in charge of setting it up.

From the beginning, IRCAM, this big brother—founded by the will of the president of the republic Georges Pompidou and his wife—almost completely overshadowed the GRM, which owed its survival only to its small size and to the fact that it was not funded from the same source. The INA, under which the GRM operated, was an institution from within the audiovisual sector, while the IRCAM was under the jurisdiction of the Ministry of Culture. The GRM, with its historical attachment to radio and the media, was able to forge ahead on its own path, which naturally incorporated the unprecedented technological evolutions of the period—first the entry of electronics into the studio in 1970, then the arrival of powerful computing using delayed time in 1974-75, and finally real time, one not excluding the other.

IRCAM was installed in the heart of Paris in underground facilities at Place Igor Stravinsky, located just next to the Centre Pompidou, the museum of art and culture. Like the GRM, IRCAM brought together engineers, technicians, and creators in a single location to promote synergies; but unlike the GRM, IRCAM had a very large budget, six times that of the GRM. The conductor and composer Pierre Boulez, then at the height of his fame, became for over fifteen years the director of this enormous structure, financed directly by the French Ministry of Culture. As soon as it opened, IRCAM embraced computer science, transferring software from the United States, from the laboratories of the Bell Telephone Company (New Jersey) and Stanford University (California), notably the famous Music V software developed by Max Mathews with the collaboration of Joan Miller and Richard Moore. Several American researchers came to work in France, under the supervision of Jean-Claude Risset, who had been approached by Pierre Boulez in 1972 to direct the computer department, and Gerald Bennett, who was in charge of the interdisciplinary "Diagonal" department. Other departments were entrusted to Vinko Globokar (instruments and voice), Luciano Berio (electroacoustics), and Michel Decoust (pedagogy). In 1976 the 4A digital sound processor was developed by Giuseppe di Giugno's team; it was the ancestor of the 4X, which was finalized in 1981 and inaugurated with Pierre Boulez's *Répons*. Since its opening in 1977, IRCAM has organized large-scale public concerts. The first cycle, "Passage du XXe siècle," included seventy musical events in 1977 alone. In 1978 IRCAM offered its first training session for composers. That same year, it inaugurated the Espace de Projection, a 375-square-meter concert hall with modular walls and ceiling that could vary reverberation times from half a second to more than four seconds.

End of the Cologne Studio's Golden Age

The Cologne Studio's golden age is generally considered to have come to a close around 1972. From that time on, Karlheinz Stockhausen was faced with slow but certain deterioration in his working conditions at the WDR. To be sure, he remained artistic director of the studio until 1980, but in 1972 two of his three sound engineers were removed from the studio, and in 1977 the contracts of his three musical assistants, Mesías Maiguashca, Peter Eötvös, and David Johnson, would not be renewed. The administration argued that National Radio did not need an electronic music studio and that it would be better suited to the conservatory in Cologne. Last but not least, the "socialist" administration of the WDR decided to reduce funding for the renewal of equipment, choosing to favor "more popular" productions.[8] The studio thus lost its position as the leader in innovation at the WDR.

Despite everything, the residency program continued: Mesías Maiguashca composed *Hör zu* (1970); Mauricio Kagel *Acustica für experimentelle Klangerzeuger und Laustprecher* (1970); Peter-Michael Braun *Essay für Oboe und Tonband* 1972; York Höller *Horizont* (1972); Henri Pousseur *Système des paraboles A* (1973); Jean-Claude Eloy *Shanti* (1974); James Whitman *The Dance of Shiva* (1975); Luc Ferrari *Allo, ici la Terre* (1974); Karlheinz Stockhausen *Sirius* (1975); and Iannis Xenakis *La légende d'Eer* (1978).

The End of Domaine Musical

After May 1968, as with the GRM, all the other actors of the learned musical milieu saw their audiences melt away. In the light of newly acquired liberties, everything seemed outdated and pale even where before listeners had perceived nothing but invention, avant-garde, and even subversion. The great concerts of contemporary music offered by foundations and other nonprofit organizations such as Domaine Musical no longer attracted attention. Domaine Musical, which had been directed by Gilbert Amy since 1967, following its founder Pierre Boulez, came to an end in 1973. Jesus Aguila shared his version of the end of the organization:

> Openly opposed to the nomination of Marcel Landowski, Domaine Musical seemed to have pursued the policy of "the begging bowl and the Molotov cocktail," with one hand asking for subsidies, while with the other hand writing bitter attacks against the Direction de la Musique, as did a good part of the avant-garde circles proudly calling themselves "leftist." . . . Caught up in these contradictions, it was with a feeling of bitterness that Gilbert Amy decided to announce the end of Domaine. . . . The bitterness stemmed from the fact that, faced with the irremediable erosion of the patronage from the aristocracy and big business bourgeoisie, the State, the only possible substitute patron, had refused to reserve for Domaine Musical the largest slice of its support for new music. Domaine could therefore no longer assume the role of trailblazer for contemporary musical creation that Pierre Boulez had always wanted to occupy alone.[9]

The Spectral School

Two years after the end of Domaine Musical, the Spectral School emerged. The name was coined by its theorist, Hugues Dufourt.[10] It was not composed of former members of Domaine Musical but of another, slightly younger group, who had seen themselves as excluded, including Gérard Grisey (*Dérives* 1974, *Périodes* 1974, *Partiels* 1975, *Prologue* 1976, *Modulations* 1977) and Tristan Murail (*Sables* 1975, *Mémoire/Érosion* 1976). Along with Michaël Levinas and Roger Tessier, they formed Ensemble l'Itinéraire. Their works tackled new musical criteria such as noisy sound, spectral research, and the use of the concept of parasites, making extensive use of modern electroacoustic equipment to enrich their compositions and present their "Live Electronic" performances.[11]

Music Education and the DMDTS

In 1969, Marcel Landowski, director of music since 1966 at the French Ministry of Culture, leading the DMDTS—*La direction de la musique, de la danse, du théâtre et des spectacles* (Directorate of Music, Dance, Theater and Performances)—launched a "10-Year Plan." It was a national program of government measures intended to revitalize music in France, particularly at the regional level, with the declared objective of "broadening the base"—the base of students considered for recruitment to the conservatories. In their 2000 book, Anne Veitl and Noémi Duchemin summarized this 10-Year Plan:

> Marcel Landowski implemented what could be described, both with respect to neighboring countries and to France, as the "French model," a pyramidal teaching model, where the development of music in the provinces was decided in Paris, through a policy of selective aid to establishments that met the exigencies of public education—a teaching system that was both egalitarian in its content and selective in its progression, where "broadening the base" was conceived above all as the means to reach "the highest level."[12]

The same year, 1969, Marcel Landowski also created a new national diploma, the Certificat d'Aptitude or CA, obtained through a national examination, to recognize an aptitude for teaching and/or administrating in a conservatory. This diploma later evolved, as a result of various reforms, towards greater professionalization in music education. Unfortunately, since its inception there have only been two or three examination sessions for the CA in electroacoustics, certifying for teaching only three or four candidates at each session, a notoriously insufficient number.

In 1975, halfway through the 10-Year Plan launched in 1969, Jean Maheu succeeded Marcel Landowski as director of music. As soon as he was appointed, Jean Maheu reassured the community by announcing that he would continue the 10-Year Plan. But he also decided to start thinking about reforming teaching content in the conservatories, targeting 1979. This reform aimed in particular at introducing contemporary music into the curriculum. Jean Maheu was forced to resign in 1978 and the project never saw the light of day.

Center for Documentation of Contemporary Music

The Centre de Documentation de la Musique Contemporaine is a nonprofit organization, directed by Marianne Lyon for its first thirty years. Created in 1976, and inaugurated in 1978 at the same time as the record label MFA—*Musique française d'aujourd'hui*—at the initiative of Jean Maheu, then director of the DMDTS, the CDMC's mission is to make the musical works of our time available to all. Anyone can go to the Centre to listen to one of the twelve thousand collected works and read the score. In addition to the scores, the center also maintains additional documents, biographies, articles, and photos. Among the CDMC's institutional partners, along with the Ministry of Culture and the SACEM, Radio France—the principal organizer of concerts in France—has undertaken to provide the center with copies of the recordings it has made.

MUSICAL CONTEXT

From the point of view of musical aesthetics, the simple dichotomy from the early days between musique concrète and musique électronique had disappeared. Composers had long since taken control of all the techniques of production, in all possible sound registers, from concrete sounds of acoustic origin to synthesized sounds, vocal sounds, sounds of the instruments of the orchestra, or even of the entire orchestra, as well as of all imaginable formulas of representation. The fusion of the concrete and electronic genres, as a result of the standardization of ideas and tools, along with the diversification of styles, led to a loss of identity, and consequently to abandoning labels. Composers no longer claimed to compose electroacoustic music or electronic music, but simply music. Of course, there remained vestiges of the two great original approaches to composition, the more formal approach of the German school, from the Vienna school, based on a priori construction, and the approach of the Paris school, based on an a posteriori construction, essentially guided by the composer's perceptive faculties, applying the "doing and listening" so cherished by Pierre Schaeffer. The historical nucleus of electroacoustic music on media remained, focusing more on exploring morphologies, dynamics, and space, and less on pitches, melodies, and rhythms. And mixed music, which united musical components composed in advance on media with components interpreted by live instrumentalists, became the junction between written learned classical music and unwritten learned music, incorporating all the possible methods for transforming instruments: prepared, electrified, amplified, live transformation, contact microphones...

In addition to the diversification of styles as a result of the increasing integration of electroacoustic techniques into all genres of music, a new and important direction was emerging at the GRM under the impetus of François Bayle: acousmatic music, in which the sound sources were unseen, but which sought to sculpt space in three dimensions, particularly during the concert. Through the Acousmonium, or loudspeaker orchestra, François Bayle expanded the concept of interpretation in electroacoustic works. In a decidedly proactive fashion, he decided to promote music and concerts rather than research. The Acousmonium, which was to become the GRM "label" for several decades, earned him renown in the musical world by acquiring a reputation of making "beautiful sound."

Research, and Development of New Composition Tools at the GRM

THEORETICAL RESEARCH WITH CANADA

Although modest in number of participants—François Delalande, the only full-time researcher, with Jean-Christophe Thomas and a few occasional collaborators—theoretical research at the GRM did not go unnoticed in the academic world. Thinking about music beyond the notes, whether or not it involved written music, was an original insight.

From 1977 to 1982, a "Franco-Quebec integrated research program" on perceptual analysis instituted a collaborative relationship between Jean-Jacques Nattiez, surrounded by a small Montreal team, and François Delalande and Jean-Christophe Thomas. The Canadian researcher Marcelle Guertin, with a few interns, furthered the analysis of music as we hear it, complemented by Jean-Luc Jézéquel in his study of electro-polygraphy indicators of the listening processes.

SOUND SYNTHESIS—STUDIO 54

Studio 54 was inaugurated in April 1970 in the premises that the GRM occupied in the Bourdan Center in Paris, an old private townhouse located in the sixteenth arrondissement near the Maison de la Radio. The particularity of this studio resided in its prototypical console, which combined electronic synthesis and processing modules, all designed at the Research Department by Enrico Chiarucci, Francis Coupigny, and Guy Reibel. A pin patch system made it possible to access the different modules and set them up in control relationships. In addition, there were Brüel & Kjær filters, a Moog synthesizer, and an analog echo chamber with mechanical reverberation plates.

With the introduction of this new equipment, works produced at the GRM were enriched with numerous electronic sounds, most of which were transformed intuitively after their creation, following the common practice for manipulating sounds from natural sources. Bernard Parmegiani was the first to use this new hardware, composing *L'Œil écoute* and marking the beginning of a long life for Studio 54. Dismantled and then reassembled exactly as before when the GRM moved to Maison de la Radio in 1975, it became Studio 116C, until 1992. The "console from 54" is today part of the musical instrument collection of the Musée de la Musique at Cité de la Musique in La Villette in Paris.

FIGURE 19. Studio 54 at the Bourdan Center, with the synthesizer console, during the composition of François Bayle's *Grande Polyphonie*. © INA (1974).

SPATIALIZATION—STUDIO 52

Right next to Studio 54, which was dedicated to electronic synthesis, Studio 52—comprising a booth and a studio—specialized in spatial effects. To this end, it was equipped with a sixteen-channel console, divided into four groups of four, to produce "4-track" sound. From its installation, the GRM repertoire would include numerous works already conceived for quadrophony from the outset of their initial sound recordings.

THE ACOUSMONIUM

With the events of May 1968 behind him, François Bayle, director of the GRM, realized that from the point of view of sound, electroacoustic concerts were often poor and boring. He asked himself how to make people hear the *unheard* ("often mentioned, but never encountered") and to win back an audience of music lovers who were somewhat disappointed to have to listen to electroacoustic works that were often poorly diffused on a simple pair of loudspeakers.

On the model of the grand orchestra in the style of Haydn, François Bayle devised (with the help of the engineer Jean-Claude Lallemand) a grand orchestra, consisting solely of loudspeakers, or *sound projectors*, which he christened the

Acousmonium. This orchestra of sound projectors, arranged by register and by tiered placement on stage, was officially inaugurated with his *Expérience Acoustique* on February 12, 1974, at the Espace Cardin in Paris, following an initial trial run a month earlier on January 16, 1974, during a concert at the church of Saint-Severin in Paris.

In the same way as orchestral instruments, each loudspeaker was selected and placed on the stage on the basis of its sound color and the acoustics of the hall. The eighty loudspeakers of the original version of the Acousmonium—which would continue its evolution over time—offered a very rich palette of sounds, from bass to treble and from muted to brilliant. From 1977 to 1985, the Acousmonium was accompanied by a Berliet truck, which served both as a means of transport and as a control booth for the concerts organized throughout France and the rest of Europe. These numerous outside performances durably established the prestige of the GRM as a major player in the fields of sound, electroacoustic music, and spatialization.

A COMPUTER PROJECT FOR STUDIO 123

Until 1974-75—in other words, before the creation of IRCAM—the focus at the GRM was primarily on research and technical development. Intensive studio use of computers was still pending.

In 1969, François Bayle had resolutely committed the Groupe to digital technology by commissioning Jean-Claude Risset to produce a piece "for computer." *Mutations*, a 2-track tape produced in the United States in the experimental studios of BTL (Bell Telephone Laboratories), premiered in France in 1971. Francis Régnier, the first person to be in charge of research at the GRM's Computer Music Workshop, introducing an October 21, 1970, in-house seminar, posed a question underlying the Groupe's research at that time: "Is it possible to develop a process of computer synthesis where the composer can describe to the computer 'sounds' or 'music' that he wants to hear, not only by assigning numerical values to physical parameters, but by providing the computer with details of a psychological nature?"[13]

This problematic was directly derived from Schaeffer's speech at the Stockholm Congress the same year,[14] which had brought together the most eminent representatives of musical research at the time, in particular the American pioneers of musical data processing, Herbert Brün of the University of Illinois and Max Mathews of Bell Telephone Laboratories.

At the GRM, the computer research team was quite small. Francis Régnier had a single assistant, Jean-Marie Pacqueteau. Two second-year students in electroacoustic composition from the conservatory (including Pierre-Alain Jaffrenou) bolstered the pair. Another GRM researcher, William Skyvington, also regularly contributed. Francis Régnier kept abreast of current research in other countries. In particular, he was aware of the existence of the Music V program, developed by Max Mathews in the research laboratories of the American company Bell Telephone (software that Jean-Claude Risset would bring to the GRM in 1976).[15] In July 1970, Francis Régnier launched the SYNTOM (Synthesis + TOM) project:

> The purpose of the SYNTOM project, over the short or long term, is to allow composers to specify the sounds they wish to synthesize in psycho-acoustic terms based on *Traité des Objets Musicaux*, and no longer through the exclusive intermediary of conventional physical parameters (frequencies, durations, etc.). This approach places us, it should be noted, in the domain of the computer "instrumentalist."[16]

Francis Régnier had even developed his "digital-analog" converter in 1971, constructed by Francis Coupigny of the GRT, and installed at the Issy-les-Moulineaux research center in the southern suburbs of Paris. This converter already used the Music V program to synthesize sound. The spool of sounds was then brought back to the GRM to be used in composition.

But in 1972, Francis Régnier suddenly resigned to devote himself to his passion, singing. This vacancy allowed Pierre-Alain Jaffrenou to assemble a new team,[17] notably inviting Bénédict Mailliard, a mathematician trained at the École Normale Supérieure and a second-year student in the conservatory's electroacoustic class.

But regardless, it must be admitted that the GRM had been slow to develop an interest in computer science. It is true that Xenakis's early propositions in 1962–63 had remained fruitless, confronted with a wary Schaeffer who saw the computer at most as a tool to assist in the classification of sounds—but certainly not as a tool capable of analyzing them and even less of transforming them. As for producing sound by synthesis, the early results were so mediocre that they did not even merit Schaeffer's attention. In 1970 at the Stockholm conference on music and computers, Schaeffer still stood his ground, declaring: "Not content with associating music with mathematics, Xenakis wants us, by the same token, to subjugate everything to his Queen. From sculpture to cinema. And why not History while we're at it? I refuse to enter into such a debate, throwing up my hands in disbelief."[18]

It is true that in Schaeffer's day computers were still very slow. It took about twenty-four hours to produce three seconds of sound, sorting punch cards. But in spite of these cards—still in use until about 1978—computers in France had been able to process sound signals since the early 1970s, when the first digital synthesis was implemented at the Institut d'Electronique Fondamentale d'Orsay (Charbonneau, Karatchentzeff and Risset) shortly before Xenakis's converter at the Cemamu, Centre de Mathématiques Appliquées à la Musique.

Nevertheless, the transformation from analog to digital also took place at the GRM, in its Studio 123 located at the Maison de la Radio. In 1974–75, Bénédict Mailliard, Pierre-Alain Jaffrenou, and Bernard Dürr began running small tests with the Bull punch card machines from GIRATEV, the organization in charge of accounting and management for the various audiovisual organizations that had been created following the dissolution of the ORTF. But in 1974, computing was still regarded as only being at a highly theoretical stage. François Bayle explained: "In brief, these were still the theories of Michel Philippot and Xenakis. We regarded sound space as being defined from low to high, by semitones or quarter tones. And the computer was grinding out the numbers representing notes, with the idea of creating sound beings using a few or more envelope plots."[19]

But at the beginning of 1975, two young engineers from the prestigious École Polytechnique arrived at the GRM for a six-month internship. They had chosen to pursue their studies under a "contract for complementary studies" with the École des Télécom.[20] As soon as they arrived, Xavier Nouaille and Jean-François Allouis were invited by François Bayle and François Delalande to conduct research into "automatic" graphic transcription of electroacoustic music, and on the search for computer-friendly graphic forms.[21] Until then, all that existed was the sonogram, displaying quite brief musical excerpts—a few seconds—with frequencies on the vertical axis and time on the horizontal axis. Since notes are insufficient to capture electroacoustic music, it was necessary to try to find one's way visually. This need for research was motivated both by François Delalande's pedagogical needs in terms of presentations to music teachers, and François Bayle's pedagogical needs in terms of concerts, since they were both constantly asked to explain the ABCs of electroacoustics to their audiences.

This research led in 1988 to the Acousmographe project, entrusted to the engineer Olivier Koechlin. In addition to his many original contributions, in Koechlin's CD-ROM *Les musicographies* he reworked the initial work of Nouaille and Allouis by colorizing the visuals.[22] Most of these new transcriptions were done for fun, like the transcription of an excerpt from Jimi Hendrix's performance of "The Star Spangled Banner" at Woodstock—with the distortion caused by saturation of his guitar's amplification while using his wah wah pedal—or the transcription of John Zorn's jazz trio in *News for Lulu*, with Zorn on saxophone, Bill Frisell on guitar, and Georges Lewis on trombone.[23]

And so Xavier Nouaille and Jean-François Allouis began "looking for forms that can be easily identified with the aid of computer tools. For example, trying to visually recognize the spectrum of an instrument in a sonogram and differentiate it from other formants contained in the overall flux, such as voice or concrete sounds." Their background prepared them well for frequency analysis, for identifying everything that is significant in a signal, and determining how to filter out interference frequencies. "At the time, we worked a lot on frequency analysis, and with resynthesis," recalls Xavier Nouaille. He also remembers his interest in reading Roland Barthes on the relationship between sign and meaning. Nouaille experienced this internship at the GRM, as did Allouis, as "a junction between scientific measurement tools and the world of music."[24]

And all this work, both ludic and serious, would lead a few years later to the important innovations of Jean-François Allouis and Bénédict Mailliard, notably comb filters, and everything that would follow in digital sound processing. "But I have to admit that in 1975 we didn't yet have a clear idea of the relevance of computers in signal processing and sound control," concluded Xavier Nouaille.

In contrast, the synthesis of digital sound was becoming increasingly well understood.

A TOOL FOR REAL-TIME COMPOSITION: THE SYTER PROJECT EXPERIMENT

At the end of the 1960s, early computer processing was carried out in delayed time. It took a night of calculations to obtain thirty seconds of processed sound. Moving closer to real time was the only solution for regaining the spontaneity of working directly on sound material as in the analog studio, and to counter the competition with IRCAM, omnipresent on the computer music front since its creation in 1975. So in 1976 when the brilliant engineer Jean-François Allouis suggested looking for computer solutions to access this famous *real time*, François Bayle immediately agreed, with the ulterior motive that one might be able to also fabricate improved phonogènes, improved filters, improved echo chambers, in short, all the enhancements that one had a right to expect. In addition, digital processing would eliminate background noise and other drawbacks in using analog. And to crown it all, the data would be writable on media and therefore indexable.

And so the Syter (SYstème TEmps Réel) project was born in 1977. Its first public performance took place the same year during François Bayle's *Cristal* concert. At the beginning, Syter only served to perform sound spatialization, using live digital synthesis during the concert.

The first description of the Syter project was from 1975, recommending:

—developing a digital prototype allowing real-time sound synthesis based on a model inspired by MUSIC V;

—developing hardware and software interfaces enabling use of the system by composers; and

—developing a set of control panels enabling use by "instrumentalists."[25]

Jean-François Allouis was resolutely in the vanguard of innovation in attempting to compete directly with the most advanced centers in the field at the time, Bell Laboratories and at Stanford University. The Americans led by Max Mathews, John Pierce, and John Chowning had been trying to develop real-time synthesizers, but nothing concrete had yet been achieved. Allouis was also competing with the research team at IRCAM, led by the Italian engineer Giuseppe (Peppino) di Giugno. The latter first developed the 4A in 1976, followed by the 4B, which would become the 4X in 1981, a digital system comparable to Syter.

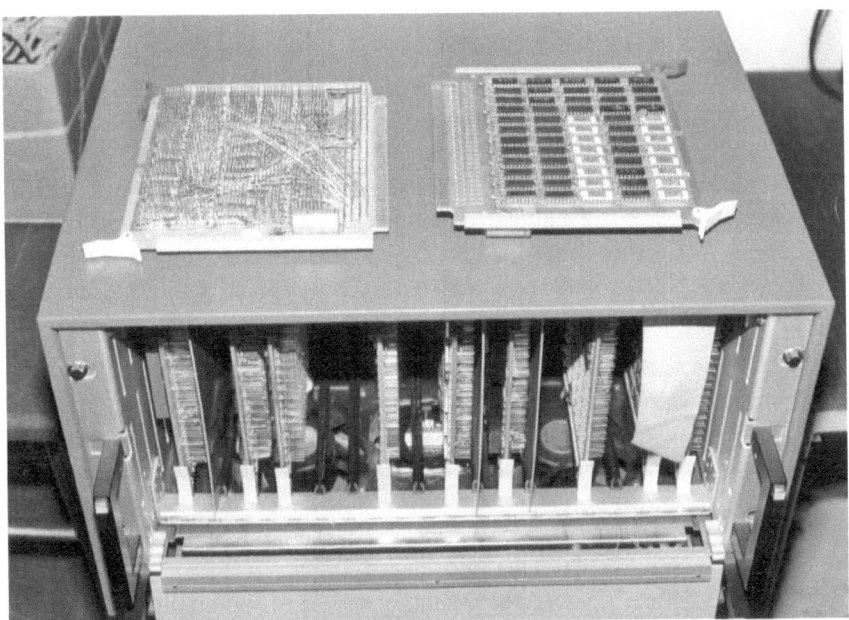

FIGURE 20. Syter 2 wired microprocessor. © Évelyne Gayou (1977).

STUDIO 123—DIGITAL MUSIC COMPOSITION STUDIO

In 1977, the small team of GRM computer researchers, Jean-Pierre Toulier, Bernard Dürr, Pierre-Alain Jaffrenou, Denis Valette, Richard Bulski, and Jean-Yves Bernier, guided by the two engineers Bénédict Mailliard and Jean-François Allouis, organized and developed a project for a digital music composition studio, which was set up in Studio 123 of the Maison de la Radio. One year later, the first computer at the GRM arrived, a PDP-11/60 from the Digital Equipment Corporation (DEC), obtained thanks to the connivance of GIRATEV, which, as the administrative management company of public broadcasting, owned computers and agreed to make one available to the GRM free of charge. The architecture of Studio 123 was that of a rather standard radio studio—two adjacent rooms, each about 300 square feet, soundproofed and separated by a double glass window.

In order to make room for computer music research, the GRM's radio production, which had been carried out until then in Studio 123, moved to Studio 116A, which would be completely reequipped for the occasion. But it would remain primarily analog for a few years: a new Studer 189Q console with sixteen inputs / sixteen outputs and eight groups, with Elison filters; four Studer A80 stereo tape recorders at 19/38 cm/sec.; an EMT digital echo chamber; and four JBL 4315 speakers.

Other Musical Initiatives in Electroacoustics

Numerous electroacoustic musical initiatives were flourishing. Centers, associations, ensembles, and many electroacoustic groups modeled on the GRM were being created in the four corners of France. Since electroacoustics had always attracted musicians from a much wider sphere than from just formal musical training, this network constituted an essential part of musical life for many creators. All these structures wove links locally with the conservatories, but also with the French educational system, universities, festivals, and cultural associations. A brief survey of some of these new centers:

The Groupe de Musique Expérimentale de Marseille (GMEM), directed in 2006 by Raphaël de Vivo, was established in 1969 in the wake of a new electroacoustic class created at the Marseille Conservatory by Pierre Barbizet and a former member of the GRM, Marcel Frémiot. They were soon joined by Georges Bœuf (who would become director in 1974), Michel Redolfi (who would later organize the Manca Festival in Nice), Lucien Bertolina, and then Jacques Diennet, Claude Colon, and Alain Fourneau. As part of its research into sound synthesis, the GMEM acquired a Synclavier 1 and then a Synclavier 2, also used to produce music for concerts and work related to dance and the visual arts. More recently, during the 2000s, the GMEM developed the sound spatialization software Holophon.

In 1970 the Groupe de Musique Expérimentale de Bourges (GMEB) was founded by Françoise Barrière and Christian Clozier, both from the GRM. They were joined by Beatriz Ferreyra, Roger Cochini, Pierre Boeswillwald, Alain Savouret, and Elsa Justel. The GMEB became famous for its international festival—founded in 1973—associated with a competition for electroacoustic works and musical software. The GMEB would become a privileged place for artists from Eastern Europe until the fall of the Communist regimes after 1989. It was also highly receptive to South American musicians. In 1994, the GMEB became the Institut International de Musique Electroacoustique de Bourges (IMEB).

One of GMEB's many technical innovations, the Gmebogosse, a small portable analog device dedicated to pedagogy, has seen multiple generations of prototypes and widespread success with users, principally elementary school children and even middle and high school students. The first model in 1972 included two cassette players/recorders, filters, controllers for speed and volume, two speakers for classroom use, and two speakers for personal listening. In the third model, which came out in 1984, sound synthesis was incorporated using a VCO and a VCA.[26] Beginning in 2004, the sixth model featured quadraphonic operation.

Among the other innovations due to the GMEB was the Gmebaphone, a loudspeaker diffusion device that predated, by one year, the GRM Acousmonium, a fact of which the GMEB takes great pride. This anecdote concerning the anteriority of the Gmebaphone over the Acousmonium provides a glimpse at the competitive atmosphere that has long prevailed in relations between the GRM and the GMEB, essentially for personal reasons, because the achievements in terms of music and research are equally impressive on both sides.

- The Ges of Vierzon, an "interdisciplinary experiment centered on music," was founded by Daniel Habault in 1972. Like the other groups, this composition studio was dedicated to hosting composers in residence and organizing training courses. Nicolas Frize, a former trainee of the GRM class at the Conservatoire, was a valued collaborator of this group.[27] The Ges also created an orchestra of loudspeakers (about forty), which it made available to concert organizers in the early 1980s. In 1985, the Ges was also the first to equip its material with a computerized system for saving playback settings, allowing for completely automated real-time playback. The Ges ceased to operate in the late 1990s.
- The Groupe de Musique Vivante de Lyon (GMVL) was founded in 1976 by Marc Favre, Xavier Garcia, and Bernard Fort. Marc Lauras quickly joined them, while Bernard Fort took over the direction of the group, which was essentially devoted to musical education, while at the same time carrying out research and composition for concerts. Aesthetically, compared to the other existing groups, pieces produced by the GMVL are certainly the closest to the problematics envisaged by the GRM.
- The Groupe de Musique Électroacoustique Albi-Tarn (GMEA), founded by Thierry Besche and Roland Ossart in 1977, has succeeded in introducing electroacoustics to an extremely large regional audience, through the organization of instructional sessions which now number in the thousands. The GMEA developed the Melisson, an electronic instrumentarium used both in its studios and in its educational programs. The GMEA also hosts composers in residence and organizes concerts.
- The Centre International de Recherche Musicale (CIRM), in Nice, founded in 1968 by Jean-Etienne Marie (1917–89), later directed by Michel Redolfi, and since 2000 by François Paris. Since 1978, its annual Manca festival has brought it considerable recognition. The CIRM's mission is to "allow composers to explore the world of micro-intervals, through reinventing instrument design and through electroacoustics." Like the other centers, its activities are structured around four axes: production, diffusion, research, and training.

At the same time, numerous performing ensembles were created. These were often small ensembles of variable size, from three to ten people, cofinanced by local authorities, interpreting and creating works primarily from the contemporary repertoire, without neglecting other repertoires in Western learned music. The oldest of these ensembles, Ars Nova, was created in 1963 by Marius Constant.

This explosion of instrumental ensembles was international. In Germany, Johannes Fritsch, Peter Eötvös, Rolf Gehlhaar, David Johnson, and Mesías Maiguashca created the Feedback Ensemble in 1970, dedicated to live electronic music. In France, the Ensemble InterContemporain (EIC) was founded in 1976 by Pierre Boulez.

The GRM waited until 1978 to found a performance ensemble for live and mixed music, the Trio GRM+ (later renamed TRIO-GRM-PLUS), featuring Laurent Cuniot, Denis Dufour, and Yann Geslin, with the aim of exploring the synthesizer repertoire. This ensemble changed its name to Trio TM+ in 1982, when it parted ways with the GRM.

In his private studio Apsome, founded in 1959, Pierre Henry composed *Apocalypse de Jean* in 1968; *Ceremony* with the pop group Spooky Tooth in 1969; *Fragments pour Artaud*, *Gymkhana*, and *Mouvement-Rythme-Etude* in 1970; the music for Maurice Béjart's production *Nijinsky clown de Dieu* in 1971; *Deuxième symphonie* in 1972; the music for *Kyldex 1*, the "spatio-lumino-dynamic and cybernetic" piece by the visual artist Nicolas Schöffer, in 1973, with choreography by Alwin Nikolais; *Enivrez-vous* in 1974; *Futuristie* in 1975; *Parcours-cosmogonie* in 1976; and *Métamorphoses*, *Dieu*, and *Instantané-Simultané* in 1977.

Meanwhile, Outside of the Field of Learned Music

The use of electricity and new tools had naturally also found its place outside of art music. An experimental attitude in composition permeated every genre. The Beatles composed *Number 9* in 1968, a piece entirely worthy of the term *experimental music*. New genres emerged. Also in Great Britain, amidst punk, rock, progressive rock, folk, and free-jazz, the Industrial genre appeared, named after the label Industrial Records. The group Throbbing Gristle (erect penis in Yorkshire slang), founded in 1976, set up its own independent label in 1977 controlled by the artists themselves, based on the model of Andy Warhol's Factory in the USA. Techno would only appear later, around 1988–90.

Another style of commercial music, disco, closer to mainstream pop, also appeared. Disco groups, with their liberated dance music in vogue in nightclubs, had a special taste for melodies sustained by a powerful bass line. The instruments of predilection were synthesizers, combined with electrified acoustic instruments. These groups, such as Hot Butter, became vectors for the popularization of electronic sounds, with their little 1972 ritornello *Pop Corn* heard around the world.

Closer to home, Jean-Michel Jarre, a former student in Schaeffer's electroacoustic class at the Conservatoire, composed *Oxygène* in 1977, published by Dreyfus Music. By becoming a commercial "hit," this piece contributed not only to popularizing electronics, but also electroacoustics, since Jean-Michel Jarre had always publicly acknowledged his time spent in Schaeffer's class in 1970.

The very original career of Edgardo Canton is also worth mentioning. A member of the GRM from 1960 to 1969, but also a tango composer, he never disavowed his Argentinean origins. At the GRM he composed a large amount of mixed music and music for film. In 1962–63, he was involved in the adventure of the *Concert Collectif* at the GRM. While continuing his professional career at UNESCO, in 1972 he founded Trottoirs de Buenos Aires, the famous Parisian tango club, a central gathering place for the French diaspora in South America in the 1970s.

EVENTS, PERFORMANCES, PUBLICATIONS, PEOPLE

The statistics reveal that from 1968 to 1978, 168 works of pure music (electroacoustic and mixed) were composed, along with 138 works of applied music (for film, television, theater, dance, mime, etc.).

During this decade, an impressive number of composers wrote their first electroacoustic piece at the GRM: Mireille Fognini, Alain Savouret, Jacques Lejeune, Joanna Bruzdowicz, Elzbieta Sikora, Micheline Coulombe Saint-Marcoux, Françoise Barrière, Vinko Globokar, Christian Clozier, Gilles Fresnais, Vlodimierz Kotonsky, Fernand Vandenbogaerde, Pierre Bernard, Pierre Boeswillwald, Roger Cochini, Claire Renard, Michel Chion, Maurice Ohana, Robert Cahen, Jean Schwarz, Jorge Antunes, Michèle Bokanowski, Bernard Dürr, Katori Makino, Pierre-Alain Jaffrenou, Roger Frima, Lionel Filippi, Denis Guilbert, Dominique Guiot, Nicolas Frize, Armand Frydman, Luciano Berio, Hinoharu Matsumoto, Rodolfo Caesar, Marcel Landowski, Michel Philippot, Denis Dufour, Eugénie Kuffler, Philippe Drogoz, Raoul Delgado, Jacques Stibler, Alain Gaussin, Ragnar Grippe, Dominique Collardey, Sergio Tulio-Marin, Charles Clapaud, Christian Villeneuve, Claire Shapira, Jean-Loup Graton, Christian Zanési, Thomas Kessler, Marc Monnet, Marie-Noëlle Moyal, Françoise Bourgoin, Yann Geslin, Laurent Cuniot, Philippe Mion, Carlos Roque-Alsina, and Marc Favre-Marinet.

All by itself, this list illustrates the eclecticism of the aesthetic sensibilities present in the GRM movement, but also the dynamism and the degree of artistic innovation in its production.

Selected Works from the GRM Catalog

1968, *Les Shadoks*, three series of fifty-two two-minute episodes each, by Jacques Rouxel, music by Robert Cohen-Solal,[28] with the voice of Claude Piéplu, and directed by René Borg.

This cartoon was fortunate enough to be broadcast at 8:30 p.m., in prime time, on Channel 1 of French television (in response to the desire of its producer, Pierre Schaeffer, to counteract the "bourgeoisization" of television).

The somewhat "British humor" featured in *Les Shadoks* immediately ensured them and their producer enormous popularity. The scenario can be understood as a metaphor for modern life's absurdity in our industrial societies and perhaps also within the Research Department itself. Indeed, the Shadoks heroes are forced to *pump* (in other words, to *work*) "stupidly" without respite, in order to conquer a better planet called Earth: in short, a paradise where there would be no more pumping. The Shadok motto, "why make it simple when you can make it complicated," was reportedly often put into practice in the Research Department. This thinly veiled derision towards our so-called advanced society provoked numerous reactions. Detractors were numerous, and *Les Shadoks* owed their survival, after the events of May 1968, only to the "official" opinion of Mrs. de Gaulle, the wife of the president of the republic, who declared that "her grandchildren liked the program a lot."[29]

The success of *Les Shadoks* was also due to the quality of its soundtrack, original concrete music by Robert Cohen-Solal. In the first series of programs, it was a model of the genre: heterogeneous sound objects, both humorous and admirably rhythmed in the editing. Who doesn't remember the musical sequence of the Shadoks on a bicycle, or the cry of the terrible Gégène, master of the planet Earth—alias Pierre Schaeffer himself in the imagination of the authors!

Owing to the reprieve provided by the cessation of broadcasting during the strikes of May 1968, the *Shadoks* team hurried to produce the second series of fifty-two episodes. But lacking sufficient time to compose—like delicate lace—concrete music for these new adventures, Robert Cohen-Solal opted for a musical romp improvised by a small modern instrumental ensemble, in the manner of emerging rock bands. The "musique concrète" touch disappeared, but the series kept its humor. The animation of the first series was produced with the assistance of the Animographe, an invention of Jacques Dejoux from the GRT, the Technical Research Group of the Research Department. This prototype made it possible to create animated drawings economically, using perforated 70 mm film strips, the drawings colorized on the gelatin and lit by transparency.

1969, *Rituel d'oubli* (30:40) by François-Bernard Mâche for tape (concrete sound), wind orchestra, and percussion. Published by Salabert. This government commission performed by the Ars Nova ensemble sought to draw orchestral sounds together with a wide variety of concrete sounds: water, wind, war, insects, amphibians, and words of men from Paraguay and women from Tierra del Fuego. In notes to the piece, F.-B. Mâche wrote, "Between the cry of a monkey and the cry of an orchestra, there is no other boundary than prejudice. Here are a few pathways leading from raw sound to the musical note."

1969, *Mutations (I)* (10:30) by Jean-Claude Risset, premiere February 25, 1971. The first work in the GRM repertoire synthesized by computer (at the Bell Labs, the research laboratories of the Bell Telephone company in the United States, using the Music V program developed by Max Mathews). The work was then finished at the GRM on 2-track tape. This commission allowed the members of the GRM, notably the researchers Enrico Chiarucci and Francis Régnier, to learn about the research work in progress at Bell Labs. Bell Telephone had fifteen thousand employees, fifteen hundred of them researchers. Its cutting-edge research, mainly focused on voice (in relationship to telecommunications), brought together—in a spirit of interdisciplinarity—engineers and artists renowned for their visionary or at least inventive abilities.

1970, *Musique en liesse* (approximately one hour in length, depending on the version) by Guy Reibel. Performance in eighteen movements, for tape, electric guitar, sound bodies, and projections of photographs by Jean-Serge Breton (images of faces), with extracts of music from India, Japan, Africa, and Europe.

1970, *L'Œil écoute* (24:24) by Bernard Parmegiani. A work for tape, in which he reused numerous excerpts from his previous sound works made for the soundtracks of short films at the Service de la Recherche. In addition to this recycling, Parmegiani was the first to employ sounds from the new analog synthesizer of Studio 54, conceived at the GRM by Francis Coupigny and Henri (Enrico) Chiarucci. The work would be associated with images in two distinct versions, one realized by the Pole Valerian Borowzyck, the other by Parmegiani himself—images of his own eyes, transformed with the help of the Research Department's image synthesizer.

At the premiere of the piece on May 9, 1970, at the Gulbenkian Festival in Lisbon, Bernard Parmegiani confided, "Perhaps in spending too much time looking, man ends up not listening. And the eye, which has become a 'solitary wanderer,' only has ears for what assaults it." Many years later, in 2002, he added, "It was research into the relationship between form and matter. From these photos of eyes, I carried out a certain number of visual manipulations with the video synthesizer in counterpoint to similar manipulations on sound in the music."[30]

Parmegiani's sounds in *L'Œil écoute* take us into a railway universe of trains hurtling along at high speed. But the author offers us "subjective" sound, the sound heard by travelers inside their compartment. Perhaps the images of eyes are also a look turned towards the interior.

1970/1973, *L'Expérience Acoustique* (130 minutes) by François Bayle. The first long program work, its composition extended over several years. Fourteen of the forty-nine projected pieces would finally be composed. François Bayle described it as an "art brut project" in which he wanted to preserve the "adventurous" character of research, "still imprisoned curiosity" to be liberated and developed.[31] The author mixed concrete sounds, often rich in harmonics, with digital sounds from synthesizers and other devices. The premiere of the entire composition took place at the 26th Festival of Avignon on August 3, 1972.

1971, *La Roue Ferris* (11:02) by Bernard Parmegiani. This work, with its repetitive character, was composed in part with synthetic sounds realized on the small Coupigny synthesizer called "The Cube." It premiered to a choreography by Vittorio Biagi.

1972, *La Divine comédie*, after Dante. François Bayle and Bernard Parmegiani joined forces, perhaps out of mutual modesty, but also with the idea of combining their efforts to give form to the all-encompassing universe that Dante had created in his work. François Bayle composed *Purgatoire* (72:15), and Bernard Parmegiani *Enfer* (65:13), with the voice, untreated, of Michel Hermon.

The following year, in 1973, the two composers finalized a "four handed" *Paradis* (39:07). This third part was not set on tape, but performed live by the two composers who each prepared a sequence, for the first time using the synthesizer to transform the music during the concert. This third section has only rarely been performed; it is not certain that its composers fully embrace it.

In his annotated catalog of François Bayle's works, Régis Renouard-Larivière described *Purgatoire* as "a new acoustic experience, initiatory, oriented towards transparency," noting that the "electronic layers, arches and arabesques are the 'blood' of the piece," while "the 'concrete' sounds bring us back to a constricted and painful present."[32] And the voice of the narrator, Michel Hermon, imparted a dramatic aspect to the piece that brought it closer to radio art.

In notes for *Enfer*, written by Bernard Parmegiani, the composer himself explained his aesthetic: "The sound pierces the skin like a slash and sucks us through this orifice towards an interior where we find the milky void belonging to the dream."

Divine comédie premiered on February 5, 1973, at Théâtre Récamier in Paris, with choreography by Vittorio Biagi. It was commissioned by the French government.

1972, *Suite pour Edgar Poe* (52:48) by Guy Reibel, to four of Edgar Allan Poe's *Nouvelles histoires extraordinaires*,* for electronic sounds, concrete sounds,

---

* Translator's note: *Nouvelles histoires extraordinaires* (New Extraordinary Tales) was the title given by Baudelaire to his collection of translations of Poe.

and vocal elements recorded by the vocal ensemble Musique Nouvelle directed by Stéphane Caillat, and text spoken by Laurent Terzieff.

1972, *Koré* (7 minutes) by Michèle Bokanowski. Work for twelve solo voices and tape, realized for the Olympic Games of Munich in 1972 with the soloists of the ORTF choirs under the direction of Michel Couraud.

1972, *L'arbre et coetera* (16 minutes) by Alain Savouret. Rondo form, i.e., refrain/verses. The initial sound recordings were quadraphonic, enabling a more differentiated mode of occupation of space (mono, stereo, quadraphonic) and drawing us closer to the sensation of "sound-sculpture."[33]

1972, *Erda* (27:02) by Jean Schwarz. Each of the eight pieces is a rhythmic study based on different source material. The last two pieces pay tribute to the jazzmen Kenny Clarke and John Coltrane, whom Jean Schwarz—himself a former jazz drummer—particularly admired.

1972, *Pour en finir avec le pouvoir d'Orphée* (PEFAPO in GRM jargon, 40 minutes) by Bernard Parmegiani. During an interview in 2002, Parmegiani explained that the title referred to his desire to "exorcise the repetitive music" that he had been composing until then, but that he had not succeeded.[34] In fact, in the early 1970's, the repetitive style from the United States had reached Europe. And even if Parmegiani has said that he had not been aware at that time of the works by Americans such as Terry Riley or Steve Reich, he still wanted to free himself from a style that was too pervasive for his taste.

In this work, I would describe Parmegiani's music as *lisse*—smooth—as in his previous works *L'Œil écoute* (1970) and *La Roue Ferris* (1971), and continuing up to *Pour en finir avec le pouvoir d'Orphée*. In the view of Deleuze and Guattari, "Lisse is continuous variation, continuous development in form, the fusion of harmony and melody in the interest of liberation from strictly rhythmic values, becoming a pure straight diagonal through the vertical and the horizontal."[35]

Despite his efforts, Parmegiani was not able to free himself from this "smooth" aesthetic in *Pour en finir avec le pouvoir d'Orphée*. That would finally be accomplished two years later in *De natura sonorum*.

1973, *Valse molle* (10:15) by Alain Savouret. A humorous and baroque work, composed of concrete, instrumental, and electronic sounds. The piece was conceived as quadraphonic from the beginning of its recording, to emphasize certain movements in space as well as certain particular states of quadraphonic sound material.

The composer, with his characteristic humor, dedicated this piece "to Saint Leger, Bishop of Autun, who, victim of the jealousy of a prominent person (another one . . .) had his eyes gouged out and then his head cut off, in the year 678."

1973, *Le Requiem* (37:20) by Michel Chion. An expressionist work first performed at Théâtre Récamier in Paris on March 19, 1973, based on texts from the funeral mass. This half-satirical, half-kitsch work aimed at "expressing the collapse of a specific religious experience." Numerous voices of children and amateur actors mechanically deliver texts in Latin and Greek. The concrete sounds that accompany them give a highly original character to the whole, somewhere between theater, radio, and music. This is one of Michel Chion's pieces most appreciated by audiences.

1973, *L'Écran transparent* (19:56) by Bernard Parmegiani. Three years after the experience of *L'Œil écoute*, Parmegiani composed this piece and codirected the visual component with the Spanish director José Montès-Baquer in the studios of Westdeutscher Rundfunk (WDR) in Cologne, Germany. The images show an actor-mime quoting a text about audiovisual perception by McLuhan. In his introduction to the concert, Parmegiani described its guiding concept: "Electronic man lives faster and faster; he is forced to see and hear everything. Every instant is information . . . And the whirlwind carousel of images and sounds builds a screen around him from which only those who know how to remain deaf or blind escape, because they listen, they look. For them the screen becomes transparent. The eye sees what the ear cannot see . . . At the crossroads of the senses is a point of non-sense, a starting point for everyday life propelled into a dream."[36]

The specificity here of Parmegiani, compared to the other GRM composers, was that he himself went behind the camera. This was the "grand era" of the ORTF's Research Department, with its exceptional vitality. Genuine research on the relationship between image and sound occurred thanks to the profound bonds that were created between musique concrète composers and film directors.

1973/1974, *Tryptique Électroacoustique* by Guy Reibel

First section, *Signal sur bruit* (27:10).

Second section, *Cinq études aux modulations* (27:30).

Third section, *Franges du signe* (37:17).

The first section (1973) explored the behaviors of perception of materials with unusual energetic profiles, the second (1973) explored morphologies generated by an instrumental constraint, in this case modulation of an electronic sound by a concrete sound, and the third (1974) explored states of unstable equilibrium, resembling phenomena bound by the logic of living organisms. These three pieces fell strictly in line with the spirit of musical research advocated by Pierre Schaeffer.

1973, *Parages* (25:20) by Jacques Lejeune. A suite of nine études built from soundscapes. The first part explored the themes of matter, space, and rhythm.

In a more dramatic tone, the second part—consisting of the final four études—explored the theme of the cycle of Icarus, successively passing through the elements of fire, air, earth, and water.

1974, *Rabelais en liesse* (first short version 50 minutes) by Guy Reibel. An instrumental and electroacoustic choral opera in nine movements based on texts by François Rabelais. First performed at the Maison de la Culture de Grenoble, then performed the same year at the 28th Festival d'Avignon in a long two-hour version with scenography by Walter Underdown, stage directed by Pierre Barrat, and conducted by Stéphane Caillat.

1974, *Thrène* (31:01) by André Boucourechliev,[37] based on an unfinished poem (which Mallarmé began writing following the death of his son) with the voices of Valérie Brière and Roland Barthes and the choirs of Radio France. Electroacoustic transformation restricted to the vocal material alone, preserving the speech that dynamically modulates the song. The author wrote in the introduction to the work that "the sound becomes ghost-word and the words, pure music." Commissioned by the ORTF for the Italia prize 1974.

1974/1975, *De natura sonorum* (28:06) by Bernard Parmegiani. In two series of six movements, each sequence brought together pairs of concrete, electronic, or instrumental sounds—each with different forms, materials, or colors. In this key work in his repertoire, as well as in the GRM repertoire itself, Parmegiani proved his virtuosity in working with sound matter.

This emblematic and often performed work in the GRM "style" encompassed practically all the known compositional processes on media. But the work should not be seen as merely a demonstrative inventory of electroacoustic techniques. The title itself, a slight twist on its source of inspiration, *De rerum natura* (On the nature of things), written by Lucretius around 50 BC, clearly revealed that all the stylistic exercises explored in the piece were underpinned by both philosophical and aesthetic reflection.

1975, *Chants Parallèles* (23:30) by Luciano Berio. Transformation of timbres by frequency modulation of the modal melodic material sung by Christiane Legrand. The engineer Bernard Dürr seconded Luciano Berio in Studio 54 of the Bourdan Center,[38] suggesting a large range of ingenuous techniques and original sound treatments, notably making use of a formant generator. Commissioned by the European Broadcasting Union (EBU) and the GRM, the piece premiered on Radio France on April 7, 1975.

1975, *le Trièdre fertile* (34:35) by Pierre Schaeffer, with the participation of Bernard Dürr. A version with computer-generated images also exists. This piece was realized in Studio 116C, using both the Coupigny and Moog synthesizers. For this final work in his repertoire, Pierre Schaeffer surprisingly agreed to work exclusively with synthetic sounds produced by equipment operated by the GRM researcher and engineer Bernard Dürr, who was deeply implicated in the piece's composition. Schaeffer listened to Dürr's audio, "just as I used to receive prepared sequences from Pierre Henry,"[39] employing his perception—in accordance with the three dimensions of the trihedron, time, pitch, and intensity—to select the best sequences, which were then mixed. The work was dedicated to Pierre Henry.

1975, *On n'arrête pas le regret* (12:40) by Michel Chion. Sound album of scenes of children, impressions and memories, in five movements.

1976, *Pierre Henry des années 50* (3 hours and 25 minutes) by Pierre Henry. Compilation of musical works and radio chronicles from the period of the 1950s. A condensed 90-minute version was broadcast on the radio, November 20 of the same year, on France Musique.

1977, *Triola ou symphonie pour moi-même* (34 minutes) by Ivo Malec. Constituted the return to pure electroacoustic composition by the composer after a four-year hiatus composing instrumental works.

1977, *Petite suite* (19:02) by Denis Dufour. Prefiguration of his more lengthy work *Bocalises* (36:43), which appeared in 1978. Denis Dufour, aged twenty-four at the time, was Guy Reibel's assistant for the CNSMP's electroacoustic class in 1976–77. All derived from acoustic sources, the music of each movement explored a different playing mode, based on the principles outlined by Guy Reibel in his teaching.

1977, *Cristal* (22 minutes) by François Bayle, for eight groups of instruments, tape, and live transformations. Premiered March 16 at the Maison de Radio France, with the Nouvel Orchestre Philharmonique conducted by Lucas Vis. The originality of the piece was that in addition to the (quite minimal) orchestral part, it exploited tape and transformations in real time, using the new Syter real-time system. One of the most spectacular transformations involved live-mixing sounds from both the instruments and the tape, and sending them to the loudspeakers, creating spatial interplay. The tape part had already been used independently for the 1973 live premiere of *Paradis* from *Divine comédie*.

EXPOSITION OF EXPERIMENTAL MUSIC, 1969

Like the four similar expositions that had taken place in previous years, the Expositions des Musiques Expérimentales of the spring of 1969 comprised five separate events in Studio 105 and in a foyer of the Maison de la Radio. Among the works presented were *Hallo d'ombre* by Guy Reibel, *Bidule en ré* by Parmegiani, *Journal* by Bayle, and *Carnaval* by Reibel.

SEMAINES MUSICALES INTERNATIONALES DE PARIS

In 1969, Maurice Fleuret, journalist at the national weekly news magazine *Nouvel Observateur* and artistic director since 1968 of the Semaines Musicales Internationales de Paris (the SMIP festival), invited the GRM to present its work for a total of thirty-six hours, including an all-nighter. All the GRM composers of the time were represented: Reibel, *Vertiges for Electric Guitar and Tape*; Bayle, *M ... E ...*; Savouret, *Kiosque*, for tape and improvising instrumentalists; Malec, *Lumina*; Parmegiani, *Thalassa*, for Ondes Martenot and tape; and Luc Ferrari, *Music Promenade*. This certainly already foreshadowed the exemplary support that Maurice Fleuret would give to musical creation (regardless of the person or genre) when he became director of music at the Ministry of Culture in 1981.[40]

SIGMA BORDEAUX—FIVE-DAY ENCOUNTER—*POURQUOI/COMMENT*

Musical production at the GRM was intensive and the public events numerous. In 1970 the GRM participated in the SIGMA festival in Bordeaux. The following year, in 1971, the Bordeaux program was repeated in Paris for "5 journées rencontres." The works presented were almost all composed in the new Studio 54, with electronic sounds produced by the recently installed synthesis modules—*Pour en finir avec le pouvoir d'Orphée* by Parmegiani and Bayle's *Jeïta* and *L'Expérience acoustique*.

At the beginning of 1975, in a public session of ARC at the Museum of Modern Art of the city of Paris, Maurice Fleuret invited the GRM for *Pourquoi/Comment* (Why/How), a series of public sessions.[41] During an interview in 2004, François Bayle recalled:

—"I don't know how to describe them, they were not lectures, nor concerts, but rather issue-based. We played music and we talked about it."

—"An intellectual and theoretical happening...?"

—"That's it. And since the room was small and there were a lot of people, the atmosphere was congenial. I asked everyone to push their ideas to the limit. There was an Allouis/Mailliard, 'computer' evening, a 'research' evening with François Delalande, and a 'Guy Reibel' evening. And for me it was an opportunity to present a lecture on acousmatics."[42]

The GRM and ACR on Radio

After a period of radio silence from roughly 1960 to 1966, which Pierre Schaeffer had wanted in order to concentrate energy on research and the writing of the *Traité des objets musicaux*, François Bayle and the various members of the GRM (Henri Chiarucci, Luc Ferrari, Marius Constant, Ivo Malec, Guy Reibel, David Rissin, Bernard Parmegiani, Agnès Tanguy, Michel Chion, and Jack Vidal) launched into regular radio production, successfully relayed by Radio France's networks

France Culture and France Musique. Annual programming would amount to about seventy hours.

The France Musique network broadcast, in extenso, works composed at the GRM and performed during the different concerts, while the France Culture network broadcast rather long programs or series of programs (about an hour each) of a didactic nature, reporting on current research, and accompanied by numerous sound examples. The subjects tackled evoked issues related not to classical musicology, but rather to what would later be called Music Sciences. These programs all reflected the new theoretical interests of the Groupe during this period, and the sign of its desire to popularize knowledge and the state of current research.

The titles of the programs speak for themselves:

La musique et l'argent (Music and money)

Comment l'entendez-vous? (How do you hear it?)

La musique du futur a-t-elle un avenir? (Does the music of the future have a future?)

Pourquoi-Comment ? (Why-How?)

Les fonctions de la musique (The functions of music)

Les dessous du sens des sons (Behind the meaning of sounds)

Le pouvoir des sons (The power of sounds)

Images, mirages acoustiques (Images, acoustic mirages)

Éveil à la musique (Awakening to music)

L'Acousmatique pour tous (Acousmatics for all)

On October 5, 1969, the Atelier de Création Radiophonique (ACR) of France Culture broadcast its first program, "Spécial Prix Italia 1969." Alain Trutat became the leader of a small team of four people focusing on this workshop project that had been proposed to management two years earlier with the support of Pierre Schaeffer.[43] Trutat wanted to create "an open space for writing sound," a bit like the Club d'Essai, where he had started as a director, notably of *Paroles dégelées* in 1952.[44] The Atelier's ambition was to regard radio not only as a means of communication but also as a means for artistic creation. Officially, the events of May 1968 had nothing to do with the inauguration of the ACR, but it is obvious that this project came at the right time in a period when any new idea could be used as an alibi for the much sought-after liberalization of the media.

When the ORTF was dissolved in 1974, the ACR was transferred to Radio France, and René Farabet, Alain Trutat's collaborator from the beginning, took over the management of the ACR until the end of his career in 1999.[45] Alain Trutat was then appointed "artistic" advisor to the management of Radio France, a position he held until the end of the 1990s. He died in 2006.

As far as radio creation is concerned, the French situation was particular compared to other countries, in particular Germany, because the affiliation of the GRM with the Institut National de l'Audiovisuel (INA) in 1974 led to a very clear differentiation at the institutional level between radio creation (at Radio France) and musical creation (at the INA). This organic separation resulted in polarizing the two activities, one towards drama and the other towards music. In contrast with French Radio, German Radio, which had never drawn such a clear distinction between music and drama, continued to develop an art form specific to radio, Hörspiel, blending poetry, drama, literature, and music.[46] In Germany this new genre had appeared under the same conditions as in France with the advent of studio techniques, but much earlier, in the 1920s.

In an interview with Olivier Corroenne in 1984, Pierre Schaeffer described the initial concept for creative production for radio in France:

> The birth of this art form, give or take a few years, dates back to 1935. With the exception of a few major premieres in various radio stations around the world in German Hörspiel and a few proponents of "directing for the airwaves,"* it was very quickly brushed aside. In pre-war France, we existed on the fringes of programming, which was dominated by advertising radio. By founding the Studio d'Essai for French state-run radio, we were the first to assert that there was such a thing as radiophonic art. This approach has been carried on with the current Ateliers.[47]

Some authors of Ateliers nevertheless attempted to establish links with electroacoustic music. Yann Paranthoën asked Philippe Mion to effectuate certain sound treatments for his program *Saigneurs de pierre* (1985). Paranthoën was even invited by the GRM to *project* some of his works in concert, on the Acousmonium. Christian Rosset creator of numerous *Ateliers*, also composed musical works that were—even when written for orchestral instruments—frequently broadcast on radio, integrated with other dramatic elements. Jean-Yves Bosseur, Lionel Marchetti, and Olivier Capparos, being composers themselves, regularly inserted electroacoustic elements into their radio creations. Several GRM composers wrote for ACR—Jean Schwarz, François Bayle, Michel Chion, Jacques Lejeune, Guy Reibel, Évelyne Gayou, and François Donato. And let us not forget their famous precursors Pierre Henry and Luc Ferrari.

Opening of the Electroacoustic Course at the Paris Conservatory

In September 1968, Pierre Schaeffer inaugurated, as an associated professor, electroacoustic classes at the Conservatoire National Supérieur de Musique de Paris (CNSMP). In this course, titled "Classe de Musique fondamentale et appliquée à

---

* Translator's note: See translator's note for *mise en ondes* in chapter 1.

l'audiovisuel," apprentice composers with classical training—or not—were invited to discover another way of listening to and making music.

The events of May 1968 were not at the origin of the course, which had already been in preparation at the Conservatoire for several months, at the suggestion of the GRM, following the model of the Martenot school's collaboration with the Conservatoire. But suddenly in May, student demands for the modernization of teaching content precipitated things. In this context, opening the course could be perceived as a concession wrenched from the faculty. The very first year, there was notably a considerable influx of applications—the Wednesday evening seminars were attended by a much wider public than the twenty or so students enrolled in the class. But after two or three years, the craze subsided. Guy Reibel, Pierre Schaeffer's assistant until 1976, attributed this disaffection to Schaeffer's excessive interest in research and correspondingly diminished interest for musical composition, while the expectation among students was the opposite.[48] This alternative approach to "making" music was not explicit with Schaeffer, or at least not explicit enough, and this shortcoming would be the root cause of all future evolution in the course. However, according to Guy Reibel,

> the creation of this course at the Conservatoire had been strongly desired by Schaeffer, then at the height of his fame and power, probably as recognition for all his efforts running from the birth of musique concrète through the writing of *Traité des Objets Musicaux*, and also probably as a way of facilitating the development of the work of the GRM, whose pedagogical "fiber" had been so plainly apparent up to that point.[49]

In 1975, the course changed its name to "Composition and Research in Electroacoustic Music." The practical sessions, made available to second-year students, left the conservatory to move to the GRM premises at the Maison de la Radio in the 210 editing room, adjacent to the famed future Studio 123, where the research team in computer music would be established in 1977.[50] The practical interest of having installations at the Maison de la Radio was considerable: the studio was open day and night, all year round, as was the entire building dedicated to radio production. Needless to say, students at the time took advantage of this to work at all hours, to the unmistakable benefit of the quality of their productions. Once again the "workshop spirit" of the past was reborn.

This proximity to radio production centers also favored opportunities, sparking the vocation of radio producer in many GRM trainees including Renaud Gagneux, Pierre Tardy, Marc Texier, Jean-Loup Graton, Philippe Mion, Christian Zanési, Gilles Racot, Françoise Bourgoin, Marc Favre, and Patrick Roudier.

The teaching staff also evolved. During the first two academic years, 1968–69 and 1969–70, Schaeffer surrounded himself for the applied sessions with GRM

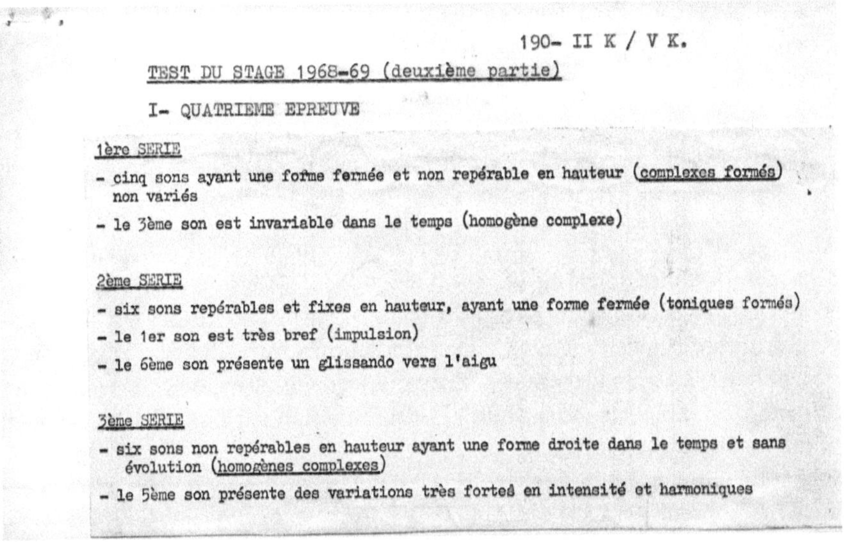

```
                              190- II K / V K.
    TEST DU STAGE 1968-69 (deuxième partie)
      I- QUATRIEME EPREUVE

  1ère SERIE
   - cinq sons ayant une forme fermée et non repérable en hauteur (complexes formés)
     non variés
   - le 3ème son est invariable dans le temps (homogène complexe)

  2ème SERIE
   - six sons repérables et fixes en hauteur, ayant une forme fermée (toniques formés)
   - le 1er son est très bref (impulsion)
   - le 6ème son présente un glissando vers l'aigu

  3ème SERIE
   - six sons non repérables en hauteur ayant une forme droite dans le temps et sans
     évolution (homogènes complexes)
   - le 5ème son présente des variations très fortes en intensité et harmoniques
```

FIGURE 21. Preparatory notes for the entrance exam for the GRM course at the Paris Conservatory. © Évelyne Gayou (1968–69).

composers and researchers. Shortly afterwards, François Bayle distanced himself from the Conservatoire to devote himself exclusively to leading the GRM, leaving his place to Guy Reibel. Other instructors then entered the picture, Michel Chion, Beatriz Ferreyra, and François Delalande. In 1976, Guy Reibel took over direction of the Conservatory course from Pierre Schaeffer, but not his official title, because Schaeffer would only retire four years later. Every year, Guy Reibel chose one student from the outgoing class as his assistant, beginning with Lionel Filippi (1972), and then one after another, Raoul Delgado (1973), Rodolfo Caesar (1974), Daniel Teruggi (1975), Denis Dufour (1976), and Philippe Mion (1977). Laurent Cuniot would become assistant in 1978, joined by Henri Kergomard in 1982, both remaining until 1991.[51]

Publications

PUBLICATIONS BY PIERRE SCHAEFFER

Le gardien du volcan (1969)

Machines à communiquer 1: Génèse des simulacres (1970)

Machines à communiquer 2: Pouvoir et communication (1972)

De l'expérience musicale à l'expérience humaine (1971)

Les antennes de Jericho (1978)

PUBLICATIONS ABOUT PIERRE SCHAEFFER

Entretiens avec Pierre Schaeffer, interviews with Schaeffer by Marc Pierret (1969)

Que sais-je ? La musique concrète, new edition revised by Michel Chion (1973)

De la musique concrète à la musique même, texts collected and commented on by Sophie Brunet (1977)

All these works can be seen as assessments and commentary on previous projects.[52] Schaeffer examined mass media and their sudden expansion, while sensing his (political) powerlessness to transform them, even though all his research converged towards universality and a quest for spirituality in the face of ever more demanding science and technique. "How can we dissimulate the personal imprint of the observer, and to what avail?,"[53] he admitted in the foreword to his book *Machines à communiquer 1*.

Another brief text by Schaeffer, delivered in Lisbon on May 11, 1970, as an introduction to his talk *De l'expérience musicale à l'expérience humaine*, was even more explicit:

> This apparently vague title masks a stubborn thought that has never wavered since the early days of musique concrète.
>
> I have been accused of empiricism, of bricolage. I vindicate—even more than I vindicate the creation of innovative works—the right to experimentation in music, just as in science. Conversely, I have never ceased denouncing the illusion that claims to base the social sciences on the model of the sciences of matter, a demarcation without originality, lacking purpose, lacking realism. I believe that I have put to the test, in the very particular field of musical research, a satisfactory methodology, susceptible of being generalized to the whole of research into man.
>
> A problem emerges in any case. If science indeed constitutes the universal panacea, the only path to knowledge, what would be the use of art? Don't we see contemporary creators tempted by the derisory and the useless, diverted from their deep vocation? Which one? I suggested this in the final lines of *Traité des Objets Musicaux*, "Art is man, described to man, in the language of things."[54]

PROGRAM-BULLETINS

In 1974, in conjunction with the January 16 concert, the GRM published the first of a series of eighteen *Programme-Bulletins*. Over the next two years these dossiers, about fifty pages each, written collectively under the direction of Michel Chion, would be distributed free of charge at the entrance of GRM concerts.[55]

From an editorial point of view, they included a first section reserved for the concert of the day, with its detailed program, accompanied by notes on the works written by the composers themselves, along with analyses of these same works authored by interns. The first *Programme-Bulletin* presented the evening concert,

Gesang der Jünglinge by Karlheinz Stockhausen, *Méditations sur le mystère de la Sainte Trinité* by Olivier Messiaen, and *Parages* by Jacques Lejeune.

In the second section of the *Programme-Bulletin*, various members of the GRM reported on current research on specific themes such as "reflections on the naming of musique concrète," "the problem of the effects of music," and "the aesthetics of the prepared piano." One of these investigations, led by Michel Chion himself, assisted by Jack Vidal, consisted in compiling a lexicon of the terms in *TOM*, ultimately resulting in a smaller companion book to accompany Pierre Schaeffer's larger volume.[56] Michel Chion associated one new chapter from this *Guide des objets sonores*, still being drafted, with each *Programme-Bulletin*. These passages were based on keywords: note; musicality/sound; form/matter; object/structure; suitable object; identification/qualification; accumulation; timbre; mass; abstract/concrete; homogeneous sounds; iterative sounds; natural/cultural; acousmatic; permanence/variation.

The third section of the *Programme-Bulletin* contained recent discographies and bibliographies concerning electroacoustics. Each issue also included a questionnaire for the audience attending the concert to collect the addresses of all those interested in this new music.

The *Programme-Bulletins* are exemplary of how the GRM worked, presenting in a single document all the activities—research, creation, and communication—and underlining their interrelationship. It is worth noting in passing the quality of the contributions and the extreme dynamism in each of the sectors. The *Programme-Bulletin* collection itself was not widely distributed, but most of the articles would later be adapted by their authors and published in books or on record sleeves.

For example, in *Programme-Bulletin No. 4* of February 6, 1974, François Delalande published the first version of one of his key articles, "Trois idées clés" (Three key ideas for childhood music pedagogy). Two years later, the article would reappear in *Cahiers recherche/musique No. 1* (1976). This series of *Cahiers* would eventually take the place of the *Programme-Bulletins*, but in book form, published by the INA-GRM and sold by mail, with six issues appearing from 1976 to 1978.

During this same period, in 1976 to be exact, another important work was published with the support of the INA, *Les musiques électroacoustiques*, cowritten by Michel Chion and Guy Reibel, with the participation of other members of the GRM. This 340-page book presented a synthesis of the history of the influence of electroacoustic music from both French and international points of view.[57]

CAHIERS RECHERCHE/MUSIQUE NO. 1 AND PEDAGOGICAL RADIO BROADCASTS

Beginning in 1973, under the guidance of François Delalande and Guy Reibel, a small team of interns—Claire Renard, Wiska Radkiewicz, Jacques Rémus, Evelyne

Gayou, Marc Favre, and Francis Wargnier—regularly visited a few French National Education kindergarten classes in Paris, working in pairs, to experiment with musical games in an effort to foster research into early musical awareness. The central axis of their exploration could be summed up as follows: how can we help children, and by extension "nonmusicians," to grasp musical concepts without their first necessarily having to learn notes? The practice of listening—"hearing," as described in Schaeffer's *TOM*—could be instrumental in developing a sense of music. The practice of "doing," during games and sound exercises, could lead everyone to "be a musician." The experiments in schools were focused on making this dialectic of "doing and listening" a reality in pedagogical practice in the classroom.

In addition to the *Cahiers recherche/musique No. 1*, these experiments in the classroom gave rise to educational radio broadcasts and numerous training seminars for music teachers in the *écoles normales*, the French teaching colleges. This activity of experimentation in the schools by the GRM "pedagogy" team was taken up by the teachers themselves in their classrooms from 1978 onwards. Their activities of musical awakening became the subject of further radio broadcasts on the network France Musique, in collaboration with the Centre National de Documentation Pédagogique (CNDP, National Center for Pedagogical Documentation).[58]

Two Former Members of the GRM

GUY REIBEL, 1936–

When twenty-seven-year-old Guy Reibel entered the GRM in 1963, he did not take part in a "training course" but instead learned electroacoustics on the job, because he arrived (at the same time as the researcher Henri Chiarucci) in the midst of Pierre Schaeffer writing *TOM*, becoming Schaeffer's assistant owing to Reibel's skills both as an engineer and as a musician—he had studied at the conservatory with Olivier Messiaen. This experience in contact with experimental research into sound matter would be a determining factor for his entire career and even in his life. In his book *L'Homme musicien*, Guy Reibel recalled:

> An entire book would not be enough to describe the consequences of this personal upheaval, in contact with this incredible material, by sheer chance at a time when it was particularly effervescent, in the throes of discovery and in a state of rapid growth.
>
> I was not only shaken by the discovery of the phenomenon itself, because in addition to these new sounds and these new gestures revolutionizing my own musical field, it was my entire prior culture that was called into question, my way of perceiving music in general that was transformed.
>
> I no longer heard Mozart as I had before. I rediscovered Debussy and Stravinsky. The dozens of musicians I thought I knew, who were part of my own intimate cultural identity, appeared to me in a new light. It was as if my listening had suddenly opened up and taken on an additional dimension.

What some people experienced as an accident or a rupture in the thread of time, sometimes even as a tragedy in the history of music, seemed to me on the contrary to be marvelous continuity, luminously obvious. I experienced this encounter with a sense of wonder, at ease, happy as a fish in water. The rupture between traditional system(s) and experimental musicalities created by new technologies (as they were not yet called) did not exist for me.[59]

In 1968 Guy Reibel assisted Pierre Schaeffer in his role as associated professor at the CNSMP. In 1976 he succeeded him as head of the electroacoustic class. His pedagogical activity also extended to Radio France where he created, on the France Musique network, programs dedicated to popularizing contemporary music, *Concerts-Lectures*, one monthly ninety-minute program from 1976 to 1991, and *Enfants d'Orphée*, a program intended for children in kindergarten classes, fifteen minutes Thursday mornings from 1973 to 1982, in collaboration with French National Education and the GRM.

Guy Reibel's first electroacoustic compositions date from 1965. *Durboth*, an eight-minute piece for tape, brass, percussion, and soprano voice, based on a Tagore poem in the Bengali language, premiered with the choral group *Choralies à Cœur Joie* in Vaison-la-Romaine, and was then recorded with the singer Suzanne Rist, under the direction of Konstantin Simonovic.

Guy Reibel acknowledges that at that time he was "under the dual influence (rather contradictory, by the way) of the intuitive method advocated by Schaeffer, and of another, more strongly linked to the cult of 'compositional processes' advocated at Domaine Musical."[60]

Variations en étoile, 1966 (15:30), for tape alone or accompanied by improvised percussion (with percussionist Jean-Pierre Drouet), was a much freer exploration of sound material, and a first step towards "this desire for gesture."[61] *Mouvances*, 1967 (11:47), was conceived for six singers and six microphones that were modified live.

In parallel with his activities as a composer and teacher, Guy Reibel directed amateur and professional choirs. He created and directed the Atelier des Chœurs de Radio France from 1976 to 1986, then directed the Groupe Vocal de France from 1986 to 1990. His works very often contain vocal parts or text: *Rumeurs* (1968), *Carnaval* (1969, revision 1971), *Suite pour Edgar Poe* (1972), *Ode à Villon* (1972), *Rabelais en liesse* (1974), *Chohina* (1979), *Langages imaginaires* (1980).

In *Langages imaginaires*, for vocal quartet, instrumental ensemble, tape, and live transformation by vocoder, Guy Reibel explored the idea that the morphologies of phonemes have a relationship with gesture and the body. In doing so, he complemented and drew upon research work into language by his friend Bernard Ucla. Here is how Guy Reibel describes the progression of his investigation:

After the "bath of basics" at the GRM, and the experience with "radiophony," came undertakings linked to childhood musical education. . . . Musical initiation, particularly through vocal games, develops musical invention on a fundamental level, for amateur beginners, children in particular, as well as for trained professional musicians.[62]

It was while developing these games that gradually the necessity became clear to me of the existence of a link between the musical idea, in the most general sense, and something fundamental, which can be called an *archetype*, in the primary sense of the term. Moreover, the territory of experiencing vocal games immediately established connections with language, inseparable from all vocal expression.[63]

Guy Reibel's other works expressed his taste for the transformation of sound materials as practiced at the GRM, but with an added tendency to explore the contribution of instrumental gesture in musical production, often resulting in a large role for improvisation by instrumentalists, *Anneaux d'ombre* (1969), *Vertiges* (1969), *Musique en liesse* (1970), the *Tryptique Électroacoustique* consisting of three parts—*Signal sur bruit* (1973), *Cinq études aux modulations* (1973), and *Franges du signe* (1974)—*Granulations sillages* (1976), *Douze inventions en 6 modes de jeu* (1979), and *Quatre études de forme* (1980).

For his piece *Douze inventions en 6 modes de jeux*, Guy Reibel collaborated with the newly formed Trio GRM+, made up of three young students from his electroacoustic class at the Conservatoire, Laurent Cuniot, Denis Dufour, and Yann Geslin. But this orientation focused on "playing," distant from the concerns and musical ambitions of François Bayle (then director of the GRM), ended up bringing the divergences between the two to the forefront. After having been a very active member of the GRM, Guy Reibel moved away from the Groupe beginning in 1981. The collaboration agreement between the GRM and the conservatory class, notably through the loan of equipment and studios, was not renewed after 1983.

The instrumental trio continued its activities until 1990, but when it broke away from the GRM it changed its name and aesthetic orientation, becoming Ensemble TM+ and devoting itself to the interpretation of works from the contemporary repertoire in the broadest sense.[64]

Guy Reibel continued his career as a composer on his own, while maintaining his teaching activities at the conservatory. In 1985, he became a professor of instrumental composition, handing over the electroacoustic class to Laurent Cuniot, Yann Geslin, and Henri Kergomard.

Starting in 1984, Guy Reibel also invested himself in the conception of new instruments, the OMNI (or Slightly Spherical Surface), and Patrice Moullet's *Corps Sonores*, both "sound bodies" for playing and morphologically transforming sounds live. Based on tactile input—the different qualities of touch emanating from gesture—these instruments piloted samplers. Guy Reibel explained:

The point in common between the Corps Sonores ... and the voice is that they can be used independently of any constituted musical system such as the tonal tempered scale." ...[65] The OMNI is like an organ which can be programmed at will, utilizing every conceivable element of music ... Resulting in a twofold examination of both the fundamental aspects of diverse "classes" of morphology, and of the gestures which can be associated with them.... A definitive assertion that gesture, in the most general sense, is indeed the essential agent of the creators' thought, the means for its implementation, and the manner of making the creators' presence visible alongside their works.[66]

Identifying the root of the musical idea, in order to develop and enrich it, appears to be the underlying principle that has driven Guy Reibel since his first encounter with sound matter in the early 1960s.

MICHEL CHION, 1947–

Michel Chion left the GRM in 1976, at the age of twenty-nine, certainly following the example of Pierre Henry (who had distanced himself in 1958), to pursue his personal career without being overshadowed by an institution or an overly imposing personality (Schaeffer). Altogether, Michel Chion was only a member of the GRM for five years, and yet his aesthetic and even moral commitment to musique concrète makes him an eminent representative of the genre, even a militant in the eyes of the musical milieu; all the more so since he never sought to deny the situation.

And so Michel Chion is "a friend of the GRM family," sometimes a bit cumbersome because of his impetuous character and his theoretical intransigence, notably concerning the definition of musical genres or the vocabulary and concepts of electroacoustic music. He often presents himself as the representative of orthodoxy by virtue of his former collaboration with (the Master) Pierre Schaeffer. Michel Chion's numerous writings on music and film are also strongly influenced by his musical perspectives. In his 2003 book *Un art sonore, le cinéma*, Michel Chion reviewed the early history of cinema and its various genres from the point of view of sound, suggesting that we should speak of "deaf film" rather than "silent film."

Chion also prides himself on penning new concepts, like the term *sons fixes* (fixed sounds), which he introduced in 1988 "to designate sounds that have been stabilized and inscribed in their concrete details on a recording medium of any kind."[67] Michel Chion himself encourages reflection concerning his new ideas and encourages questioning. On the subject of the relationship between the real and *son fixé*, he expressed his interrogations in 1998: "The expression 'fixed sound' ... is ambiguous because it is not what happened that is fixed. It is its trace that is imprinted. The footprint of someone does not fix the appearance of the person, but the trace remains and we can observe it and that is what a fixed sound is."[68]

Michel Chion entered the Conservatoire de Versailles at the rather late age of sixteen to study Music. But he had developed his auditory sensitivity from early childhood thanks to his first experiments on a Philips tape recorder that his father had brought back from Germany when he was ten years old, a time when the concrete manipulation of sound was still unknown to the general public.

From 1969 to 1971, Michel Chion studied at the GRM. His impressions:

> The first year, there were quite a few of us, about forty I think, but not everyone came to the Bourdan Center for the training course. The majority came to follow Schaeffer's theoretical seminar. After that, things settled down and in the second year, after the exam, there were only six of us.[69] I knew from Guy Reibel that I had just barely made it into the second year. I wasn't "really gifted for sounds" but I was known for knowing how to talk about them, for having a kind of intellectual vision of them and I had also written good summaries of the seminar for *Étapes*, the in-house journal of the Service de la Recherche, edited by David Rissin, and it was these capacities as a writer and intellectual that were remembered.
>
> So in the second year, there were only six or seven of us left. Among them were Robert Cahen, Roger Cochini, Jean-Michel Jarre (who would leave during the second year), Jean Leduc and Nina Coissac.[70]

Immediately after, Michel Chion was offered a research contract at the GRM, first as an assistant to Pierre Schaeffer at the Conservatoire, in 1971–72, for teaching the history of electroacoustic music. He then became the assistant of Agnès Tanguy, who was in charge of the GRM's radio programs, and then took over from her from 1973 to 1976.

Michel Chion composes musique concrète, makes videos, teaches at the university, and writes. Among his important musical works are: *Le prisonnier du son* (1972, 34:40), an electroacoustic melodrama; *Requiem* (1973); *On n'arrête pas le regret* (1975); *Tu* (1977, 60 minutes, 40 seconds), a mystery in two acts and sixteen movements; *Diktat* (1979, 95 minutes), a melodrama in seven parts; *La ronde* (1982, 24:08), in ten movements to his own poems; *La Tentation de Saint Antoine* (1984); *Samba pour un jour de pluie* (1985, 11:30), melodrama; *Dix études de musique concrète* (1988, 31:51); *24 préludes à la vie* (1989, 42:35); *Gloria* (1994, 20:25); *Perpetuum Kyrie* (1997, 28:40); and *L'île sononnante* (1998, 28:40).

Among his twenty-seven books on music, sound, and especially cinema, can be found: *Les musiques électroacoustiques* (with Guy Reibel), 1976; *Pierre Henry* (a monograph), 1980, reedition 2003; *La musique électroacoustique* in the Que Sais-je? book collection, 1982; *La voix au cinéma*, 1982; *Guide des objets sonores*, 1983; *Le son au cinéma*, 1985; *La toile trouée ou la parole au cinéma*, 1988; *L'audio-vision*, 1990; *L'Art des sons fixés ou la musique concrètement*, 1991; *Le Poème symphonique et la musique à programme*, 1993; *Le Promeneur écoutant*, 1993; *La Musique au cinéma*, 1995; *Le Son*, 1998; *Un Art sonore, le cinéma*, 2003; and *David Lynch*, 1992, reprinted in 2007.

In 2005, the GRM devoted a book to Michel Chion: *Polychrome Portrait no. 8*.[71]

At the turn of the twenty-first century Michel Chion took on a singular role by presenting himself as a defender of the concrete style, as opposed to all the variants that have emerged since Pierre Schaeffer and Pierre Henry's first efforts. By refusing to "go digital" Michel Chion was determined to remain faithful to the techniques of analog editing and mixing.[72] These techniques do not allow the infinite multiplication of sound layers and transformations without loss of quality, unlike audio sequencers, which, with all their plug-ins, encourage the creation of a multitude of sound planes and effects. As a result, Michel Chion, also a keen film lover, has settled into a melodramatic style, with a meticulous improvisational component introduced during the recording process, in order to *fix* in the best possible way his sound materials—voices, texts, atmospheres, and spaces.

But after more than a decade of resistance, he himself converted to digital.

9

1978–1988

Real and Nonreal Time

The GRM continues to operate under the aegis of the INA, in congruity with the spirit of the new situation in the music world. The philosophy of its studios and its sound processing tools remains faithful to the principles of the beginnings of musique concrète, technology at the service of people, as an extension of their thinking and their fingers. High-speed computing was emerging. The decisive turn of research towards digital audio tools in Studio 123 opened a new era of significant artistic productivity at the GRM, to which the works created give ample evidence. Competition in digital technology between the GRM, with Syter, and the IRCAM, with its 4X, was in full swing already in 1981, as well as with the Americans, who were still far ahead.

The launching of regular collaboration with Radio France in 1978 for organization of annual concert seasons, under the name of Cycle Acousmatique, strengthened the GRM's position within broadcasting and music circles. If we look at the concert programs, we also notice an active policy of inviting outside composers to work in the two studios dedicated to creation, 116C and 123.

HISTORICAL CONTEXT

With the creation of the INA in 1975, a new GRM emerged. Established in the premises of the Maison de la Radio, the GRM forged links with Radio France, both human and statutory. A list of responsibilities, voted by the French legislature in the 1975 audiovisual law, described the reciprocal services to be provided by the INA and Radio France. In exchange for access to the audiovisual archives held and maintained by the INA, Radio France paid a fixed sum each year and gave the INA access to its airwaves for forty hours of creative and research radio programming annually, entrusted to the GRM.

In 1984 a collective labor agreement for audiovisual personnel came into force. It also concerned GRM's collaborators, now hired in long-term contracts. The position of researcher-composer disappeared, replaced by levels and rungs, associated with a seniority grid and salaries harmonized with the entire audiovisual sector. The members of the GRM were now divided into researchers, administrative staff, and production staff. The activity of composing was no longer officially recognized. Collaborators were expected to compose on their own time, even though their musical skills were a requirement for joining the Groupe. In order to have enough free time to create, most composers opted for part-time contracts. The GRM followed a policy of commissioning new works for its concerts from composers of its choice. Research and development work was differentiated from the rest; engineers were hired for specific research projects, which became all the more important as computers became omnipresent.

Lastly the INA, with its status of EPIC (*établissement public à caractère industriel et commercial*, industrial and commercial public establishment),* had an obligation to generate revenue. The GRM participated in this commercial strategy, publishing records and books that were distributed and sold to the public.

This deep evolution in the audiovisual landscape was coupled with an important change in the French government's cultural policy, beginning in 1981 with the arrival of the Left to power. The first Fête de la Musique (World Music Day), held June 21, 1981, on the initiative of Jack Lang, the incoming minister of culture, reflected this change. Lang appointed Maurice Fleuret to the post of director of music. In a mere four years, Fleuret would impose a lasting influence on the renovation of the French music sector.[1] From the beginning of his mandate, Maurice Fleuret—journalist, music critic, festival organizer, and an aficionado of contemporary and non-European music—strove to promote amateur music, evoking a 1980 poll revealing that one of every three French households owned a musical instrument and that one out of two people under eighteen years of age was a musician.[2] He also supported the renovation of musical education; the first three Centres de Formation des Musiciens Intervenants (CFMI) were created in 1983 at the universities of Aix-Marseille, Lille III, and Toulouse-Le-Mirail.[3] In this effervescent context, electroacoustics naturally found its place.

Numerous centers for research and musical creation, inspired by the GRM model, opened in France, financially backed by the French government. This was a far cry from the wave of creation of the major national historical centers attached

---

* Translator's note: The EPIC is a special French legal status for what would normally be a private corporation open to competition, but given exceptional circumstances is considered to best serve the public interest as a governmental entity.

to radio institutions during the 1950s–60s—NHK in Japan, WDR in Germany, and RAI in Italy. The structures emerging in the 1970s and 1980s, both in France and abroad, were often at the initiative of composers trained at the GRM. They constituted a second generation. With the benefit of hindsight, it could be argued that institutional electroacoustic music was at its peak.

THE MUSICAL CONTEXT
Catching Up with the Americans

Ever since his arrival at the GRM in 1973, first as an intern and then as a full-time engineer from 1975 onwards, Jean-François Allouis had kept abreast of the technical progress in computing in the country most advanced in the field, the United States. He monitored the latest microprocessors invented by firms such as Intel; he read professional journals; he knew Max Mathews, the inventor of the Music V program, which the GRM (like the IRCAM) had had at its disposal since 1971 thanks to permission from Mathews himself.[4] Mathews worked with John Pierce at Bell Labs in New Jersey, they were in contact with researchers in the scientific community, and ideas circulated. John Chowning, another American, from Stanford University in California, devised frequency modulation sound synthesis in 1967. In 1980 the patent for this discovery was bought by the Japanese firm Yamaha, which developed and in 1983 marketed the famous DX7 synthesizer, followed by all its descendants.

All of this progress made by "the Americans" was transmitted to the IRCAM through the intermediary of Jean-Claude Risset, a French musician and engineer, graduate of the prestigious École Normale Supérieure, whom Pierre Boulez called upon to set up the IRCAM computer project and who had the idea of inviting some of the American pioneers of computer science to come and work in Paris. The arrival of these American researchers, whom Risset had met during his studies at Bell Laboratories, would allow France to catch up in the field of computer music.

The American advance in computer science was considerable. The first milestone cited in the American saga of computer music is always the work of Lejaren Hiller and Leonard Isaacson at the University of Illinois, giving birth in 1957 to the first piece of automatic composition "on a computer," *Illiac Suite for String Quartet*. The Music V program developed at Bell Labs by Mathews dated from 1967, its precursors Music IV from 1961, Music III (the true breakthrough) from 1959, Music II (four-channel mono), and Music I with its single channel of mono sound from 1957.

In a 2003 radio interview, Marc Battier also insisted on the essential role of Jean-Claude Risset: "Jean-Claude Risset revealed that the problem was not one of technology, but one of knowledge of sounds. It is with Risset that timbres were enhanced. This led to research in psychoacoustics, and the work of John Chowning, etc."[5]

John Chowning was indeed able to crystallize and direct cutting-edge research into computer music.[6] After his breakthrough discovery of sound synthesis by frequency modulation, patented in 1967, he never abandoned his interest in the concept of timbre. In 1970 the Japanese firm Yamaha bought the rights to this patent, later producing a range of synthesizers, including the DX7, offering musicians the ability to create magnificent timbres. Chowning also explored the idea of the illusory sound spaces created in our brains by our perception of finely calculated timbral trajectories.

In addition to his personal projects, Chowning founded a major U.S. center for musical research and innovation at Stanford University in California, in 1973 opening the Center for Computer Research in Music and Acoustics (CCRMA—pronounced Karma), which not only trains composers but also acts as a bridge between technical and artistic research and the industrial world of Silicon Valley.[7]

Flashback to 1978: The American researchers who came to Paris, sometimes under contract for several years with IRCAM,[8] also seized upon the occasion to visit the GRM and get to know this singular site, living witness to the history of the invention of musique concrète. And over the years warm human relationships grew between the members of the GRM and "the Americans": Max Mathews, John Chowning, Jon Appleton, David Wessel, Miller Puckette, Eric Lindemann, Andy Moorer, and Gerald Bennett.

Studio 123, Nonreal Time at GRM

Another outstanding researcher, Bénédict Mailliard, had been working in the field of computer science at the GRM since 1975, at the same time as Jean-François Allouis, but with a more fundamental approach, concentrating on pure computation and algorithmic research. Bénédict Mailliard studied mathematics at the École Normale Supérieure.

Following the first workshop in 1973 during which the GRM computer team had tried, without success, to interest composers in Music V programming, a second one-week workshop took place in 1979. In the meantime, Bénédict Mailliard had thoroughly reexamined the possibilities of the studio. He developed simplified access to the tools, at last enabling composers to become autonomous, working without the assistance of an engineer. New one-week workshops, each bringing together four or five composers, were then regularly organized through 1985. But still too often when the computer was used for composition, Mailliard's technical expertise was called upon for assistance. This aid brought by engineers to composers fueled a controversy between the two professions. Should engineers be considered as contributing authors of the music when they are involved in a major proportion of the work in the studio? Jean-François Allouis always refused to "hold

FIGURE 22. Computer training course at Studio 123 led by Bénédict Mailliard. From left to right: Augusto Mannis, Bénédict Mailliard, Christian Zanési, Jacques Lejeune, Alain Savouret, Philippe Mion, Guy Reibel, Jean-Loup Graton, Yann Geslin. © Guy Vivien (1982).

the hands" of composers; Bénédict Mailliard did so only in hopes of eventually seeing things change. This is how the search for more ergonomic user-friendly software tools became central to the development of the fledgling Syter system.

But after collaborating in launching Studio 123, the two founders gradually came into conflict, as early as 1977.[9] And in 1986, a sudden turn of events took place. Bénédict Mailliard resigned because his project had been blocked. For one thing, the equipment was old (ten years old), leading to frequent technical malfunction, but above all he was frustrated by the repeated refusals from François Bayle and Jean-François Allouis to finance his "deferred time" research, in favor of research on "real time." The computer was definitively shut down in 1988.

Was this strategic choice a mistake or not? Several years later, François Bayle explained his viewpoint in an interview: "In the end, the future would prove that real time was the best path, especially in acousmatics where one absolutely needs to hear what one is doing. [Favoring deferred time] would have meant returning to the predictive approach that already had its home: the Ircam."[10]

After his resignation, Bénédict Mailliard suffered greatly from his separation from musical research, even more so because his high level of intelligence, recognized by all, certainly needed the stimulation of the group to function well.[11]

Syter: SYstème TEmps Réel (Real-Time System)

In Studio 123 of the GRM, real-time research, directed by Jean-François Allouis, had been carried out in parallel with research on deferred time since 1975.

But composers, in spite of the first introductory workshop in 1973, did not easily adapt to the logic of computer codes, leading to the idea of Jean-François Allouis, assisted by Bernard Dürr, of attempting to approach the ideal of real-time sound processing while developing an interface that would be simple enough to not discourage composers.[12] In the previous chapter, we saw that Syter 1 in 1977 was simply a computer that controlled an analog synthesizer, the Syter 1 card running on the PDP-11/60 computer. After an intermediate stage using wafer microprocessors for Syter 2 (1980), a third Syter appeared in 1981, the final version of the initial project. The system had been purchased in 1978 and the converters built over the next year by the technician Richard Bulski, who would continue to be affiliated with the GRM for the rest of his career. He spent hundreds of hours "wrapping" the circuits, elements of four bits each, to optimize computing speeds. Microprocessors at the time were still too slow. Then a new computer from the English manufacturer SMS was acquired by the GRM in 1982, a clone of the DEC PDP-11/23. And then came the 11/73, more advanced, less expensive, and more compact.[13]

SYstème TEmps Réel (Syter, the real-time system) was developed at the same time as Peppino Di Giugno's 4X at IRCAM in 1981. The two systems produced practically the same results, but by following opposite paths. Jean-François Allouis explained:

> I began with the concept of Music V, which contended that with a programmable object one could do anything one wanted. But in reality, since microprocessors at the time did not offer enough computing power, I chose to return to digital wiring; it was the only way to obtain computing speed. Di Giugno, on the other hand, had started with microprocessors, but they were quite unsophisticated in the beginning. For the 4A in 1976 they were simple oscillators. In the end, with Syter at the GRM, we achieved the same results as the 4X at the Ircam.[14]

IRCAM's 4X made working in real time possible but required the assistance of an engineer or a tutor competent in computer programming to carry out the operations dictated by the composer. In his book *L'homme musicien* published in 2000, Guy Reibel explained very clearly the nature of this fundamental divergence between IRCAM and GRM:

> If one had to summarize it, one could say, in a schematic way, that for the IRCAM new technologies constitute an enlargement of the traditional musical field, progressive improvement of savoir-faire in the field of instrumental composition. A "bonus" in a way, which does not fundamentally modify a certain way of thinking about music, nor the musical institution as a whole.

FIGURE 23. Jean-François Allouis in Studio 123 in front of Syter 3. © Guy Vivien (1985).

On the other hand, for the GRM musique concrète is an irruption, a fracture in the way of listening, of thinking, of "being" musicians, questioning our attitudes and our musical behavior, and it is for this reason that Schaeffer was more interested in the "musician as human" than in the works of composers, judged by him as often being enthralling. But more as being "experimental" works, open doors to research into listening, just as much as they were artistic objects in themselves.[15]

Once the Syter system had been built, Jean-François Allouis began writing signal processing software that made the conversation between the composer and the machine easier, without requiring any particular knowledge of computers. The success of this feature is undeniable. The final version of Syter became available in 1985, consisting of a host computer housing a specialized modular processor with graphical control panels, input/output interfaces and programming software for interactive control of sound processing. It also featured synthesis, mixing and looping. Between 1985 and 1987, the Digitone company sold a dozen of the 1985 version, the most advanced Syter. But the retail price was about 700,000 francs, as much as the purchase price for a two-room apartment in Paris.

In hindsight, we can only underline the foresight of Jean-François Allouis, who declared in 1986: "Syter can be seen as a hardware precedent to design the prefigura-

FIGURE 24. Évelyne Gayou in Studio 116 A, devoted to radio production. © Éric Bourbotte (1987).

tion of what will, in my opinion, be the studio of the future, namely a 'software studio.' In this studio, all the functions (recording, synthesis, processing, manipulation/editing/mixing) will be performed by software, in an integrated environment."[16]

And yet the Syter experience would have never seen the light of day without two unconditional and essential supporters (psychologically and financially): the director of the GRM, François Bayle, and the director of the Digitone company, Jean-Louis Lapeyre.

The Introduction of Digital Technology into Analog Production Studios

The GRM always defended the philosophy of the composer's autonomy in relation to tools, not out of vanity but to reduce as much as possible the distance between doing and listening. The simplest solution lay in making composers the direct authors and actors of their music by putting the tools directly in their hands. More trivially, this also reduced problems of ego and interpersonal relations between composers and their assistants, technicians, or engineers. So the ergonomics of composition studios were adapted to new technologies as they became available.

Digital recording arrived on the market in 1980, in the form of videotape-based systems, but it was not immediately adopted at the GRM. Analog recording on 6.25mm (¼-inch) tape would continue until 1998. In 1977, a 1-inch 8-track tape recorder was acquired to complete the equipment of Studio 116 C, which already had four ¼-inch 2-track stereo Studer A80 tape recorders. In 1987, Hugues Vinet, technical director of the GRM since 1985, bought a Studer 2-inch 16-track analog tape recorder; it was installed in Studio 116A, which already possessed as standard equipment four ¼-inch 2-track analog tape recorders and a DDA console with sixteen inputs, four outputs, eight groups, and JBL speakers with Studer amplifiers. In 1982, Studio C, geared more towards composition than Studio A, was equipped with a Publison sampler ("the French infernal machine") from the DHM company. Studio A was primarily devoted to radio production. Afterwards, many devices incorporating digital hardware arrived, including a DX7 synthesizer in 1983 and a Casio sampler in 1983.

MIDI equipment was gradually introduced after creation of the MIDI standard in 1983. An in-house workshop was organized, without much success, for members of the GRM wanting to familiarize themselves with the new standard.

The CD player was introduced in 1984. In only three years, it dominated the studio, pushing aside the black vinyl records. Finally, the personal computer—the Macintosh system—made its entrance at the GRM in 1988, initially exclusively in the research studio.

The IRCAM during This Same Period

Since 1980, IRCAM had been the target of considerable criticism concerning its real usefulness and its cost. Several people resigned, notably Jean-Claude Risset, who had founded the computer department. In an interview with Olivier Meston in 2001, Risset explained: "I left Ircam in 1979 for personal reasons. It was obviously a very exciting place to be, but at the same time difficult to deal with, because it was necessary to carry out both multidisciplinary research as well as to hold public presentations in the Palais de la Découverte style."[17]

The five original departments were eliminated in order to decompartmentalize and rejuvenate the teams. Then the teams were composed of reception staff and technicians who guided and assisted guest composers. Tod Machover was leading musical research, Jean Kott, computer research. Giuseppe Di Giugno was designing a real-time digital synthesizer, the prototype of which was completed in 1981.[18] The first version of Pierre Boulez's *Répons*—a work for six soloists, ensemble, and 4X computer—premiered that same year in the German town Donaueschingen.[19]

Conservatories: CNSMP, CNSML, and CNR in Lyon

Beginning in 1982, the trajectories of the electroacoustic program at the Paris Conservatory and the GRM began to diverge. The CNSMP course, created in 1968 by Pierre Schaeffer, was still active, with Guy Reibel as director since 1976. But with Schaeffer's effective retirement in 1980, the relationship between the GRM and "the class" changed. In 1982, the working studio (of the conservatory) at the GRM moved back to the Paris Conservatory. It was modernized, but still "in analog." In

1984 the schism had become complete when the partnership agreement between the two institutions was not renewed. Then, with the arrival of Marc Bleuse as director of the conservatory in 1985, the organization of studies was reformed. In particular, the instrumental composition class absorbed the electroacoustic class, it becoming merely an option during the first two years of composition.

Since 1978, Guy Reibel's two assistants in the electroacoustic class had no longer been replaced every year. Laurent Cuniot and Henri Kergomard, the two most recent assistants, remained at the conservatory and in 1982 even took over teaching the class when Guy Reibel himself accepted a teaching position in an instrumental writing class.

The Conservatoire National Supérieur de Musique de Lyon (CNSML) was created in 1980. Pierre Cochereau was its first director, Gilbert Amy succeeding him in 1984. In spite of the creation of this national conservatory with its own electroacoustic course just as in Paris,[20] the situation of conservatory education in general was in dire need of modernization. According to Anne Veitl and Noemi Duchemin: "The situation of musical education in 1981 can be summarized as follows: since 1795, French music policy has been focused on developing centralized music education focused on specialized music training."[21]

Still in Lyon and still in 1980, but at the *conservatoire national de région* (CNR), Denis Dufour, a former student of Guy Reibel at the Paris Conservatory and a member of the GRM since 1976, inaugurated an "acousmatics class."

Workshops with the City of Paris

In 1978 the city of Paris created the Association pour le développement de l'action culturelle (ADAC, Association for the Development of Cultural Action), appointing Francis Balagna as its director.[22] In 1979, as part of the Ateliers Nouvelles Technologies, the GRM was entrusted with the Atelier d'initiation à la musique électroacoustique, which would train about ten people every year in analogical composition techniques, under the direction of two composers associated with the GRM, Philippe Mion and Jacques Lejeune. The GRM equipped the studio, located on the premises of Théâtre Présent,[23] and the city of Paris partially remunerated the teachers.[24] The workshop attracted a varied public, ranging from young amateurs curious about new technologies to composers hoping to become professionals.

Grame in Lyon

Among the new sites for research and musical creation, the Grame stands out as a model. Created and directed by James Giroudon and Pierre-Alain Jaffrenou, Grame was established in Lyon in 1981.[25] Thanks to the support from the Ministry of Culture, this center has developed a full range of activities from musical creation to the diffusion of works in concert and research. Going beyond electroacoustics, Grame focuses on contemporary music, even founding the Ensemble Orchestral

Contemporain (EOC), an ensemble of performers of varying geometry. Grame also organizes educational sessions with various regional partners, as well as the Musiques en Scène Festival once a year, and an international competition for musical creation. Yann Orlarey, its scientific director, created the Sinfonie in 1984, together with Pierre-Alain Jaffrenou. This was a diffusion device, without potentiometers, for recorded or live music, which enabled "diffusion patterns" to be played.

From the point of view of research, Grame sees itself

> at the heart of the Art-Science debate and at the crossroads of many disciplines; research in computer music constitutes one of the essential poles of its activity. . . . Research projects are articulated around two main axes, on one hand . . . development of collaborative real-time music applications, and on the other, developing formal languages for symbolic manipulation in computer music writing.[26]

Grame is also notably recognized for development of the programing language Faust (Functional Audio Stream).

Music and Research in Belgium

In 1982, Annette Vande Gorne, a former student of the electroacoustic class at the Paris Conservatory, created the nonprofit organization Musiques & Recherches in Ohain near Brussels, Belgium. The following year, the organization acquired a diffusion system and began organizing regular concerts of acousmatic music. From 1988 onwards, Annette Vande Gorne also offered training courses in composition. Spatialization courses using the organization's fifty-four-loudspeaker Acousmonium were an important innovation in 1999. In September 2004, Musiques & Recherches merged its teaching activities with those of the Belge Royal Conservatory of Mons, as part of the reform of advanced conservatories in Europe. From that date on, conservatories in Belgium have functioned like universities, issuing comparable diplomas—bachelor, masters, and doctoral degrees. Education in electroacoustics has become quite comprehensive, with twenty hours of classes per week, six electives, and numerous hours of studio work.

Francis Dhomont in Canada

Another example of the spread of electroacoustics comes from Canada, where the French composer Francis Dhomont emigrated in 1978. He had discovered in France the philosophy of working with concrete sound matter based on listening, and brought it with him across the Atlantic. His teaching at the University of Montreal from 1980 onwards led to the creation of a Canadian electroacoustic aesthetic movement, the "Montreal School." He grafted this new approach onto a Canadian musical environment whose almost unchallenged masters until then had been John Cage and Erik Satie, even if a few Canadian artists had already opened the door to this new approach. Maurice Blackburn, composer and collaborator of the filmmaker Norman McLaren, had attended the GRMC in 1954–55; composer and director Pierre Mercure had done the same in 1957–58, Micheline Coulombe Saint-Marcoux in 1968, Marcelle Deschênes in 1968–71, and Yves Daoust in 1976–78. Among the best-known composers of the Montreal School trained by Dhomont are Robert Normandeau, Stéphane Roy, Christian Calon, Roxanne Turcotte, Gilles Gobeil, Ned Bouhalassa, Claudia Tamayo, Claude Shryer, Monique Jean, and Michel Frigon.

The Birth of Electro

In so-called commercial music, alongside hip-hop and rap, electro appeared around 1982–83. The term *techno* originated in part in connection with the German group Kraftwerk and their 1978 song *The Robots*. I evoke this example because it clearly illustrates the dialectic of interaction between aesthetic influences. In the beginning, a group, Kraftwerk, claimed to be influenced by Stockhausen's music. But after the period from about 1985 to 1995 in the sphere of commercial music producing numerous aesthetic variant forms, in turn in the early 2000s one of them in particular, electro music, became a source of inspiration for "serious" composers.

EVENTS, PERFORMANCES, PUBLICATIONS, PEOPLE

The First "Acousmatic Cycle"

The first concert season of the GRM in coproduction with Radio France was held in 1978. This "Cycle Acousmatique," organized by Ivo Malec, François Bayle, and Geneviève Mâche,[27] took place in the Grand Auditorium of the Maison de la Radio, Studio 104, which would be renamed Salle Olivier Messiaen in 1992 in homage to the composer who died that same year. Each season, numerous works premiered over the course of six or seven evenings, each usually including two concerts, resulting in fifteen to twenty new pieces each year. In addition to purely electroacoustic pieces, there were also mixed works and live electronics, and sometimes multimedia. Most of these works were commissioned from composers who would come and work in the GRM studios to compose them. The concerts—especially the mixed works—were recorded by Radio France sound engineers and then broadcast on the radio. Cycle Acousmatique continues to exist, following the same formula, simply renamed Son-Mu in 1992, and then Multiphonies in 1998. Since 1986, it has also sometimes been accompanied by other events, such as public panels exploring research topics.

A Selection of Musical Works from 1978–1988

Musical production reflected the new institutional and artistic context of the decade. In addition to the new stimulating collaboration between the GRM and Radio

France for the Cycle Acousmatique concerts, the major technological changes from the introduction of high-speed computing in Studio 123 had a crucial impact. Finally, the founding of GRM PLUS Trio would lead to a new repertoire of mixed works.

1978–80, *Erosphère*—cycle of three pieces by François Bayle.
    La fin du bruit (26 minutes) has become the classic example of successful use of the resonant filters developed by Bénédict Mailliard at Studio 123, the computer studio, but the work also explored the possibilities of other nonreal-time processing.
    Tremblement de terre très doux (28:11). The title is an homage to the surrealist artist Max Ernst, who painted a canvas by the same name, *A Very Gentle Earthquake*.
    Toupie dans le ciel (22 minutes). The piece opens with the "real" sound of a spinning top.
    Just before joining the IRCAM team, Marc Battier assisted François Bayle for the three interleaved movements *Eros bleu*, *Eros rouge*, and *Eros noir*.

1979, *Pacific Tubular Waves* (25 minutes) by Michel Redolfi was entirely composed of electronic sounds from a Synclavier digital synthesizer.

1979, *Une saison en enfer* (44:10), 8 tracks, by Gilbert Amy, based on Rimbaud's text. Electronic sounds and sounds of human origin realized with the assistance of Yann Geslin. Voices: Nelly Borgeaud, Eweda Malapa, Michel Hermon, and Jean Tibaudeau.

1979, *Douze inventions en six modes de jeu* (40:30) by Guy Reibel, with TRIO GRM PLUS (Laurent Cuniot, Denis Dufour, Yann Geslin). In this piece for 16-track tape, each "invention" was developed from an electronic element treated as an instrument with a specific playing mode. In movements 3, 6, 8, and 11 the TRIO plays its own inventions solo, following the composer's graphic score.

1979, *Quand est-ce qu'?* (12:22). The only piece by Nicolas Frize in the GRM repertoire, composed during his studies at the Conservatory (1972–74). Since then, Nicolas Frize has worked as an independent composer, often drawing attention for his left-wing political and social commitment, through exceptional creations in unexpected locations with atypical participants—workers in Renault factories, prisoners, *Nuits blanches* of Paris.

1979, *Bilude* (2:18). Final piece composed by Pierre Schaeffer. Minka Roustcheva, piano. This short piece with its ironic character, composed in homage to René Daumal, was inspired by the first piece in Johann Sebastian Bach's *Well-Tempered Harpsichord*.

1980, *Verbes comme cueillir* (12:20) by Marc Battier. Digital transformation of words using the Studio 123 software.

- 1980, *La Galerie* (46:30) by Denis Dufour, for three "electronic" musicians (TRIO-GRM-PLUS) and three "concrete" musicians (Marie-Noëlle Moyal, Philippe Mion, and Roland Petit) playing diverse microphoned bodies.
- 1980, *Immersion* (24:10) by Michel Redolfi. This piece was composed in part at the GMEM, from concrete sounds transformed with the Synclavier digital synthesizer.
- 1980, *Quatre études de forme* (32:30) by Guy Reibel. Commissioned by INA-GRM for piano solo (movements 1 and 4) and for tape and piano (movements 2 and 3).
- 1980–83, *Son vitesse-lumière* by François Bayle. I. *Grandeur nature* (32:40); II. *Paysage, personnage, nuage* (24 minutes); III. *Voyage au centre de la tête* (20 minutes); IV. *Le sommeil d'Euclide* (30:30); V. *Lumière ralentie* (34 minutes). This work is haunted by the universe of Jules Verne.[28] In the program notes, the author indicated: sound recording, Christian Zanési; digital audio processing, Bénédict Mailliard, Yann Geslin, and Jean-Yves Bernier. In 1986, in collaboration with visual artists, François Bayle took part in the F.A.U.S.T. festival in Toulouse, performing *Son-Vitesse-Lumière* on the Acousmonium, accompanied by lasers, smoke effects, and a very complex lighting system designed by Jacques Rouveyrollis, a specialist in lighting shows for superstars.
- 1981, *Don Quichotte corporation* (22:42) by Alain Savouret. In this piece, analog transformation and digital transmutation intermingle. Bénédict Mailliard contributed significantly to this work with his assistance for the nonreal-time processing, notably in the extreme use of the possibilities of Studio 123.
- 1982, *Variations didactiques* (10 minutes) by Yann Geslin. To Mallarmé's poem "Un coup de dés jamais n'abolira le hasard" (A roll of the dice will never abolish chance) recited by Michael Lonsdale. The author started with a single material, the voice, subjecting it to multiple variations using only a computer (the one in Studio 123 of the GRM). Some transformations are so powerful that the voice is no longer recognizable.
- 1982, *Hinterland* (30 minutes) by Carlos Roque-Alsina, for tape, piano, and percussion. Carlos Roque-Alsina, piano, and Gaston Sylvestre, percussion.
- 1982, *Gaku-no-michi* (*Les Voies de la musique*—The way of music, 3 hours) by Jean-Claude Eloy, version for tape and synthesizers. Film without images for electronic and concrete sounds. This work is divided into four parts. *I–Tokyo* is composed of sounds from daily life in the city (subway, bus, doorbells, elevators . . .) mixed with other sounds from Japanese life (bells of the Kyoto temples, announcements of sumo fights . . .). *II–Fushiki-e* (*La Voie des sons de meditation*—The way of meditation sounds) evolves in a universe of electronic timbres, contrasting with the first part. *III–Banbutsu-no-ryudo* (*Le Flot*

incessant de toutes les choses—The unceasing flow of all things) returns to everyday sounds manipulated in such a way as to become something altogether different. *IV–Kaiso* (Reminiscence) concludes with movement towards abstraction.

1982, Monday, March 22, *Hymnen*, Regions I and IV, by Karlheinz Stockhausen, version premiered by TRIO-GRM-PLUS.

1982, Friday, April 23 concert "Hommage à Pierre Schaeffer." The Grand Auditorium of Radio France was packed. The evening, introduced by Pierre Schaeffer himself, included three new works: Pierre Henry's *Pierres réfléchies*, François Bayle's *Voyage au centre de la tête*, and Iannis Xenakis's *Pour la paix*.

Concerning *Pierres réfléchies*, its composer Pierre Henry clarified in the program:

> I started with just five notes, B on the bassoon, F sharp on the oboe, C on the flute, C sharp on the tuba, and B flat on the contrabassoon, all recorded at 76cm/s, transposed, assembled, multiplied, and agglomerated by the hundreds. A sort of geological accretion.[29]

François Bayle stated in the program concerning his own piece:

> I will be brief concerning *Voyage au centre de la tête*—because the flaw with words is that they anchor concepts—while here we are dealing with an exercise in continuous logic, engendering subjectivity and variability.

Iannis Xenakis wrote regarding *Pour la paix*:

> The thread is text from *Écoute* and *Et alors les morts pleureront* by Françoise Xenakis. The music was essentially composed on the UPIC at the CEMAMU. The choral interventions were recorded at the Maison de la Radio. The final tape was edited and produced at the GRM.[30]

1982, *Week-end* (13 minutes) by Ivo Malec. Use of nonreal-time processing in Studio 123. A twenty-three-minute version was performed as mixed music, with the synthesizer trio TRIO-GRM-PLUS, May 3, 1982, at the Maison de Radio France.

1982, *Le Lis vert* (37 minutes), Denis Dufour's second "musique concrète" suite. The piece premiered during concerts coproduced with the IRCAM at the Forum de la Création on March 15–16, 1983, in the IRCAM's projection space. Also on the program were Arnaud Petit, *Espace de pleurs*; Stephen Srawley, *Cette langue si simple*; François Bayle, *Les couleurs de la nuit*; Bertrand Dubedout, *Aux lampions*; Philippe Leroux, *Hommage à Andreï Roublev*; and Guy Reibel, *Les quatre éléments*.

1982, *Les couleurs de la nuit* (38 minutes) by François Bayle. 4-track. First performance on March 15, 1983, at the Grand Auditorium of the Maison de Radio France. François Bayle explained:

> My equipment loaded (with deformable segments of meaning) . . . , I built the sound image of a large forest in which "I entered into listening," where through the effect of highly structured spatiality capable of shaping within myself—after days of attentive listening—what I will call, for lack of a better term, "a veritable-sentiment-of-absolute-danger."

1982–83, *La Ronde* (24:09) in ten movements, by Michel Chion, voices of Lanie Goodman and Ellen Larsen, texts by the composer. First performance on February 28, 1983, at the Grand Auditorium of the Maison de Radio France. Chion described it as "brief contrasting movements forming one single curve stretching from one end of the work to the other. . . . Preference was given to microphone sound sources." (M. C.)

1983, *L'invitation au départ* (44:54) by Jacques Lejeune, pictographic succession in nine episodes.

1983, *Lumière noire* (18 minutes) by Bernard Parmegiani. "The absence of tonality characterizing a large portion of the sound material . . . is due to the use of 'white noise,' defined as sounds whose mass in theory contains every frequency, accumulated statistically."[31] This white noise was processed by Parmegiani using the 123's computer instruments, and then mixed in analog.

1983, *Sous le regard d'un soleil noir* (50:41) by Francis Dhomont. New version. Texts by Ronald D. Laing, voices of Arthur Bergeron, Marthe Robert, and Pierre Louet. First performance on March 28, 1983, at the Grand Auditorium of the Maison de Radio France. The work broached the difficult subject of schizophrenia. According to the composer, it was "the story of a shipwreck: hallucinatory drifting through an obsessional landscape where the B note is the haunting figure, the tonic axis cementing the eight sections together." (F. D.)

1983, *Tempo Lontano* (Temps jadis, 19 minutes) by Daniel Teruggi. First performance on May 30, 1983, at the Grand Auditorium of the Maison de Radio France. In this work, "there is a deliberate use of consonances, symbolizing the consonance between time spent dawdling and time spent living." (D. T.)

1983, *Vier jahreszeiten* (The four seasons, 65:10) by Jean Schwarz. After Goethe, with the artistic collaboration of Petrika Ionesco. Jorge Chaminé, baritone. First performance on May 30, 1983, at the Grand Auditorium of the Maison de Radio France. The work uses sounds realized on the computer of Studio 123 and mixed in an analog studio. "The music of Goethe's verses, coupled with the music of the German language, was more influential in my choice of texts for the vocal part than the meaning of the words . . . At the risk of sounding decadent, I wrote the score for the soloist in a traditional manner, avoiding acrobatics and performative touches." (J. S.)

1983, *Organimal* (31:30) by Gilles Racot. First performance on May 30, 1983, at the Grand Auditorium of the Maison de Radio France. "*Organimal* is in its entirety a 'composition' of dance actions (2 dancers), visual projections and music. These three levels interpenetrate one another." (G. R.)

1984, *Stop! L'Horizon* (18 minutes) by Christian Zanési. First performance on March 26, 1984, at the Grand Auditorium of the Maison de Radio France. Composed during three ten-day periods spread out over six months. The 8-track Studer tape recorder was used throughout the piece.

1984, *Tides* (30:08) by Denis Smalley. First performance on April 30, 1984, at the Grand Auditorium of the Maison de Radio France. Processing of the different sources was finalized in the GRM digital Studio 123. "The raw matter was produced in 1981 using a digital synthesizer at the University of Toronto Computer Systems Research Group. In April and May 1983, this material was transformed with analog and digital systems at the Finnish Radio Experimental Studio. Aquatic sounds were created using the Synclavier II." (D. S.)

1984, *Éléments mécaniques* (30:30) by Christian Eloy. First performance on March 26, 1984, at the Grand Auditorium of the Maison de Radio France. Tribute to Fernand Léger and New Realism, "as personal research into the force of the mechanical element in the world of sound."

1984, *L'Image éconduite* (70 minutes) by Philippe Mion. Four tracks. Voice of Marin de Charette and texts by Henri Michaux. A significant portion of this piece was produced in the GRM digital audio studio.

1984, *E cosi via* (14 minutes) for piano and tape, by Daniel Teruggi. The author indicated that this work was the first piece composed on Syter, "at a time when it was closer to a prototype than to a studio tool. The core instruments were used, in a very dense tape serving as a backdrop for composing the piano part."

1984, *La Tentation de saint Antoine* (95 minutes) by Michel Chion. Concrete melodrama with Pierre Schaeffer as Saint Anthony, Michèle Bokanowski as Narrator, Art Leroi Bibbs as the "Preacher" (in English in the original), Korinna Ralphs-Frisius as Queen of Saba, and Michel Chion, the Harbinger.

1984, *Janek Wisniewski décembre–Pologne* (18:20) by Elzbieta Sikora. Tribute to the Polish workers at the Gdansk factories in 1970.

1984, *La création du monde* (103 minutes), complete version, by Bernard Parmegiani. First period: *Lumière noire* (Black light); Second period:

Métamorphose du vide (Metamorphosis of the void); Third period: *Signes de vie* (Signs of life). Work composed in part at the GMEM of Marseille and at the GRM (sounds realized at Studio 123).

1984, *Pli de Perversion II* by Denis Dufour, for the electroacoustic instrumental TRIO-GRM-PLUS (Laurent Cuniot, Denis Dufour, Yann Geslin) and Syter (Jean-François Allouis). This concert was the first time the Syter system was used in public. Musically, the acoustic instruments and Syter are treated on an equal footing, while in François Bayle's earlier *Cristal* in 1977, Syter had instead been exploited essentially as a spatialization tool.

During rehearsals for this piece, a judicious remark by Denis Dufour was to be at the origin of a major Syter development. Dufour complained that he was limited to a mouse controlling a maximum of two parameters at a time and that he could not memorize his settings. Jean-François Allouis responded to this complaint by developing a parameter control page, ancestor of the famous "interpolation screen." This function of interpolation would be maintained in all future software developments at the GRM.

1985, *Germinal*, a group project bringing together fifteen composers from the GRM or closely affiliated with it. Their compositions all used the nonreal-time Studio 123.

Feux d'eau (5:53), François Bayle
Le Labyrinthe de l'amour (3:34), Denis Dufour
Stand by (5:10), Patrick Fleury
Gwelands argents (3:30), Évelyne Gayou
Rebours (4:52), Yann Geslin
Impromptu nuage (5:10), Jacques Lejeune
Éloge du zarb (5:09), Denis Levaillant
Affleurements (4:06), Bénédict Mailliard
Cotillon 2000 (6:11), Philippe Mion
Furientanz (5:40), Arnaud Petit
Anamorphées (7:30), Gilles Racot
Étude numérique, aux syllabes (6:25), Alain Savouret
Grimenal (5:11), Jean Schwarz
Léo le jour (5:15), Daniel Teruggi
Zéro un (4:11), Christian Zanési

The fifteen composers of *Germinal* came together, led by Bénédict Mailliard, in a group project where each composed a short three- to five-minute work using the new computer sound processing tools developed by

Bénédict Mailliard for Studio 123. Each work was based on a very short sound kernel (a concrete sound a few seconds long), using only this initial sound to create sound material through nonreal-time transformations. In contrast, final mixing was carried out in an analog studio because digital audio sequencers were not yet mature at the time.

Gilles Racot described his experience:

> For me, *Germinal* was an ideal apprenticeship for learning to use the tool, even if I was already quite familiar with it. When dealing from the beginning with sequences that are longer, as you usually do for a work, you just add reverberation, perform spectacular filtering, etc., using the 123 as an instrument for adding color. It becomes an instrument for synthesis. I'd be happy to begin with just a metronome "click," still achieving much the same thing. A simple impulse can be sustained, reverberated, filtered until it becomes a continuous tonic sound. The poorest and noisiest of sprouts leads where one wants, as long as one knows how to define the path—I start here, I go there. Reducing the initial material, as imposed by the *Germinal* project, pushed me to see and use all the programs. The merit of this constraint was to develop and reveal genesis in arborescence, through a series of transformations. Creating traces and cross-references, following very powerful logic, that of an unfolding tree.[32]

- 1985, *Sud* (24 minutes) by Jean-Claude Risset. Commissioned by the French government, for 4-track tape, produced at the GRM from sounds synthesized by computer in Marseille with the Music V program and then processed at Studio 123. This work was also emblematic of the search for convergence between concrete sounds recorded in nature—in this case the sea and insects, captured in the Marseilles inlets, so dear to the composer—and functional anamorphosis generated by pitch grids preestablished by computer.[33]
- 1985, *Eterea* (26:45) by Daniel Teruggi. 4-track. Most of the sounds in this work were generated from white noise.
- 1985, *Château de sable* (20 minutes) by Gilles Grand. Realized at the nonprofit organization Canope in Lyon and at the INA-GRM.
- 1985 *Exultitudes* (23 minutes) by Gilles Racot. Mixed work for tape, with sounds processed in nonreal time in 123, and saxophones processed in real time by Syter. First performance on April 28, 1986, at the Maison de Radio France, with Daniel Kientzy on soprano and double bass saxophones. Gilles Racot explained, "the exultation evoked in the title alludes to the feeling of jubilation provoked by working on the computer at the time—that of 123—with the powerful possibilities for sound mutations combined with the absence of deterioration in signal quality," which made it possible to go very far with sound.[34] It is necessary to stress the dreaded deterioration in signal-noise ratio with analog sound: with the third generation of copying a sound, the hiss

associated with the medium became unbearable to the ear, and all the more so when a sound was faint.

1979–1986, *Nuit noire* (24:30) by Michel Chion. First performance March 24, 1986, at the Grand Auditorium of the Maison de Radio France. This piece, drawn from the final movement of Chion's *Diktat*, was dedicated to the memory of Philip K. Dick. "*Nuit noire* is a nightmare, composed of a series of falls into different depths of dream, discomfort and terror." (M. C.)

1986, *Chantakoa* (30 minutes) by Jean Schwarz, for 2-track tape, clarinetist, and Syter system. Michel Portal, clarinets. Certain sequences were transformed live by the Syter: spatialization, reinjection, comb filtering, harmonization, delay.

1986, *Messe à l'usage des enfants* (30:30) by Denis Dufour, 4-track. Text by Thibault d'Orion, alias Thomas Brando.

1986, *Le vide et le vague* (20 minutes) by Philippe Leroux. Based on five sequences of analog sounds, processed and hijacked by the computer of Studio 123, the author built his piece by linking analogies of musical ideas.

1986, *Instantanés* (10 minutes) by Philippe Manoury. Live electronic version for percussions by Florent Jodelet, and synthesizer trio, the instrumental ensemble TM PLUS: Laurent Cuniot, Denis Dufour, and Yann Geslin.

1986–1987, *Noctuel* (15:50) by Gilles Racot. First performance on June 20, 1988, at the Grand Auditorium of the Maison de Radio France. Alexandre Ouzounoff, bassoon. Homage to the pictorial work of Pierre Soulages. Gilles Racot explained that "the electroacoustic work is exclusively concrete, where the sound of the bassoon, harmonically and spectrally reduplicated upon itself, creates harmonic curtains that are more or less opaque, that the superpositions and juxtapositions animate, deny, reveal."

1987, *La complainte du Bossué* (14 minutes) by Alain Savouret. For double bassist-speaker and Syter real-time digital processing instrument. Humorous work in which the double bass player Frédéric Stohl assumed a key role by "dialoguing" with the Syter system played live by Daniel Teruggi, assisted by Richard Bulski. Commissioned by the GRM.

1987, *Polyphonie polychrome*, second tableau, *Valeurs d'ombre* (26:45) by Patrick Ascione. The piece plays on increases in dramatic tension by stages.

1987, *Jardin secret II* (11 minutes) by Kaija Saariaho, for harpsichord and tape. Elisabeth Chojnacka, harpsichord. The sequel to *Secret Garden I* (1984). The tape consists of concrete harpsichord sounds and the author's voice, processed on the Syter system. The rhythmic materials were elaborated with the FORMES program at IRCAM.

1987, *Éloge de l'eau* (23:40) by Denis Levaillant. The original sounds were recorded by Madeleine Sola and Daniel Deshays. The work was realized at the digital audio Studio 123 with the assistance of Bénédict Mailliard.

1987, *Huit chansons précieuses* (14:30) by Philippe Mion, for tape and soprano. Liliane Mazeron, soprano. Dual confrontation between lyrical and electroacoustic composition.

1987, *De l'autre côté du miroir* (13:45) by Bernard Fort. Musical presentation of some possibilities of sound processing offered by the digital Studio 123.

1987, *Suite pour violoncelle et bande* (24:30) by Pierre Alain Jaffrenou. Christophe Roy, cello. The tape plays the role of another cello whose expression of timbre and technique are extended and warped, in relationship with the reference instrument. Processing was done in the digital Studio 123.

1987, *Aquatica* (22 minutes) by Daniel Teruggi. 4-track. *Aquatica* follows *Eterea* and precedes *Terra* and *Focolaria* in the work *Sphœra*. A large part of the sounds originated from Syter. *Aquatica* explores the materiality of water, from the drop to the ocean. The 1993 CD presented a short 13:35 version in 2-track stereo.

1988, *Courir* (18:51) by Christian Zanési. First performance on February 27, 1988, at the Grand Auditorium of the Maison de Radio France. "I remember the initial performance: a single sound capture, the microphone right in my mouth. Running as fast as possible. Soliciting the entire body machine to reach a state where the mind is in short-circuit." (C. Z.)

1988, *Desert Tracks* (about 35 minutes) by Michel Redolfi. Commissioned by the National Endowment for the Arts. The sounds were recorded in the California desert.

1988, *Douze mélodies acousmatiques* (25:50) by Denis Dufour. First performance on June 20, 1988, at the Grand Auditorium of the Maison de Radio France. "Neither text, nor dramatic support, twelve short forms, melodies without words (just a few words) taken in some cases from other of my measures," explained the author in his introduction.

1988, *10 études de Musique concrète* (27 minutes) by Michel Chion. First performance on January 25, 1988, at the Grand Auditorium of the Maison de Radio France. In the program notes, the composer described the aesthetic plan for his piece: "Physical work on sounds, their definition, their shape, their inscription on the support, bringing into play concrete techniques experimented for the occasion, such as micro-editing without scissors by applying successive layers to the tape or manually varying the contact between the tape and the tape recorder playback heads."

Public Forums

The 1987–88 season saw the inauguration of public forums where the GRM presented its current research in the form of encounters and mini-concerts at the Maison de la Radio on the eve of concert evenings.

—January 20, 1987, François Delalande organized an evening titled *Vues sur l'écoute*, with the participation of Jean-Luc Jézéquel on the theme of aesthetic analysis and electropolygraphy and Jean-Christophe Thomas for the study of verbalization. During a second part, Philippe Mion demonstrated the interpretation of diffusion over loudspeakers.

—February 24, 1987, François-Bernard Mâche: *Temps réel et temps irréel en musique* (Real time and nonreal time in music)

—March 24, 1987, François Bayle: *Écouter et comprendre* (Listening and understanding)

—April 28, 1987, Jean-François Allouis: *Temps: réel—différé—composé* (Time: real—deferred—composed)

—June 16, 1987, Michel Chion: *Du son à la chose* (From sound to the thing itself)

The Noroît Acousmatic Exhibitions

The acousmatic exhibitions in the north of France at the Noroît center in Arras, organized by François Bayle at the request of the great admirer of contemporary art Léonce Petitot, presented a different composer each time, in collaboration with the GRM. Several monographic concerts accompanied an exhibition of photos and archival documents, together with a richly illustrated catalog on the invited composer and the history of electroacoustics.

Bernard Parmegiani, February 26 to March 6, 1983

Ivo Malec, May 7–14, 1983

Pierre Henry, April 13–28, 1984

François Bayle, March 15–26, 1985

Michel Chion, February 26 to March 15, 1986

Jacques Lejeune, in the form of five concert encounters, May 26–27 and June 9–11, 1986

After the death of Léonce Petitot, founder and principal patron of the Noroît center, the final exhibition from March 13 to April 3, 1987, the Acousmatic Exhibition, recapitulated the previous ones and prefigured the project of the Acousmathèque dear to François Bayle.

In 1988, Bernard Petitot, son of Léonce Petitot, relaunched the initiative, setting up a biennial international competition of electroacoustic works, accompanied by the release of a CD gathering together the works of winners. The fifth session of this competition in 1996 marked the end of the competition.

Publications

VINYL RECORDS

In 1978 François Bayle launched a collection of 33 rpm vinyl records, INA Collection GRM, followed by an economical collection, the Gramme series, with more than thirty often award-winning releases through 1985. These record pressings present the panorama of the Groupe's musical production, from its beginnings to more recent works: those by Schaeffer, Bayle, Amy, Schwarz, Reibel, Parmegiani, TM+, and others.

CDS, COMPACT DISCS

In 1984, the GRM released the first contemporary music CD in France, *Le concert imaginaire*, a medley of excerpts from GRM works by Parmegiani, Schaeffer, Henry, Chion, Lejeune, Malec, Schwarz, Zanési, Dufour, Mion, and Bayle. Long considered as the GRM "calling card," this CD would be re-released five times. Beginning in 1988, it was followed by a collection of electroacoustic music CDs. In 1997, this collection included fifty titles. In 2005, it reached the number one hundred.

BOOKS

Le Répertoire Acousmatique 1948–1980, documented by Geneviève Mâche and Annette Vande Gorne, edited by François Bayle, is a catalog of the GRM's musical works and scores, presented in chronological order of their composition. Unfortunately, as the title indicates, this repertoire does not go beyond the year 1980.

In 1982, François Delalande published his first work, in collaboration with Bernadette Céleste and Elisabeth Dumaurier, *L'Enfant du sonore au musical*, in the collection Bibliothèque de Recherche Musicale, which he edited at Buchet-Chastel in Paris. This publishing activity received the support of the INA-Publications.

In the same collection other works were published:

L'Envers d'une œuvre, in collaboration with the French-Canadian musicologist Jean-Jacques Nattiez in 1982. In this work, François Delalande, Jean-Jacques Nattiez, Philippe Mion, and Jean-Christophe Thomas analyzed Bernard Parmegiani's *De natura sonorum*.

In 1983, Michel Chion's *Guide des objets sonores* was published as a complement to Pierre Schaeffer's *Traité des objets musicaux*, presented in the form of an annotated index of the key terms established by Schaeffer in his *TOM*.

In 1984, François Delalande published *La musique est un jeu d'enfant* (Music is child's play). Later translated in Argentina and Italy, it laid the foundations of pedagogy based on a concrete attitude by the child towards sound, leading to creation.

In 1986, a quadruple issue of *La Revue musicale*, no. 394, 395, 396, and 397, "Recherche musicale au GRM," and a separate issue of the *Catalogue GRM 1948–1986, sélection d'œuvres du GRM*, a selection of works by the GRM, were published. The latter is a succinct extension of *Le répertoire acousmatique 1948–1980*.

François Delalande's Books on Early Childhood Musical Awakening. After the experiments in preschools, a question arose: since three-year-old children are capable of inventing a sequence based on a sound they have discovered, how about a child only a few months old? Pedagogical research turned towards the youngest children, focusing on fundamental rather than directly applied research, studying sonic exploration in young concrete musicians and the emergence of musical behavior.

These observations in nurseries, first in Paris with Bernadette Céleste (*L'Enfant du sonore au musical*, 1982) and then Jean-Luc Jézéquel, continued in Italy in Florence (1991–92) and in Lecco (2002–6). They would have a profound impact on the career of François Delalande.

RADIO

After the departure of Agnès Tanguy in 1973 and Michel Chion in 1975, a new radio team was formed under Jack Vidal, himself already at the GRM since 1973.[35] Omar Bouffaïd, who arrived in 1975, left the GRM in 1978, followed by Jack Vidal in 1983. Eventually the team stabilized around Évelyne Gayou, who arrived in 1975, and Christian Zanési in 1977.

French audiovisual law changed in 1986, but it retained the terms defining relations between the INA and Radio France—the GRM would continue to produce about fifty hours of programs annually. Programs were part of the schedules of the France Musique and France Culture networks, in the form of weekly series lasting from half an hour to an hour. Some of them were renewed for several consecutive seasons; all of them are preserved in the INA's archives:

1978, *Catalogue électroacoustique illustré*—45-minute series. Hosted by Évelyne Gayou, programming by Jack Vidal, directed by Christian Zanési. This series, on France Musique, broadcast works from the electroacoustic repertoire, alongside recent music from a wider repertoire of jazz, improvised music, film music, and experimental work.

1978–79, *L'Art des bruits*—45-minutes series, later replaced by the series *Qui dit quoi à qui?* on France Musique, hosted by Jack Vidal and directed by Christian Zanési. Weekly listening show featuring electroacoustic music along with commentary.

Beginning in 1980, each month the works performed at the *Acousmatic Cycle* concerts were broadcast on France Musique, in the form of programs redesigned to fit the 75-minute format of the show. The name of the series changed to reflect the titles of the GRM concert seasons at Radio France: *Son/Mu* in 1992, then *Multiphonie* in 1998.

On France Culture, the programs tackled research topics, almost always involving outside guests—composers, musicologists, researchers, engineers. Listening excerpts were chosen from all musical styles. These programs, which required lengthy documental preparation, were rarely broadcast live. They were produced in turn by Jack Vidal, Évelyne Gayou, and Christian Zanési. But after Jack Vidal's departure in 1983, Christian Zanési gradually began specializing in programs for France Musique and Évelyne Gayou in programs for France Culture.

From 1985 to 1990 on France Musique, Christian Zanési directed the long-lived series *Acousmathèques*, programmed in groups of four half-hour shows, with a different guest producer each time. The series numbered 206 episodes.

From 1978 to 1988 on France Culture, Évelyne Gayou directed programs for various series: *Recherche musique*, *Points d'écoute sur . . .* , *Miroirs*. The subjects explored were evoked in their titles:

Animer la musique (Animating music)
Images, mirages acoustiques (Images, acoustic mirages)
Classiques des idées et du répertoire GRM (Classics of ideas and of the GRM repertoire)
De la musique à l'ordinateur et réciproquement (From music to computer and vice versa)
Électricité à tous les étages (Electricity on all floors)
Recherche et création sonore à la radio (Research and sound creation on radio)
Miroirs, photos–musique (Mirrors, photos–music)
Tendances de la sémiotique musicale (Trends in musical semiotics)
Écouter et comprendre (Listen and understand)
Le Sens de la musique (The meaning of music)
Sur l'écoute musicale (On musical listening)

In addition to all these regular series, special programming was carried out in collaboration with Radio France on the occasion of technical experimentation, multicasting throughout Europe, or in duplex with foreign countries, the GRM being considered at the forefront of musical creation, and Radio France at the forefront of production and broadcasting. It should also be noted that Radio France often called on the GRM for its expertise in putting together live music events with electronics and spatial projection of sound. A strong partnership with the technical services of Radio France was established on a long-term basis.

"*Paris, impressions*"—*a Group Radio Production.* Finally, let us mention a group enterprise, an approach that often occurred at the GRM. Alongside *Germinal*, a series of short musical compositions produced with the new digital audio tools of Studio 123 from a single sound, the collection *Paris, impressions en 6,25* was conceived first and foremost as a vehicle for radio. This collection of ten creative works for radio, realized in analog on standard ¼-inch tape (6.25 mm in the metric system), had the city of Paris as its theme, and a rule of using only the ordinary technical means of the radio studio:

> Jacques Lejeune, *L'Enfant et l'oiseau* (11:40), evocation of the bird market;
> Évelyne Gayou, *À l'aube du jour des morts* (16:10), early morning streets;
> Jean Schwarz *Interurbain* (14:30), the metro;
> Bernard Parmegiani, *Ici la radio* (13:44), sound evocation of the Maison de la Radio;
> Denis Dufour, *Entre dames* (19:30), Notre Dame de Paris;
> Philippe Mion, *Suite fauve* (16:40), the zoo at the Jardin des Plantes;
> Christian Zanési, *La Traversée* (9:25), the city;
> Daniel Teruggi, *Montparnasse la nuit* (14:25), the Montparnasse district;
> Bénédict Mailliard, *D'un pays l'autre* (26:28), cosmopolitan Paris;
> Laurent Cuniot, *Diabolus Games* (13:50); the atmosphere of a Parisian café.

GRM Members

THE TRIO-GRM-PLUS, OR THE SYNTHESIZER TRIO

At the end of 1977, one year after the experience of the *Cristal* concert,[36] where the prototype of Syter (Syter 1) was used live in public for the first time to transform sounds in concert, François Bayle decided to establish, within the context of the GRM, a small instrumental ensemble capable of manipulating Syter (Syter 2) live during concerts, because using the machine required considerable skill. Thus was born the Trio GRM+, then TRIO-GRM-PLUS, composed of three young instrumentalists—Denis Dufour, violist, Laurent Cuniot, violinist, and Yann Geslin, pianist—all three from Guy Reibel's electroacoustic class at the Paris Conservatory. In François Bayle's mind, this small ensemble could also perform mixed music and even live electronic music later, after becoming acquainted with the synthesizers recently introduced on the market.

In 1985 Laurent Cuniot took over as director of the trio, which had been renamed TRIO-GRM-PLUS since late 1981. The aesthetic and personal discord between François Bayle and Guy Reibel ended up isolating the trio, and in 1990, after financial support from the GRM was withdrawn, the TRIO-GRM-PLUS became the electroacoustic instrumental ensemble TM PLUS. Yann Geslin resigned, followed a

FIGURE 25. Trio GRM+: Laurent Cuniot, Denis Dufour, Yann Geslin. © Éric Bourbotte (1980).

year later by Denis Dufour. Once alone, Laurent Cuniot, a proponent of Guy Reibel's theories, after first strongly redirecting the ensemble towards research into "musical gesture,"[37] widened the repertoire in correlation with the orientation of his personal career towards orchestra conducting. Since 1986, the TM PLUS ensemble, which he still directs, has a varied geometry. TM PLUS adapts to the works to be performed from its extensive repertoire of contemporary music, now extending throughout the entire twentieth century and sometimes even well before.

From the early period of the trio, I will only mention a few works that are still accessible today, since unfortunately many were never recorded:

Pli de perversion II (1984) by Denis Dufour, for violin, synthesizer and Syter (Syter live for the first time in concert).

Hamlet by Laurent Cuniot, orchestral piece with two synthesizers, recorded on CD Salabert.

12 inventions en 6 modes de jeux, by Guy Reibel, exploiting the capabilities of the trio.

Stries (28 minutes, 1980) by Bernard Parmegiani.

Also notable are *Gamma plus* by Jean Schwarz, and *Symphonie au bord d'un paysage* by Jacques Lejeune, recorded on vinyl. Finally, found in the GRM's sound library, there are previously unreleased recordings of "situations de jeux": improvisations on synthesizers, by the trio.

DEFECTING ENGINEERS AT THE GRM AND IRCAM

While composers rarely moved between the GRM and IRCAM, in contrast, researcher-engineers alternated their engagement with one or the other institution rather easily. The first to have maintained constant contact with the GRM was Risset, both during and after his time at IRCAM from 1975 to 1979, on leave of absence from his position at the CNRS of Marseille, where he would return in 1980. GRM member Bernard Dürr had been in contact with the IRCAM since 1975, introduced to Pierre Boulez by Luciano Berio.[38] Jean-François Allouis left the GRM to become technical director of the IRCAM from 1987 to 1991 at the request of Boulez. In doing so, he was careful to keep open the possibility of returning to the INA with a formal leave of absence agreement. During Jean-François Allouis' absence, his successor Hugues Vinet, an engineer with the École Nationale Supérieure des Télécommunications (ENST), presided over the destiny of GRM technical research from 1986 to 1994, before himself joining the IRCAM. After his experience at the IRCAM, Jean-François Allouis returned to the INA to direct research into image synthesis and processing, subsequently resigning in 2001 when this field ceased to be a priority for the INA. A third engineer, Emmanuel Favreau, who started out in Di Giugno's team at the IRCAM as a signal processing specialist, completed this strange triangle. He succeeded Hugues Vinet at the GRM in 1994, when the latter joined the IRCAM. In the meantime, Emmanuel Favreau had followed di Giugno to Italy at IRIS, the research and development center of the Bontempi company. More recently, in 2006, the GRM recruited a new engineer, Adrien Lefèvre, in charge of development of the Acousmographe, who had previously worked at the IRCAM from 1995 to 2001, notably on the Diphone in the Analysis/Synthesis team directed by Xavier Rodet.

10

1988–1998
Innovation

To create is divine, to reproduce is human.
—MAN RAY

This decade, which coincides with the end of François Bayle's career, mirrors the GRM that he helped to build and preserve throughout his thirty-six years of presence. The Groupe was firmly rooted in the INA while remaining in permanent contact with Radio France. The relationship with the musical milieu was not forgotten and creative and production activities converged towards the systematic search for contact with the public, always with an underlying ambition for popularization. All of these facets continue to define the GRM today. The overall picture could not be more positive, the GRM having been able to adapt to new technologies. GRM Tools was internationally recognized and the Acousmographe was gently embarking on its path.

HISTORICAL CONTEXT

The decade was scarred by a somber note. The GRM lost its great Mentor. On August 19, 1995, Pierre Schaeffer passed away.

The decade 1988–98 was a time of commemorations, beginning with the anniversary concerts. It began with the "30, 40 ans du GRM" concerts for the thirtieth anniversary of the GRM and the fortieth anniversary of musique concrète in 1988 at the Théâtre National de Chaillot and at the Grand Auditorium of the Maison de Radio France.[1] That same year, the GRM began its tenth season of Cycle Acousmatique concerts, still under the artistic direction of Ivo Malec and still in partnership with Radio France, whose first edition of the Présences festival took place in 1991. At the other end of the decade, in 1998, the jubilee of musique concrète was celebrated with several important public events.

Inside the GRM, the Acousmathèque was officially inaugurated on February 10, 1993, with a ceremony in Studio 116 of the GRM in the presence of Jean-Noël Jeanneney, French secretary of state for communication; Georges Fillioud, president of the INA; and Thierry Leroy, director of music at the Direction de la musique, de la danse, du théâtre et des spectacles (DMDTS). Acousmathèque was the title given by François Bayle to the activity of preserving the five thousand tapes in the GRM collection, including fifteen hundred pieces composed since 1948 by more than two hundred composers. A plaque, affixed in Studio 116B of the Maison de Radio France, commemorates the event. In François Bayle's plan for the project, this site was also intended to be used for auditory consultation and seminars.

Numerous events of the period softly presaged the great upheavals of the 2000s. The National Audiovisual Institute, INA, evolved towards concentrating more on its vocation of curating the French audiovisual heritage, following the vote of the law of June 20, 1992, by the French Parliament, instituting a legal digital repository for all French radio and television media.

At the GRM, the initial partnership from 1988 with the French navy was extended for three more years, with the technological impact of accelerating development of Syter and digital audio research. Since the early 1990s, the advent of personal computers had given the general public access to computer tools. GRM Tools appeared in 1991 and immediately found an audience. In parallel, a network-based approach was introduced with the INA launching its first website in 1994, and in the same year the GRM acquiring computers for office work with its first 20GB local server installed in 1997.[2]

In 1996, taking over direction of research and development at the INA for two years, the philosopher Bernard Stiegler initiated his term with a highly symbolic act, replacing the word "development" in the name of his department with "innovation."[3]

So, when François Bayle retired in 1997, Jean-Pierre Teyssier, the INA president, entrusted Daniel Teruggi with managing the Groupe, which was integrated into the new Innovation Department, headed by Bernard Stiegler.

Commemorations for the anniversary of musique concrète as well as the series of "firsts"—the first Electro CD Forum, the first participation in the Présences festival—gave a general impression of tranquil strength. All the GRM lacked was more dynamic engagement on the international level. Daniel Teruggi would attack the issue as soon as he took over the destiny of the Groupe.

MUSICAL CONTEXT

Let us take the 1989–90 season as an example representative of the philosophy of the concert seasons given by the GRM in coproduction with Radio France. Concerts took place in the Maison de la Radio building in the Grand Auditorium or

Studio 104, and after 1992 in Olivier Messiaen Hall, which seated six hundred people below and three hundred in the balcony. Acoustics in Olivier Messiaen Hall were excellent, and concerts there were recorded directly from one of the adjoining technical booths by Radio France sound engineers to be broadcast on the France Musiques network in the following weeks.[4] Attendance numbers for GRM concerts ranged from a minimum of three hundred people to an upper average of six hundred. Some exceptional events, such as the Pierre Schaeffer concert of April 25, 1988, sold out the nine hundred seats.

Another important observation concerns the average age of audiences. Since the emergence of techno in the 1990s, the GRM audience has been in part rejuvenated, attracting a new, young audience. The term *techno* was coined in Detroit in 1988 to describe the sound of a new musical genre that then spread to the rest of the world. Different directions included "the French Touch," represented since 1996 by groups such as Daft Punk. Ephemeral groups like Stardust coexisted with others that were much more resistant, including Daft Punk and Chemical Brothers. Even well-established rock bands like U2 have explored this new genre. One of techno's most distinctive musical characteristics is its rhythm fixated on 120 bpm (beats per minute). The first French techno parade took place in Paris in 1998.

Syter, Acousmographe, MIDI Formers, GRM Tools

Throughout this period, technological development continued within the GRM. After the resignation of Bénédict Mailliard in 1986, activity at the "nonreal time" Studio 123 declined in favor of focusing completely on the Syter project.

The accomplishments of the Syter venture are particularly impressive. Over the period of exploitation of the system from 1985 to 1993, 125 people—including 8 women—received training on the system in the form of one-week courses conducted by Daniel Teruggi.[5] About three hundred works of music, incorporated—to varying extents—sounds or treatments utilizing Syter.

Beginning in 1994, maintenance of Syter had become problematic, both because it was too costly and because of the unavailability of replacement components. The maintenance contracts were terminated. Malfunctioning units were used for spare parts for machines still working. The Digitone company had as early as 1985 expressed the desire to extend the Syter venture by developing a new version of the processor (its development had been frozen since 1983), but nothing happened.[6] The last Syter in working order is at the Conservatoire de Paris. Since Hugues Vinet's departure in 1994, Yann Geslin alone is capable of servicing Syter. He is its guardian, scrupulously running it once a year, each year with a bit more anxiety.

From 1988 to 1990 the engineer Olivier Koechlin worked on development of the Acousmographe software. Denis Dufour, as part of his salaried activity at the GRM, was the first to produce "listening scores" of some of the classics from the GRM repertoire: Pierre Schaeffer's *Étude aux objets*, François Bayle's *Vibrations*

FIGURE 26. Syter workshop under the direction of Daniel Teruggi. © BBR (1990).

composées, Michel Chion's *Requiem*, Ivo Malec's *Luminétudes*, and Francis Dhomont's *Novars*. During an interview in 1997, Denis Dufour explained that he "drew the scores with a small drawing program on the tiny MSX computer" that he owned. "Only Dhomont's [*Novars*] score was done on the Acousmographe, which was still in its infancy." Denis Dufour declared that "it was difficult, even discouraging, to continue working on this software [the Acousmographe] whose incessant bugs sometimes made me lose entire days of work!"[7] Denis Dufour's initial work was later taken up and completed by Dominique Besson.

In 1990 the first Acousmographe, a tool for the graphic representation of music, was presented at the MIM in Marseille, along with demonstrations of acousmatic music transcription. An interactive CD-ROM, *Les Musicographies*, realized by Dominique Besson and Stéphan Dunkelmann in coproduction with Festival Les 38è Rugissants,[8] presented ten graphic and multimedia transcriptions of music, from all kinds of repertoires. It was released noncommercially in 1995.

MIDI Formers, developed at the GRM by Serge de Laubier between 1988 and 1998 under the name of MacSoutiLs, functioned in two ways, either in the studio, generating sound fluxes by recreating known synthesis software effects, or connected to a Mars workstation, allowing composers to control the flux of events synthesized by the station. In 1992 MIDI Formers was released in international distribution. Unfortunately, Serge de Laubier's research contract ended in 1998, for lack of sufficient funding to adapt these tools to new computer platforms and to continue development.

Beginning in 1990, the engineer Hugues Vinet, who had joined the GRM in 1985 to continue digital audio development work on Syter, began developing GRM Tools, creative sound processing software for transforming and synthesizing sound in real time. This software consisted of diverse treatment processes, some of which followed models implemented in the tools developed by Bénédict Mailliard and Yann Geslin for Studio 123. Some of these tools used the Music V program internally, with a user interface corresponding to transformation operations used in "concrète" music. In 1991 Hugues Vinet presented his first version of GRM Tools under the name DSP Station at the ICMC in Montreal. In 1992 an English-language version was released. The engineer Emmanuel Favreau joined the GRM as head of research and development in 1994, succeeding Hugues Vinet. He launched the new GRM Tools, which received an international award in 1997, the American magazine *Electronic Musician*'s Editor's Choice Winner.

The Composition Studios—116A, B, C, and 123

Mixing control consoles never stop getting bigger and bigger, with more and more features. On the new 116A DDA, sixteen channel settings could now be stored as digital data. Digital technology gradually entered the studios, at first inside newly purchased machines such as the Lexicon reverb, because the GRM no longer fabricated its own sound processing tools. Fostex D20 DAT (digital audio tape) recorders were introduced into all the GRM studios in 1990, and by 1992 they had brutally replaced analog recording. The first digital multitrack recorders, the 8-track ADAT, and the first (4-track) Pro Tools system were acquired in 1992. The 8-track Pro Tools arrived in 1994.

From an aesthetic point of view, these new tools enabled composers to multiply the number of audio layers without loss of quality. The resulting compositions explored the concept of internal space, with numerous superimposed planes of presence moving and unfolding. This gave works an increasingly baroque musical style, where the sculptural aspect predominated in an infinity of layers and details.

At IRCAM

The Station d'Informatique Musicale, a digital audio processing platform for research, production, and musical creation, was launched in 1991.

In 1992, Laurent Bayle (not to be confused with François Bayle, director of the GRM) succeeded Pierre Boulez at the head of the Institute, promoting a policy of communication and diversification of activities.[9]

In 1994–95, several software programs were finalized: Audiosculpt, software for the graphical representation of sound, enabling sound "sculpting"; Spatialisateur, software for controlling room acoustics; and Patchwork, software written in Common Lisp handling symbolic musical structures in the form of time/frequency representations.

In 1996, IRCAM began to distribute FTS, a software version of the Station d'Informatique Musicale.

Among the composers invited to work at IRCAM in recent years are notably Alejandro Viñao, York Höller, Michaël Levinas, Ichiro Nodaïra, George Benjamin, François Bayle, Pierre Henry, Philippe Manoury, Gilles Racot, Barry Anderson, Kaija Saariaho, Trevor Wishart, Jonathan Harvey, Philippe Leroux, and Martin Matalon.

The Decline and Demise of the Cologne Studio

York Höller had overseen the studio since 1992, but in 1998 he became involved in the financial reforms of the management, which was unwilling to continue costly production at the electronic music studio. The historic studio was dismantled and moved to a building on the outskirts of Cologne, under the benevolent eye of Volker Müller, the last remaining technician. There are plans to turn it into a museum. The WDR continues to host a few composers in its modern studios, but a page has turned.[10]

Following the timid festivities of its fiftieth anniversary, this is a tragic outcome for the Cologne Studio, which sees no future for itself, even while remaining a mythical landmark in the minds of all those who remember it.

The ADAC Workshop in Paris

Beginning in 1988, in addition to the music composition workshop equipped with analog equipment, the GRM launched a second "computer-assisted electroacoustic music workshop" at 3 rue Amyot, in the fifth arrondissement of Paris, in partnership with the City of Paris's Association pour le développement de l'action culturelle (ADAC, Association for the Development of Cultural Action), first under the direction of Philippe Mion, and later Christian Eloy, and then Régis Renouard-Larivière—who himself had studied in the Workshop.[11] The presentation brochure of the second workshop stated:

> —This second electroacoustic music workshop provides an overview of digital studio techniques (microcomputer—MIDI system—synthesizer—sampler) and their musical applications.[12]
> —It provides theoretical instruction and practical realization work.
> —Preference will be given to those with knowledge of this musical orientation.
> —Experience working in a studio or with a synthesizer is desirable.

The workshop offered a nine-month course consisting of three three-month sessions. The first session of initiation in the basic techniques of electroacoustic composition was mandatory before the second session, geared more towards live music and MIDI techniques. Finally, the third level focused on the compositional process.

The workshop was not free of charge, but it still attracted numerous candidates interested in composing. At the end of each session, critical listening sessions were organized at the GRM, showcasing works by participants.[13]

The Emergence of Electroacoustics Classes in France

New classes in electroacoustic music composition opened in some municipal "conservatories": in Chalon-sur-Saône (Nicolas Vérin), in Pantin (Christine Groult), in Gennevilliers (Jean Schwarz), etc. All of these courses would expand, and thanks to the students trained, a network would be created.

Among Jean Schwarz's students in Gennevilliers was Gino Favotti, who would later establish the class at the Conservatory of the Twentieth Arrondissement of Paris, until 2006 the only electroacoustic class in Paris, apart from the CNSMDP and the Adac.

In Pantin, the electroacoustic composition class of the conservatory of the arrondissement had enjoyed a long-standing reputation, attracting people from all over France and abroad. The electroacoustic studio had been founded in 1972 by Fernand Vandenbogaerde as part of the experimental conservatory of Pantin created by Michel Decoust. In 1980, under the direction of Sergio Ortega, the studio passed into the hands of Stephen Srawley, and was later restructured in 1990 by Christine Groult.

The studio accepted an average of sixteen students studying music composition for three levels and then an advanced course. The course was open to everyone, not only musicians, but also to visual artists, dancers, organizers, and technicians. This diversity was an important factor in the vitality of the class. A space equipped with an Acousmonium allowed the organization of master classes throughout the year, as well as concerts by students and internationally renowned composers. Thanks to Sergio Ortega, director until 2003, a particular connection had been established with South America and Spain. Teaching centered on composition, but also offered a framework for reflection on the very conditions of music: its social, cultural, and economic aspects. In 2007, a few former students founded the KM Pantin association, with the aim of encouraging the creation and dissemination of electroacoustic music and creating a true platform for the exchange of knowledge within this community.

Pierre Schaeffer's Archives

The nonprofit Centre d'Études et de Recherche Pierre Schaeffer (CERPS, Pierre Schaeffer Center for Study and Research) was founded in November 1995 under the presidency of Jacqueline Schaeffer, Pierre Schaeffer's widow. The historian

Sylvie Dallet, who had assisted Pierre Schaeffer in classifying his personal archives, became its director.

The CERPS was located near Paris in a small room in Montreuil-sous-Bois, where its documentary collections were available to members. It organized concerts and participated in colloquiums and other events centered on the person and the work of Pierre Schaeffer. But the organization faced financial difficulties. In 2003, Sylvie Dallet found a more stable position at the University of Marne-la-Vallée and resigned, leaving the organization in debt. It was formally dissolved a few months later. In 2005 the Schaeffer archives were rehoused at the Institut Mémoire de l'Édition Contemporaine (IMEC) in Caen, according to the wishes of Pierre Schaeffer's widow. Since then, negotiations with the INA have been underway to return the collection to the INA, thus increasing its accessibility and visibility.

François Bayle's Private Studio

After his retirement in 1997, François Bayle devoted himself entirely to his private studio and his label Magison. Founded in 1990, the Magison nonprofit organization supports seminars and publications as well as composition and interpretation prizes open to composers in the field of acousmatics, including Prix Noroît, Prix Musiques et Recherches, and the Electroacoustic Music Foundation (EMF).

Magison's objective is to promote the concept of acousmatics, based on François Bayle's theoretical work and repertoire. The Magison label has released all of François Bayle's works in updated versions, totaling eighteen volumes as of 2006.

The Influence of GRM Theoretical Research

The GRM has played an important role in the development of music analysis in Europe. In 1985, François Delalande was one of the founding members of the Société Française d'Analyse Musicale (SFAM) and of the journal *Analyse Musicale* (and later of *Musurgia*, its successor), and one of the main organizers of the first European Congress of Music Analysis in Colmar in 1989. He organized, with the help of Jean-Jacques Nattiez, the inaugural plenary session where ten European and American researchers compared, in terms of method, their analyses of a single composition.[14]

A European current of musical semiotics found its roots at the GRM, following the meeting organized by François Delalande in 1986 between European researchers (Eero Tarasti from Finland, Gino Stefani from Italy, Costin Miereanu and Daniel Charles from France) with François Bayle, and Marcello Castellana, a semiotician under contract at the GRM, in order to launch a European laboratory for research into musical meaning. The project was finally taken in hand by Eero Tarasti, resulting in a series of European conferences under the title International Conference on Musical Signification, where Delalande continued to play a significant role.

EVENTS, PERFORMANCES, PUBLICATIONS, PEOPLE
The Radio France Présences Festival, Multiphonic Concert, Noroît Competition

In 1991, Radio France launched the Présences festival, featuring a different contemporary music composer each year. Taking place at the end of January, with about twenty free concerts attracting fifteen to eighteen thousand listeners in total, this major event has sometimes been described as the "FIAC of contemporary music."[15] (The FIAC is the world-famous annual Foire internationale d'art contemporain—International Contemporary Art Fair in Paris.)

Each year, the GRM is regularly invited to collaborate for one of the festival's concerts, presenting either mixed works or live electronics. The GRM's skill in the technical implementation of this type of concert was widely and highly recognized, both in terms of the installation of diverse equipment and for its Acousmonium sound projection system.

In 1992, following the retirement of its programmer Ivo Malec, the Acousmatic Cycle changed its name to Son/Mu and programming was assumed by a committee known as BATZ, named after its members, Bayle, Teruggi, and Zanési. Jean Schwarz had also been part of the committee in the early days. Over time, a formula of two concerts in a single evening was established, one at 6:30 p.m. and the second at 8:30 p.m. following a half-hour intermission. In general, the 6:30 p.m. concert either featured works by young artists or was educational in focus, with the 8:30 p.m. concert reserved for more established composers. Each season, held in partnership with Radio France, consists of an average of six or seven evenings, adding up to a dozen concerts in all. But the GRM also participates in other events—festivals, Fête de la Musique, concerts abroad—bringing the annual number of concerts to about thirty, not counting the seminars, salons, forums, exhibitions, and workshops to which it is often invited.

A small statistical survey evaluating the 1992–96 concert seasons demonstrated that over these five years musical creation at the GRM amounted to 101 new titles, including 26 mixed pieces and one applied work. Of these 101 titles, 86 were commissioned by the GRM, 10 were commissioned by the French government and 5 were coproduced with Radio France.

On March 16, 1997, the GRM presented its first multiphonic concert to the public, in the Salle Olivier Messiaen of the Maison de Radio France, using its new 8-track portable sound projection device, linked to the Acousmonium. The program included eight short pieces composed for the occasion by composers either from the GRM or closely related with it. These eight pieces were played in succession, without interruption: *Navigation à l'ouïe*, Francis Dhomont; *Arc* and *En Ciel*, François Bayle; *L'Église oubliée*, Jacques Lejeune; *Étude S*, François Donato; *Fanfare*, Denis Dufour; *Echoi*, Daniel Teruggi; *Jardin Public*, Christian Zanési; *Octosax*, Jean Schwarz.

In 1989, the First Noroît-Léonce Petitot International Acousmatic Composition Competition was held at the Noroît Center in Arras. For the first time the GRM was involved in organizing a competition, even if it remained modest compared to the Bourges Competition that had been created in 1973.

A Selection from the GRM Catalog, 1988–1998

Note: passages in quotation marks are taken directly from program notes authored by the composers themselves.

1989, *Rouge-Mort* (16:50) by Bernard Parmegiani. "At the origin of this musical piece, an assumed theme: *Carmen*! No more and no less. I abandoned the narrative to pursue a linear sequence whose sequential organization and formal structure provided me with the framework for a musical drama broken down into five movements." (B. P.)

1989, *Novars* (19:07) by Francis Dhomont. According to Dhomont, "*Novars* celebrates the birth of Musique Concrète, the *Ars Nova* of our century." The piece includes some borrowed sound material from Pierre Schaeffer (with his kind permission). Sound processing was executed at the GRM in Studio 123, following the nonreal-time sound processing workshop Dhomont attended in the same studio.

1989, *Focolaria Terra* (26 minutes) by Daniel Teruggi. Piece for 4-track tape. Spatial projection created with the Acousmonium and Syter.

1989, *Espaces-Paradoxes* (31 minutes) by Patrick Ascione. Prix Ars Electronica 1994. A piece entirely composed and formatted in the studio on a large number of independent tracks, with a simultaneous stereo version. The author worked mainly on the GRM's Studer A 800 2-inch 16-track analog tape recorder. Premiered February 16, 1990, at the Grand Auditorium of the Maison de Radio France.

1989–90, *Grand bruit* (20:45) by Christian Zanési. Work realized on the 16-track analogical Studer A 800 tape recorder recently bought by the GRM (for the considerable sum of 500,000 francs). The "big noise" mentioned in the title is the RER train the composer took every day from home to studio. The twenty-one-minute journey with its eight stations gave the piece its length and form.

1990, *Tournoi* (15:30) by Philippe Leroux. "This work is essentially informed by periodic elements that are self-generating (spinning on their axis, hence the title, '*Tournament*'), forming different strata grouped according to analogues of movement, harmonic timbre, and matter." (P. L.)

1990, *Immersion* (15 minutes) by Costin Miereanu. Composed on the Syter system at the instigation of Ivo Malec, to whom the piece is dedicated. "Immersion mixes multiple layers of acoustic sources modeled with Syter's transformative instruments with glossy and limpid sonorities of metallic-colored electronic sounds, in 'shifted' discourse that is unceasingly 'transitive.'" (C. M.)

1990, *Ash* (16:10) by Horacio Vaggione. Composed from a repertoire of small sound objects (two to three seconds long) of instrumental origin—flute, clarinet, bassoon, piano, saxophone, and percussion. The morphologies of these sounds were then transformed in Syter.

1990, *Mobilis in mobile* (30:40) by Jean-Marc Duchenne. Here the composer explores three scales of space and time, in other words, three levels of complexity in the sound matter, in three movements: *Abyssa*, *Les météores*, and *Globular-Star-Clusters*.

1990, *Ys* (20 minutes) by Cécile Le Prado. "Omnipresence of water in the city, its topography, its kinetic forms." (C. L. P.)

1990, *Pomort 1: La disparition de Randolf Carter* (19 minutes) by Roland Cahen. Inspired by the French collection of Lovecraft short stories *Demons et Merveilles*. Exploring the acousmatic listening process, the author situates his work between theater and concert and asks listeners to "use their ears to excavate stacked layers and depths of sounds."

1991, *L'Ange ébloui* (20 minutes) by François Donato. As introduction to listening to his piece, the author quotes Anaya: "What does the wind do when it passes over us, silent and bluish with incomparable wings? . . . What do our spirits breathe, now awakened beyond the stars?"

1991, *Les pentacles* (21 minutes) by Ricardo Mandolini. Quadraphonic piece created with Syter. "According to the ancient Kabbalah, pentacles are geometric representations of absolute principles, diverse manifestations of a veiled and implicit unity. My piece, created based on the concept of recreating pentacular forms, constructs its perceptual and multi-parametric space through an interplay between displacement of the sound source, systematic variation in the depth of the musical fabric, and abrupt modification (amplification/reduction) of density." (R. M.)

1991, *Ixchel, La lune—Une histoire du temps* (20 minutes) by Georges Gabriel. "Opening and launching the sound—Freezing space, forgetting time—Cracking it, breaking it. By inlaying it with everything. The formal aspect of the work also functions to disrupt the spiral where all sound tends to be neutralized and set at point zero; affirming the sonorous, its presence, its sensory reality, revealing its intimacy." (G. G.)

1991, *Le présent composé* (23:48) by Bernard Parmegiani. This piece is the first in the *Plain-temps* series. "The *present* that this title evokes is in reality the instant which Bachelard (again and always) describes to us as the only reality of time. 'I' compose the instant, it composes me." (B. P.) The sounds are taken from everyday reality.

1991, *Appel d'air* (21:04) by Michel Redolfi. Commissioned by the French government. Pierre-Yves Artaud and Lanie Goodman, sampled flutes. Processing on the Syter system; mixing on the DYAXIS system at the Cirm in Nice, France. "This title, *Appel d'air*, an *indraught*, because the work, nourished by the breath of the flautists, unabashedly indulges in old-fashioned musical pleasures (arias, orchestrations, etc.), a sort of breath of oxygen within the work."

1991, *Fabulae* (53 minutes) by François Bayle. Four movements: *Fabula*, *Onoma*, *Nota*, and *Sonora*.
  Fabula: in an enlarged space, "bird song brings order, through a series of episodes, to manifold rustlings. Built on the directions of intervals and neumes, it contrasts distances (ascending, cancelled, descending, increased, tightened) with clashes, irregularities, and morphologies in the landscape." (F. B.) Processing on the Syter system.
  Onoma: "Five musical postcards. Five snapshots—in every sense of the word—five small scenes captured in the colors of the sound Ektachrome." (F. B.) Sounds of the gurgling of boiling water, harmonica, ukulele guitar, electric guitar, rainstorm, percussive chimes, car horns, the squeaking of a swing.
  "With *Nota* I offer something which, starting with the note taken as a complex object of brief, terse, and concise value, will embrace space to the point of tracing direction, the path of an arrow towards the goal, or ricochets off the goal." (F. B.) Use of MacSoutiLs, developed by Serge de Laubier.
  Sonora: reprise of an organ motif from *Nota*, which reappears five times as a refrain, framing traces of a nursery rhyme played in the background.

1991, *Chants d'ailleurs* (16:40) by Alejandro Viñao, for soprano and computer sounds (Syter). Soprano, Frances Lynch. "Songs from Elsewhere" is a series of three songs—Song I, Song II, and Song III—songs from an imaginary society, a culture highly developed technologically while remaining rural. This unlikely intersection explains the ritual and sometimes monodic nature of the singing, with the computer-generated element aimed not at harmonizing or orchestrating the songs but rather at extending the phrasing and timbre of the voice beyond its natural acoustic possibilities." (A. V.)

1991, *Objets obscurs* (14:43) by Åke Parmerud, with the voice of Dominique Andrieux. "The *Obscure Objects* are a set of short enigmas describing in part the original sound material and in part an aspect of the musical content. The solution to these enigmas can be found in each of the four movements. A short text introduces each enigma, followed immediately by its sound equivalent. Each movement is based on sounds produced by one single common object (a chair, a glass . . .). One movement = one object. At the end of the final movement, the 'solution' of all the enigmas is given. However, these solutions create a new 'inner' enigma, to which no answer is given." (A. P.)

1992 *Syrcus* (18 minutes) by Daniel Teruggi, for percussions, Syter, and recorded sounds. Percussions, Florent Jodelet. The title comes from merging the words *Syter* and *percussion*. Daniel Teruggi, who from 1985 to 1993 worked with composers, training them in using the Syter, perfectly mastered the tool. In composing his piece *Syrcus*, he believed he had exhausted practically all the resources of the instrument:

—pseudo-electronic sounds caused by filters whose frequency varies periodically;
—real time processing;
—recorded sounds, read from the hard disk; and
—half-second fragments of instrumental sources played-back in loops.

Several months of preparation were required to refine the effects and sound processing. After having been performed live four times with Syter, *Syrcus* was recorded on disk, after which the composer transposed the Syter manipulations of the piece to GRM Tools. The work was performed some ten times using this new methodology.

1992, *Volta redonda* (19:35) by Rodolfo Caesar. The title was borrowed from the name of a Brazilian steel-producing city. The composer explained, "This is the third piece I have composed based on ideas inspired by Milton Machado's installation sculptures *Heavy Metal* and *Hi-Fi*. The most striking idea is that of creating meaning from circumstances. In the present case, the circumstances move propelled by spiraling dynamism, itself formed of helixes and whorls, in steellike, if not sidereal, space. Against a dense background—a blend of chords and timbres—'objects' and events loop, in panoramic and somewhat contortionist trajectories."

1992, *Simulacre* (33 minutes) by Jean-François Minjard, with the voices of Nicole Toche, Noël Toche, and Christiane Mataix. Piece belonging to a series of studies of matter, where the author questions his work as a composer. "Is our place to let the (immanent) meaning of sounds guide us, or rather to seek to create new planes of (transcendent) meaning?" (J. F. M.)

1992, *Message* (11 minutes) by Thomas Kessler. "The source material is a completely incomprehensible message captured on a tape recorder, and the challenge in passing it on is so great that neither the origin nor even the language can be deciphered. But the more the true meaning of the message remains hidden from us, the more important the words seem to become." (T.K.)

1992, *De la distance* (17 minutes) by Ramon Gonzales-Arroyo, for tape and double bass (Jean-Pierre Robert). "Distances in space, localization, and depth; distances in the character of expression; interplay of distance within the sound itself, whose great potentiality remains evident thanks to synthesis. Distances, finally, in the language and sound material. The work is divided into four movements, linked two by two." (R.G.-A.)

1992, *Tangram* (28 minutes) by Robert Normandeau. The piece exists in two versions, the first, composed between 1989 and 1992, was reworked in 1993 for a second version. The author dedicated this "feature-length sound film" to the composer Francis Dhomont. The formal structure of the piece was inspired by the Tangram, a Chinese construction game made up of seven geometrically shaped pieces with which one can reassemble shapes. Here the author used two times seven sound tracks that he played back on the same number of loudspeakers. Compositional work in the studio was carried out on a playback system similar to that used in concert.

1992, *Alpensymphonie* (13:10) by Dieter Kaufmann: I, *Mehr Luft*, based on brass sounds recorded in the '70s with the Kärtner Blasmusikkapelle Rudi Platzer. II, *Chanson*, a mixture of sounds and words from Ernst Jandl's poetry collection, *Laut und Luise*.

1992, *L'impatience des limites* (18 minutes) by Bernard Fort. A piece dedicated to his wife, Mady, who died prematurely, "departed to the garden of sleep," and for whom "past, present, future are only one state and who, nevertheless, watches her children grow." (B.F.)

1993, *Les corps éblouis* (20 minutes) by Christian Calon. "Just a few words, a short sentence which, by its insistent recurrence, served me throughout the compositional work to channel the energy contained in the sounds. Here it is: *Then came the spring like drapery, Upon our dazzled bodies*. (C.C.)

1993, *Nout* (16 minutes) by Miao-Wen Wang. In the program notes, the female composer Miao-Wen Wang explained, "Nout is the Egyptian goddess of the sky, mother of all the gods. During the day, the solar boat travels through the sky; at night, the gods rest on Nout's body, who gives birth to them at dawn. The piece is divided into four sections: *The Solar Boat*; *Divine Night*; *Dawn*; *Rebirth of the Gods*."

1993, *Tchakona* (32 minutes) by Yuri Kasparov. Work for bassoon, Alexandre Ouzounoff; cello, Christophe Roy; and electronic accompaniment, François Donato. Commissioned by Radio France and INA-GRM. "The foundation for *Tchakona* is the theme introduced by the bassoon and the cello. The structure of the theme, made up of several sections, defines the form of the work as a whole. Each section is the source of the next variation." (Y. K.)

1993, *Epiphonies* (21 minutes) by Gilles Racot. As in his preceding piece, *Subgestuel*, Racot here makes "use of the same process of harmonic propagation, realized by mutual and combined multiplications of nine fundamental chords of four harmonizers, generating complexes of four, eight, sixteen, thirty-two, sixty-four . . . components." The materiology of *Epiphonia* is mainly the voice, "which partially explains the title: 'epi-phônê,' 'over-voice' (and also: 'over-sound')." (G. R.)

1993, *Sonneries des orgues* (21 minutes) by Serge de Laubier. Méta-instruments, Rémi Dury and Serge de Laubier. "I believe I've somewhat deviated from the original concept, which was to compose music on tape using only sounds from the organ in the Olivier Messiaen Hall. The tape indeed exists, but I have added two pairs of Meta Instruments to it." (S. d. L.)

1993, *D'un souffle retrouvé* (13 minutes) by Carlos Grätzer, for flute and tape. Gilles Burgos, flute. "From a breath, from a solitary sound material, from a single gesture, the instrumental score and the tape were composed. At the beginning there were five chords; they were developed in echelons and then underwent several transformations necessary for their propagation. The tape was created from sampled flute sounds, transformed with Syter using the intervals of the preceding chords as a filtering framework." (C. G.)

1994, *Echanges de la lumière* (20 minutes) by Christian Rosset. For seven-string bass viol (Matthieu Lusson) and tape. "The left hand holds the microphone, the right hand slowly slides a piece of wood, opens a door, moves a piece of furniture . . . Later, the studio work condenses this material, setting its contours . . . Between these phases of recording and transformation, the instrumental composition takes shape . . . *Échanges de la lumière* aims at maintaining the idea of mobile form in a field where hardware tends to bind the material, and at a time when excess of precision in composition risks rigidifying instrumental playing." (C. R.)

1994, *Feuillage de silence* (14 minutes) by Elsa Justel, for tape, flute (Vincent Touzet), and oboe (Hélène Devilleneuve). "Drawing a parallel between the composer and the gardener has always seemed to me something quite natural. Having lived in the country as a child, I observed my father carefully monitoring the harmonious growth of his plants. Looking at the imperfection of nature—be it vegetal or human—we experience this sensation of dissatisfaction, inciting us to constantly renewed searching. This piece tries to combine natural sounds from two instruments, in a certain idyllic fusion, with sounds manipulated through technology." (E. J.)

1994, *Portraits de femmes* by Luc Ferrari, for vocalist, five instrumentalists, and memorized sounds. Elise Caron, vocalist; Carol Mundinger, clarinetist; Sylvain Frydman, clarinetist; Christine Lagniel, percussionist; Michel Maurer, pianist; Michel Musseau, synthesizer. Concert comprising two works intertwined in time and space: *L'escalier des aveugles* (1994), a set of purely electroacoustic pieces performed in sound projection, and *Chansons pour le corps* (1988–94), a suite of songs with instrumental musical interludes performed live onstage.

1994, *Forêt profonde* (33 minutes) by Francis Dhomont. Acousmatic melodrama in six sections, based on Bruno Bettelheim's essay on the psychoanalysis of fairy tales. "*Deep Forest* is the first part of *Cycle des profondeurs*, a diptych inspired by psychoanalytical reflection, an adult interpretation of children's tales oscillating between remembering my naive wonder and discovering their secret mechanisms." (F. D.)

1994, *Drama Symphonie* (24 minutes) by Denis Levaillant. "The source sounds for *Drama Symphonie* are excerpts from my orchestral piece *Les couleurs de la parole*, sampled, then processed and mixed as concrete material. Today, the back and forth between instrumental writing and studio work is extremely fruitful for me. Throughout this process, I have been thinking about Edgar Varèse and his symphonic utopia, which is why this work is dedicated to his memory." (D. L.)

1994, *Lady gosse* (17 minutes) by Gino Favotti. "I aimed at 'toy-object' music. A nod at Raymond Queneau and his *Cent Mille Milliard de poèmes*...—like those books we offer children, made of different fabrics and colors, which are attached and detached to the tune of the children's play, constantly recomposing new images." (G. F.)

1995, *Arkheion Stockhausen* (16:24) by Christian Zanési. First work at the GRM realized on the 8-track Pro Tools system that Zanési was proud of having the GRM acquire after seeing it in use in Sweden. *Arkheion* means *archive* in Greek. Here the archives were recordings of Stockhausen's voice, drawn from the INA archives.

1995, *Renaissance* (16 minutes) by Åke Parmerud. In program notes for the concert, the composer commented, "In this age of computers, hard disk recorders, and digital synthesizers, it is unusual to compose using 'old' analog systems.... The last time I worked with analog systems was in 1986, almost ten years ago. The invitation from the GRM to compose a piece allowed me to carry out the idea of exploring the possibilities of an analog renaissance within the digital domain. The piece is built on sounds from a Serge modular synthesizer, with some additional sounds from Renaissance instruments such as tambours, crumhorns, and viola da gamba."

1995, *L'estran* (19 minutes) by Christian Eloy. "The dictionary, with its obligation of brevity, explains to us that the '*estran*' is the portion of the littoral between the highest and the lowest seas. Everything in this definition (*portion—littoral—between—the highest and the lowest*) indicates the in-between, the uncertain, opposites, flesh or fish, land or sea, who knows?" (C. E.)

1995, *Hot Air* (23 minutes) by Jonty Harrison. "One of the primary sound sources for this piece—balloons at children's parties—is at the root of a succession of ideas that allowed me to associate 'toy' balloons with 'hot air' balloons. Then numerous other concepts related to air came into play (breathing, expression, natural phenomena), as well as concepts related to heat (energy, action, danger)." (J. H.) Sound material developed with the GRM Syter system and the software GRM Tools, mixed in the University of Birmingham Electroacoustic Music studios.

1995, *Annam Sarvam* (30 minutes) by François Donato. This work builds on *Annam*, premiered two years earlier. Each of the seven movements is similar to an étude with its own specific territory even while embracing the same harmonic model as the others.

1995, *Exercices de style* (18:30) by Francis Larvor. Michel Musseau, voice; Dominique Auber, sound effects; Marie-Hélène Bernard, captions. Based on Raymond Queneau's *Exercices de style* (Gallimard).

1995, *Pandemonium Architecture* (11 minutes) by Tim Brady. Electric guitar and electric guitar samples by Tim Brady, processed in GRM Tools and mixed in Protools. "This composition is the fourth movement of *Strange Attractors*, a six movement work for electric guitar. *Pandemonium Architecture* exploits the sound of the electric guitar in the context of computer music. The live performance component attempts to establish a dialogue of textures and colors with the tape." (T. B.)

1996, *On eût dit des coups d'ailes* (22 minutes) by Guy Reibel. Slightly spherical surface: Jean Geoffroy; clarinet: Renaud Desbazeille; voices: Anne-Marie Heleau, Marie-Claude Patou, Daniel Durand, and Patrick Radelet. An autobiographical work, "in which a few closed grooves are scattered, in homage to Pierre Schaeffer, the inspiration for Corps Sonores and for my own musical journey." (G. R.)

1996, *Futaie* (14 minutes) by Régis Renouard-Larivière. "In the manner in which it unfolds over time, like a long, slow phrase where only the punctuation remains—in contrast with the sort of flatness situated between immobile masses and pointless flare-ups—*Futaie* has something to do with a time frame depicting a specific object: it is difficult to find in it representations of rural groves or little bunnies." (R. R.-L.)

1996, *Mariposa clavada que medita su vuelo* (17:20) by Nicolas Vérin. Flute, Cécile Daroux. Work created in the GRM studios and in the Lygis studio. "The title is a verse (in parentheses in the poem, which reinforces its suspended, timeless aspect) from Federico Garcia Lorca's *Ode to Salvador Dali*. 'Pinned butterfly meditating its flight'—the ephemeral agitation of fluttering, flickering life is nevertheless our only possibility of flying." (N. V.)

1996, *Récit sur la vielle roue* (25 minutes) by Shuya Xu. Commissioned by the French government. Soprano, Rufeng Xing. "The central concept for this work, inspired by a poem by Li Bai (701–62), is the story of a painful separation between lovers. The three primary vocal materials are singing, declamation, and recitation." The tape part is composed essentially from sampled Chinese instrumental sounds, sheng (mouth-blown polyphonic reed), xiao (end-blown flute), and guqin (seven-string plucked instrument). Mixed in the GRM studios.

1997, *Des jambes de femmes tout le temps* (30 minutes) by Philippe Mion. Commissioned by the French government, with support from Muse en Circuit.

1997, *Fugitives voix* (17 minutes) by Daniel Teruggi. "Sound processing applied to the voices of Jean Bollery and Anna-Maria Kieffer, with four generations of GRM digital instruments: the 123 (delayed time), Syter (real time), GRM Tools 1.51, and GRM Tools TDM Plug-Ins volumes 1 and 2. The voices are structured into seven linked episodes—fugitive, because barely perceptible. Voices generating musical drama transcending their initial narrative." (D. T.)

1997, *Avel* (13:15) by Jean-Claude Risset. Commissioned by the French government. "*Avel* means *wind* in Breton. The sounds, recorded or synthesized with the program Music V, were modified and assembled with Pro Tools, Syter and GRM Tools. *Avel*, an 'aeolian' piece, is governed by a logic of flux. It incorporates a few phrases sung by Irène Jarsky and a few notes of Celtic harp played by Denise Mégevand. The swaying at the end is reminiscent of François Bayle's celestial spinning top." (J.-C. R.)

1997, *Empty vessels* (7:40) by Denis Smalley. "The 'empty vessels' are large garden pots from Crete and a Turkish olive jar. The resonances of these containers were recorded as a starting point for the piece. Since the recordings were done in a garden, the microphones placed inside the pots also picked up sounds from the environment . . . The piece explores the relationships between recognizable phenomena in the environment and their more abstract qualities." (D. S.)

Seminaries

The 1989 colloquium "Le son de la musique," organized by the GRM in coproduction with the music division of the network France Culture, gave rise to radio broadcasts and above all to a book by François Delalande. Published in 2001 as *Le son des musiques entre technologie et esthétique*, the book included a transcription of interviews carried out at the colloquium, expanded with extensive commentary by François Delalande.

Public debates preceded each of the five concerts in the Son-Mu 1992 season under the title *Réel/Virtuel*. They focused on the dialectic between *fixed sound* and "Son-Mu."[16] The speakers—François Bayle, Michel Chion, François Delalande, Denis Dufour, Xavier Garcia, Daniel Teruggi, and Jean-Christophe Thomas—addressed five themes: support, narrativity, morphologies, controlled space, and simulated gesture.

The research seminar of the 1992–93 season, consisting of six three-hour sessions between December and May, was titled "Du temps à l'œuvre" (From time to the musical work"). François Delalande introduced it:

> A large number of the styles of music in the second half of the twentieth century, and in particular electroacoustic music, have abandoned earlier models for organizing duration, whether at the level of the general plan of the work (grand form) or at the level of *writing*—rhythm, meter. A theory of time and form remains to be constructed. On this level, music produced on a support undoubtedly represents an extreme case which can shed light in an exemplary manner on a more general issue. The objective of this research seminar is to contribute collectively to such theoretical formulation, starting from the compositions and from compositional praxis.[17]

The following year, the research seminar of the 1994–95 season was titled "Le champ morphologique, outils, concepts, écriture" "The morphological sphere—tools, concepts, writing." François Delalande opened the debate:

> Ever since music has been made with machines, a new conceptualization of sound has developed in parallel. In particular, the notion of sound "morphology," defined and studied by Schaeffer since the earliest years of musique concrète, is a tool for thinking that is increasingly relevant for composition, for at least two reasons. First because in a much-needed way it goes beyond the concept of timbre, which is too limited because of its origin with instrumental sounds. This step forward is in line with the history of musical thought, but also because—and this is related—studio composition tools have been conceived and developed, especially at the GRM, to act on the morphology of sounds. This has resulted in fertile reciprocal interaction between morphological operations, constituting a permanent axis for musical research.
>
> The object of this 1994–1995 seminar is to become aware of this back-and-forth adaptation mechanism between thinking and tools, identifying its aesthetic implications and drawing lessons from them for musical analysis and for the design of composition software.[18]

Over both the 1995–96 and 1996–97 school years, a seminar explored the theme of defining the classification of so-called electroacoustic works and of drafting a typology based on specific examples.

The "Empreintes" Sessions

From 1993 to 1997, a research activity of the GRM, the monthly "Empreintes" events, took the form of recorded public talks, with a different composer each time. Daniel Teruggi, moderator of these sessions at the Acousmathèque, described the problematic on the invitation card: "The 'Empreintes' sessions invite composers whose research is 'concretely' inspired by sound to describe their trajectory, to articulate their chosen approach, illustrated with sound or graphic examples."[19]

Numerous composers have lent themselves to the exercise of these sessions: Jean-Marc Duchenne, Gérard Grisey, Marc Favre-Marinet, Michel Pascal, Denis Smalley, Thierry Lancino, Xavier Garcia, Michel Chion, Rodolfo Caesar, Michaël Levinas, Alain Savouret, Philippe Mion, Francis Dhomont, Kaija Saariaho, Luc Ferrari, Horacio Vaggione, Michel Redolfi, Gilbert Amy, Jean-Claude Eloy, Philippe Manoury, Enrico Corregia, Nicola Bernardini and Alvise Vidolin, Pierre Henry, Bernard Parmegiani, and Javier Alvarez.

Editorial Activity

Most of the GRM's research activities have resulted in editorial spin-offs in the form of radio programs, books, and articles.

RADIO

The national cultural network of Radio France, France Culture, with its specific musical programming, is the favored place for broadcasting these works. With the GRM, the France Musique network primarily broadcasts musical programs with commentary, in particular electroacoustic works premiering in concert. The GRM delivers "turnkey" programs—already recorded, edited, and mixed in its studios—to the broadcasters, in consultation with network programming departments. It is important to note, however, that from 1987 onwards, Radio France's program directors, both on France Culture and France Musique, began standardizing the professional contracts of the "delegated producers"—who had before always held an independent status. In order to ensure producers more regular income, Radio France took over the assignment of programs to them, including those at the GRM. From that time on, the GRM collaborated with producers who, while designated by the network, were selected for their affinity with new approaches to music creation on support. GRM programs thus became coproductions between INA, which contributed by directing and providing archives and documents, and Radio France, which paid the producer. Among these producers, Philip de la Croix, Christian Rosset, Jean-Yves Bosseur, and David Jisse have worked, or still work, with the GRM on programs or series:

—The *Opus* series, a weekly ninety-minute evening program on France Culture, featuring portraits of composers. The GRM contributes to this segment of programming by producing portraits of electroacoustic composers two or three times a year. After the portrait of Pierre Schaeffer in 1988 followed those of François Bayle, Francis Dhomont, Jean Schwarz, Konrad Boehmer, Luc Ferrari, Bernard Parmegiani, Michel Chion, Horacio Vaggione, François-Bernard Mâche, Bernard Lortat-Jacob, Christian Zanési, Beatriz Ferreyra, Denis Smalley, and Guy Reibel.

—The series *Multipistes*, inaugurated in 1987 with Philip de la Croix, ended in 1991 after 193 programs. This weekly thirty-minute series provided a panorama of the activities of electroacoustic music centers and studios in France and some neighboring countries.

Other series on France Culture also featured the GRM in their programs:

—The series *Le rythme et la raison* (Rhythm and reason, half an hour, five consecutive mornings, on a single theme);
—*Euphonia* (one hour, five consecutive days in the early afternoon);
—*Jeux de l'ouïe* (Listening games, ten minutes in the morning, five days a week);
—*Les chemins de la musique* (Music pathways, half an hour, in the morning, five days in a row); and
—*Coda* (ten minutes in the early evening).

On the France Musique network, in addition to regular broadcasting of works created at GRM concerts, the *Akousma* series ran for ten years, from 1994 to 2004.

BOOKS AND JOURNALS

In 1993, the Bibliothèque de Recherche Musicale (collection edited by François Delalande at Éditions Buchet-Chastel) published a book by François Bayle, *Musique acousmatique, propositions . . . . . . positions*.[20] Bayle outlined his musical theories, in particular the theory of *i-sounds* (or images of sounds), offering an original listening model based on concepts stemming from the work of Klee, Bachelard, René Thom, Charles S. Peirce, Husserl, and Whitehead. Bayle also evoked his theories concerning sound projection in concert.

In June 1995 the first issue of the review *Ars Sonora* was published, edited by a collective including Christian Zanési, Régis Renouard-Larivière, and Christine Groult, under the direction of Jean-Yves Bosseur, with the collaboration of the Centre de Documentation de la Musique Contemporaine (CDMC) represented by Marianne Lyon and Catherine Vayne. Nine issues followed, until publication was suspended in October 1999.

In 1997, François Delalande's book *Il faut être constamment un immigré* (It is essential to be constantly an immigrant) was published by Buchet-Chastel, consisting of a lengthy interview with Iannis Xenakis about his music.

ELECTRO-CD FORUMS

In 1993 and 1995, the first and second Electro-CD Forums were held. These events, which brought together twenty electroacoustic music CD labels, were held three evenings in a row in Salle Olivier Messiaen at the Maison de Radio France, in the form of three public exchanges and six concerts. The objective was to bring together various actors in the electroacoustic music production sector, from composers to record distributors, in an attempt to gain insight into the future of this art and its economics. Already noted in 1993 were the tendency towards proliferation of small labels and their difficulties in gaining public recognition.

New and Former GRM Members

FRANÇOIS DONATO

A new collaborator-composer joined the GRM. François Donato arrived on July 11, 1989, to oversee concert production and coordinate composers during their work in the studio. He would resign for personnal reasons on September 8, 2005, but continued to be musically active in the southwest of France.

ALAIN SAVOURET'S CLASS AT THE CNSMDP

In 1997 on the occasion of *Journées Musique et Quotidien Sonore* in Albi in the French department of Tarn, Alain Savouret wrote in an introductory brochure, "The old Bescherelle dictionary of 1846 is going to bail me out. I will call myself an 'inductor of music,' based on the definition it gives (and that I embellish) of the 'inductor' in physics."[21] Savouret, former student in the GRM's electroacoustic class at the Paris Conservatory in 1968 and research fellow at the GRM from 1968 to 1972, was above all a devotee of live music. In 1969 one of his first compositions, *Kiosque*, was conceived for tape and instrumental improvisation. In 1991, along with Jean Pallandre, Savouret organized "loudspeaker vigils" during the course of an entire winter in the homes of local residents of a working-class neighborhood of the city of Calais, literally giving voice to the residents. Since 1993, he has been teaching the Generative Improvisation class at the CNSM in Paris, combining his expertise in typo-morphological description of sounds based on Schaefferian criteria, with vivid hands-on music-making, using not only classical orchestral instruments but also a wide range of contemporary electronic instruments, especially synthesizer keyboards.[22] Alain Savouret insisted on the necessity of all contemporary music practitioners of receiving "complementary ear training" based on the concepts developed by Pierre Schaeffer in *Solfège de l'objet sonore*. Over time he increasingly shifted from what he no longer called "solfège of the sound object" to instead "solfège of the audible," calling into question the concept of sound object, too narrow for his tastes.

Following Rainer Bœsch's retirement in 2003, Alain Savouret was joined by Alexandros Markeas. Savouret himself would retire in 2007.

TEACHERS AT THE PARIS CONSERVATORY ELECTROACOUSTIC CLASS

When the CNSMP moved from rue de Rome in 1991 to new premises at the Cité de la Musique at La Villette, three new studios and the auditorium were completely equipped with digital technology,[23] thanks to a comfortable budget of 7 million francs from the Ministry of Culture.[24] After the departure of Henri Kergomard in 1991, Laurent Cuniot took over the electroacoustics teaching position, assisted by Luis Naon. Yann Geslin, present since 1987, continued in his area of expertise in computer music. Luis Naon succeeded Laurent Cuniot in 2002.

11

1998 and Beyond

The fundamental rule of a network is that its usefulness is proportional to the square of the number of people who use it.
—METCALFE'S LAW, AFTER ROBERT METCALFE

For the GRM, at first glance, the years since 1998 would seem to be a continuation of previous ones—the same institutional structure, the same partners, the same activities. However, two essential elements of change have appeared. First, the arrival of a new director, Daniel Teruggi, who succeeded François Bayle after he retired in 1998. From the very beginning, Daniel Teruggi sought to give a new face to the GRM by taking advantage of the exceptional circumstance of the commemoration of the fifty-year anniversary of musique concrète. He encouraged the members of the group to work more collectively, fostering new initiatives. The second innovative element in the landscape stemmed from the INA, which was redirecting its activities primarily towards preservation of the audiovisual heritage for which it is responsible.

The GRM's activities for the period following 1998 reveal a fever pitch of software development as well as musical, radio, and publishing production. The new technological context with the arrival of internet and multimedia, coupled with the generalization of personal music production systems, accelerated the transformation of the Groupe towards the international scene and a wider public. It became a crossroads where artists, researchers, producers, and educators met, with the French national education system increasingly becoming an important collaborator. The renewed confidence of Radio France in the GRM as specialized in "electronic" music, in the framework of the "Présences *électronique*" festival from 2005 onwards, tended to reinforce the role of the Groupe in a musical milieu undergoing a process of soul-searching. The act of composing on media was no longer limited to members trained by the group, but to all those willing to tackle sound material with software available on the market, GRM Tools, and the like.

Taking note of this opportunity, the GRM invited aesthetically compatible composers from all horizons to perform in its concerts.

In the midst of all this turbulence, the GRM has once again managed to find renewed momentum by adapting to the demands of the music world, the public, and its parent institution, the INA.

HISTORICAL CONTEXT
Reinforcing the Preservation of Cultural Heritage at the INA

Under the presidency of Emmanuel Hoog, who succeeded Francis Beck in 2001, the INA has resolutely focused on its mission of preserving the archives of French public broadcasting. Every effort has been engaged to strengthen the heritage sector. The Research and Experimentation Department (DRE) reoriented its activities towards this objective. The GRM, part of the DRE, continued ongoing projects, but also contributed to this new objective by participating in designing filtering systems for the restoration of archives and algorithms to suppress background noise on magnetic tapes. The Acousmographe, the software tool for computer-assisted graphic transcription of sound, whose first version dated back to 1988, naturally found its place in this new challenge. In April 2004, the INA launched an online service, Inamédia, for professional researchers, who could now consult and purchase online from the two hundred thousand hours of archives in the national television collection.[1] The 1.6 million associated records are accessible in French and English. From 1995, when conservation began, through 2002, the first video files were compressed in the Mpeg 1 format at 1.2 Mbps. Since then, they have been compressed in Mpeg 2 at 8 Mbps. Inamédia is the world's largest bank of digitized audiovisual archives. The particularity of this system is that it offers an interactive view, allowing direct production of time-coded extracts. Since 2002 the audiovisual Dépôt Légal, the French national repository system managed by the INA, has been extended to cable and satellite channels, and since 2005, to the web. The audio counterpart of Inamédia is titled Inasound, working on the same principle of access restricted to professionals. A version for the general public, the Offre Grand Public (OGP), was launched in 2006. It is a worldwide success, the site Ina.fr receiving one million visitors per day. GRM records and even musical extracts can be purchased online, in addition to material from the television and radio archives.

The orientation chosen at the INA to reinforce the activity of indexing and automated recognition of archives for industrial-scaled exploitation, although remote in principle from musical preoccupations, took on relevance for the GRM. The question of the conservation and exploitation of recording media had always been a source of concern for electroacoustic composers. Since the early 1960s, the GRM had had its own system for indexing its archives, an adaptation of older classification systems. But today, when the earliest recordings are all more than sixty years old, the question of the durability of the heritage becomes crucial. Practically all GRM activities make use of the archives, first and foremost the concerts that regularly feature works from the repertoire, in tandem with recent works. Radio broadcasts and research seminars also make extensive use of the archives. Each new use of material from the archive sheds new light, and at the same time contributes to enriching the collection. The digitization of the collection became a priority, as throughout the INA. At the GRM, digitization began in 1998 for the older collections on tape. Since 2002, new works and their metadata have been systematically deposited in a specific online database, Acousmaline.

New Editorial Choices at Radio France

Radio France, the GRM's historical partner, was undergoing profound restructuring. Beginning in 2000, the new director of Radio France—the well-known journalist Jean-Marie Cavada—developed a programming policy radically different from his predecessors.[2] Time slots reserved for music on the France Culture network decreased significantly, replaced by live programs and magazines in the form of debates and commentaries on nonmusical current cultural affairs. Faced with this decrease in the volume of production on France Culture, the GRM reoriented its collaborations towards creative programming, notably coproducing with the Atelier de Création Radiophonique, which had a regular eighty-minute broadcasting slot on Sunday evenings.[3]

But radio creation also faced difficulties. In France, Europe, and across the planet, broadcasting slots for creative radio programs were being cut at a dramatic pace, along with the elimination of many creative centers such as in Italy, Germany (Berlin), Australia, Slovenia, and Canada. The internet became a refuge for creators (Kunstradio in Austria and AudioHyperspace in Germany, Denmark, and the Netherlands). The arguments put forward by terrestrial (traditional) broadcasters to justify this decline were ratings, aging and shrinking audiences, and the need to economize. In 2004, Ars Acustica, a member of the European Broadcasting Union (EBU), decided to launch a European website and satellite broadcasting, on a frequency having been provided free of charge by the EBU. This international organization brings together the world's leading artists and producers of creative radio programming.

In another signal of the depth of change, in 2004 Pierre Boulez refused, at the last minute, to serve as the guest composer at the Radio France Présences festival, denouncing Radio France's cultural policies on the national radio networks.

But in contrast, on the network France Musique the volume of broadcasting of musical works presented by the GRM remained stable, even if the time slots were becoming later and later in the evening.

Meanwhile, the audience for terrestrial radio stations was in decline. The production of DVDs (cinema) was exploding. In 2011, 120 million DVDs were sold in

France but only 32 million in 2021. The sale and exchange of music on the internet was disrupting the entire sector of production and distribution. Record publishers began to complain about piracy, which was reducing their revenues. And DVD manufacturers were proposing new formulas such as the disposable DVD and pairing of audio CDs with DVDs. The fusion of audiovisual, telecoms, and computing was becoming a reality.

As far as the music world is concerned, and in particular at the GRM, two worlds coexist, becoming increasingly polarized. The year 2002 was a milestone on the road to globalization. The strong symbol of the passage to a single common currency in Europe on January 1, 2002, also appeared to signify the unleashing of a new cultural order of things. The old "Franco-French" system was breaking down in the face of a European—and even international—network of musical production and distribution in full development.

Alongside the Paysage Audiovisuel Français (PAF, French Audiovisual Landscape), the Paysage Internet Français (PIF, French Internet Landscape) was born. At the GRM, the Acousmaline server, multimedia publication and the *Portraits polychromes* book series, as well as web radio, all contribute to its global outreach.

Technological innovations in internet communication and computer developments are at the heart of these upheavals.

Relocation to the Maison de la Radio

Following the 9/11, 2001, terrorist attacks in New York, the central tower of the Maison de la Radio building in Paris was designated as a "high-rise building," subject to very strict new safety standards. The requirement to bring the tower into compliance with these standards, ordered by the Paris Prefecture of Police, meant that all the staff and equipment there had to be moved. Two entire months were necessary, at the end of 2004, to completely relocate the various libraries installed on the eighteen floors of the central tower, INA's sound library, the record library, and the Radio France library. Radio France's vinyl records would be stored in premises rented from the Calberson corporation, north of Paris, and INA's magnetic tapes and 78 rpm recordings were stored at the Essarts Center, some 50 km south of Paris. Every day couriers had to shuttle, delivering material requested by program producers. This relocation was experienced as a "loss of memory," especially for the longest-serving employees, who were well-versed in the collection, using it regularly. Many archives, still not indexed—particularly those dating from before 1975—were only familiar and therefore accessible to certain specialists.[4]

To compensate for this physical transfer of the archives, Jean-Paul Cluzel, president of the Radio France group, decided to accelerate the plan to digitize radio production in order to have the sound documents available directly online as soon as possible. The renovation of the Maison de la Radio building was planned for a period of seven years extending through 2012. The project also included replacing the auditoriums open to the public and the Salle Olivier Messiaen (nine hundred seats) with a new fifteen-hundred-seat auditorium, later around 2013, since the initial work had already fallen behind schedule.[5]

The GRM also needed to comply with the demands of the relocation. Its historical Studio 123 closed permanently in December 2004. But the GRM, once threatened with having to leave the Maison de la Radio completely, saw its lease renewed thanks in particular to public letters of support from prestigious composers Pierre Boulez, Karlheinz Stockhausen, Pierre Henry, and François Bayle. In the end the GRM would remain in the circular "Maison," near the radio and concert production centers. In compensation for losing Studio 123, the GRM was granted an equivalent amount of space on the sixth floor of the Maison de la Radio, and its analog archives would be stored at the INA Center in Essarts.

The Vinegar Test—the State of the Archives

When the GRM was forced to move its archives from the thirteenth floor of the tower of the Maison de la Radio in December 2004, it undertook testing the state of preservation of its magnetic tapes—four thousand tapes filling 129 linear meters, for a total of three thousand hours of sound documents.[6] The vinegar test consisted of placing a small strip of reactive paper in each tape box and reading it forty-eight hours later, verifying the level of contamination from a microscopic fungus that causes an odor smelling strongly of vinegar. At the GRM, it was discovered that 20 percent of the four thousand reels were vinegared at level 5—the highest. To save this collection from obliteration, the only solution was to immediately digitize it—at a cost of €82 per digitized hour.

Towards a Halt in Downsizing at the GRM

Since the early 1990s, the GRM had seen its staff stagnate and even decrease as older collaborators retired without being replaced. Between 1991 and 2007, the GRM lost seven of its composer collaborators: Ivo Malec (1991), Bernard Parmegiani (1992), François Bayle (1997), Jean Schwarz (1999), Jacques Lejeune (2000), Denis Dufour (2001), François Donato (2005). Other important collaborators also retired: Nicole Rouzoul (1993), secretary of the research team who had previously been Michel Philippot's secretary; Geneviève Mâche (1992) and Jocelyne Curjol (1999), documentalists in charge of the sound library; Jean-Paul Bourget (2004), technical supervisor in charge of studio maintenance; François Delalande (2006), researcher in music sciences; Bernard Bruges-Renard (2007), administrator; Jacques Darnis (2007), technician in charge of the Acousmonium; and Jean-Christophe Thomas (2008), researcher in music sciences. All of these departing collaborators, almost all of them having spent the majority of their careers at the GRM, represent "the memory of the GRM" disappearing.[7] A few people were hired to fill the gaps, but in smaller numbers: Diego Losa, in charge of digitizing

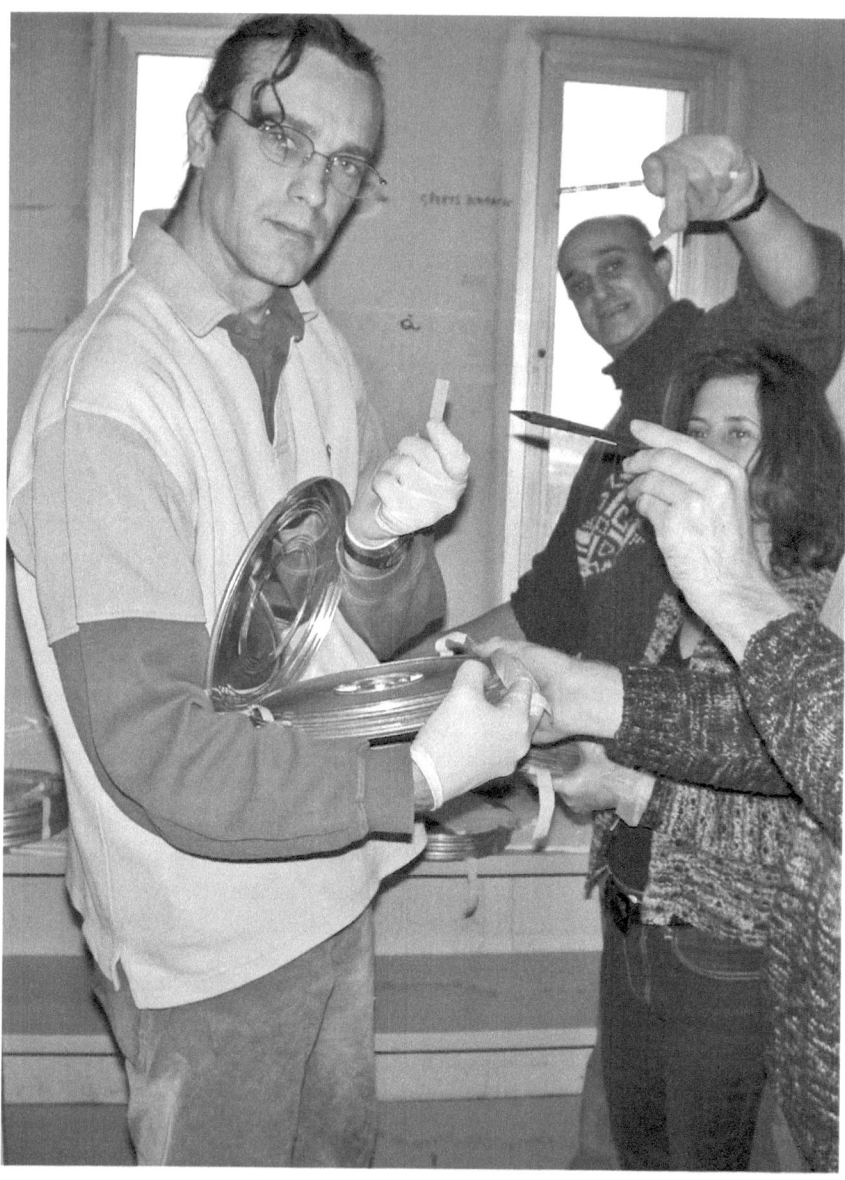

FIGURE 27. Vinegar test by François Donato, Diego Losa, and Solange Barrachina using testing strips on the GRM archive tapes. Blue indicates good health; yellow indicates the presence of fungus, giving off the smell of vinegar. © Évelyne Gayou (2004).

the collection, joined in 2001; Adrien Lefèvre, a computer engineer in charge of developing the Acousmographe, in 2006; and François Bonnet in 2007, as a sound engineer in charge of production alongside Philippe Dao. In 2007, only fifteen people remained at the GRM, compared to twenty-two ten years earlier, and the age pyramid of those remaining had a significant proportion of people over fifty years old. This situation has raised concerns among the remaining employees about the Groupe's sustainability. For a while, the GRM had relied on temporary workers and interns. Finally in 2008 two new people joined the group with long-term contracts: Alexandre Bazin, in charge of the creation of radio programs, and Jean-Baptiste Garcia, in charge of concert production and the management of the archives. Then, in 2019 and 2020, two new researchers were hired, Nicolas Debade and Mathias Puech. The group continues on its way.

The Transition to All-Digital

In 1998, the GRM presented a prototype of a digital program in Dolby Surround to the program director of France Culture. The program, produced by Évelyne Gayou, met with great interest from management, but proved impossible to broadcast on the air because Radio France's transmitters were not yet equipped for multichannel. Broadcasts could only be transmitted in analog, and only in stereo. The first broadcasts in Dolby Surround only began in 2001, via satellite on the music channel Hector (superseded by Vivace in 2005), at night from 1 to 3 a.m. Once again, GRM was first, Christian Zanési and David Jisse inaugurating the genre with their program *Fins de mois difficiles*. To enjoy the sound spatialization effect, listeners first needed to connect their tuner or decoder to the satellite.

Beginning in January 2004, Radio France launched its 5+1 multiphonic gateway on web radio, along the lines of what was already being done in Sweden. Production on analog tape was rapidly receding. In May 2004, Yann Paranthoën, chief sound operator at Radio France and above all a leading artist in radio creation and winner of numerous Italia prizes,[8] announced that EMTec, the last French manufacturer of magnetic tape, had officially decided to stop production, its staff of 120 people dismissed. EMTec was Radio France's supplier.[9] That was the end of analog.

In July 2004, the France Inter network went all-digital overnight. The acceleration of the transition to digital production was real and brutal, even though the first digital editing units and postproduction studios dated back to 1996. Radio France opted for the Sadi audio sequencer developed by the BBC. Production managers attended training courses in small groups.

In March 2005, Yann Paranthoën passed away. A program manager awkwardly paid tribute to him on France Culture, declaring, "it is the end of an era."

As for the GRM, it had been working digitally with Pro Tools software on Macintosh computers since 1992. Digital musical and radio production had been standard since 1995. By the end of 2003, all GRM studios were digital, using hard disks. Even DAT (digital audio tape) had already disappeared.

In 1998 the GRM started operating on a network, both internally for administrative tasks and externally with the internet. A new collaborator, Dominique Saint-Martin, who had arrived in 1997, first set up a dedicated server for the GRM. From the year 2000 onwards, he oversaw monitoring the functioning of the GRM's internet pages on the INA website. Some recurrent sections were quickly set up and improved year after year, in particular *Web Radio*, which after its initial broadcasts on terrestrial networks made GRM programs available to Internet users, as well as *Studio Reports* (short presentations of composers invited to the GRM to perform one of their pieces in concert) and monthly public announcements of programming.

Acousmaline: The GRM Server

But Dominique Saint-Martin had a more ambitious project in mind, developing a large server that would contain the entire GRM archives, once digitized, as well as all new sound or written works, when possible from their earliest stages of creation. In his view, this server would make it possible to "pool sources and share knowledge" both within the GRM and with the public, as Pierre Schaeffer had done with the shared sound library in the '60s.

In 2003, the computers in the three composition studios (116A, B, and C) and in all GRM offices and other technical workspaces—in particular the digitization unit for the archives collection—were interconnected to the server known as Acousmaline. This meant just over twenty computers could access the new database at any time. Administrative services were directly connected by a high-speed link to the Bry-sur-Marne center in the eastern suburbs of Paris, where most of the INA headquarters were located, including management and financial services.

Development of the Acousmaline server by the Direction des Services Informatiques (DSI) and an external service provider (for a sum of 250,000 FF) was an ambitious challenge, but necessary for the survival of the GRM archives, threatened by aging media.[10] Phase one was carried out 2002–3, phase two in 2005. This server, a prototype, consisted of two components, an interface, and a media server, initially 500 GB, both hosted at the Bry-sur-Marne center. Acousmaline allowed researchers and members of the GRM to access, through multiple points, the GRM's digitized archive. But work was slow because each document had to be manually indexed, and the GRM was cruelly short of manpower. Diego Losa had been hired by the INA in 2001 to continue digitization of the Acousmathèque, the GRM's music collection, in 24 bit / 96 kHz. As of 2005, 280 GB had been digitized, including twenty-seven hundred sequences or sound works accompanied by their documentation. In 2004, Solange Barrachina, responsible for monitoring Acousmaline, assigned the chronological order number one thousand to a CD in the database, which included all sources together: music, interviews, seminars, radio broadcasts, and external records.[11]

In the meantime Inamédia was created, an INA server accessible to the general public. The Acousmaline project was incorporated into the larger Inamédia project, launched in 2004 and renamed Ina Mediapro in 2007, with three hundred thousand hours of digitized television archives and seventy thousand hours of radio.

New Directions in Research: The "Big Projects"

The emergence of major European projects (with substantial European funding) in 2002–3 (following more modest projects, such as Epicure, which came to an end in 2005) should in theory have compensated for the lack of direct investment in basic research by companies and institutions.[12] But funding durations were often notoriously short (usually a single year, occasionally three years), hampering the continuity of investigations and inhibiting a return to fundamental research, even if things appeared to be evolving towards increased continuity from one project to the next.

The INA and its Département de recherche et d'expérimentation (DRE, Research and Experimentation Department) adapted to this new way of doing research, and in particular to its new sources of funding. Let's take an example of a research project in which the GRM, through the DRE, invested heavily. "PrestoSpace" was officially launched on February 1, 2004.

Initiated in 2003, the European project PrestoSpace consisted in federating, for a period of four years, research by thirty-seven European partners in preservation, restoration, and storage of audiovisual archives. In Europe these audiovisual documents amounted to one hundred million hours.[13] In addition, the project aimed at facilitating long-term access to these documents, including through integrating metadata. PrestoSpace was a joint project led by INA (France), the BBC (Great Britain), and RAI (Italy), involving industrial companies such as Philips (Netherlands), Thomson (France), and Elsacom (Italy), as well as public institutions, notably research centers, INRIA (France), CNRS (France), Teseo and EMF (Belgium), Fraunhofer (Germany), broadcasters such as ARTE (France and Germany) and SWR (Germany), and universities in Poland, Hungary, Ireland, Austria, and France. This list is far from exhaustive, given the number of partners and affiliates, but it clearly indicates the new form research was taking at the INA, and consequently at the GRM. The GRM and the DRE had already hosted and steered university interns and PhD students on research topics directly related to this extensive project—research into ontology, metadata, and the restoration of archives.[14]

This research was of interest to the GRM because it generated synergy with other institutions also interested in sound, financial synergy (the overall budget was approximately 11 million euros at the outset, including 2.3 million for the INA), as well as research synergy. But at times the administrative burden of such projects (numerous meetings, intermediate reports to be written, consultations, legal problems, all complicated by the multiplicity of languages) tended to monopolize the energy of participants, to the detriment of research itself. Another potential danger lay in the risk of fragmentation of the work and atomization of results, because of the multiplication of participants and the use of numerous interns and temporary contracts.

Another example of a research project was the French project Écrins—Environnement de classification et de recherche intelligente des sons (Intelligent Sound Classification and Retrieval Environment). This eighteen-month 2001–2 project brought together three partners: the IRCAM, the INA-GRM, and the company Digigram. The GRM created textual descriptions of the sound samples (perspective, morphological, instrumental, causal, etc.), the IRCAM modeled the data and integrated the results of this research into a system for automatic analysis of signals deposited in the database, and Digigram looked for ways to industrialize the results. The results were presented at the ICMC in Gothenburg in 2002.

MUSICAL CONTEXT
Equipping the Studios

Newly appointed director of the GRM in 1998, Daniel Teruggi decided to renovate the composition studios in function of the budgetary means at his disposal, progressively moving them towards "all digital" and multichannel, 5+1 and 8-channel. In 1998 the studios were equipped with the first generations of digital devices, DAT, samplers, filters, Macintosh G3 computers, and hard drives, but the consoles were still analog. The renovations lasted several years, from 1999 for Studio 116A to 2004 for Studio 116C.

In 1999, Studio 116A was returned to service after one year of transformations, the analogical DDA console disappearing in favor of a Pro-Control console associated with the Pro Tools TDM system and a G4 computer. In one corner of the studio a Studer A80 analog 2-track tape recorder was preserved to play back tapes, in case of need. In another corner of the studio, we kept the Studer 16-track analog tape recorder, which could also be used to play back old archival tapes. The JBL and Genelec loudspeakers made it possible to play back in two to eight channels. In 2003 the Focus Right Green preamplifiers were replaced by Avalon AD2022s with excellent class A transistors. In 2007, the Pro-Control console was replaced by an analog SSL AWS 900+ console, with a digital driver allowing it to be used as a control surface.

In 2002, following the definitive shutdown of Syter, Studio 116B had been left unused. It was renovated and equipped with a Pro Tools system installed on a Macintosh G4. The Tascam TMD4000 digital console allowed routing sound to a two- to eight-channel speaker system. Studio 116B became a full-fledged digital workstation, like Studio 116A.

FIGURE 28. The all-digital Studio 116 C; in production with François Bayle as "guest" composer. © Évelyne Gayou (2004).

In 2004, the third GRM studio, 116C, was "finally" renovated. Increasing budget restrictions considerably delayed work, but patience paid off. The studio was completely redone—electricity, fiber optic cabling, decoration of the floor (light wood color parquet), acoustic treatment of the ceiling, and new equipment. The result was successful both aesthetically and technically, thanks to the judicious choices of Philippe Dao, François Donato's assistant. Here too it became all digital—Pro Tools TDM system on Macintosh G4, Yamaha D2000 console, and dedicated 5+1 listening, reducible to stereo.

The three 116 Studios—A, B, and C—thus became more or less equivalent in equipment, allowing composers to move from one to the other, even while working on the same material, without any difficulty in adapting.

It shouldn't be forgotten that since 1999 the GRM Tools software had been accessible to PC platforms in its VST version.[15] Having become an all-purpose tool, GRM Tools was sold everywhere, for uses ranging from home studios to major postproduction film studios. And the Acousmographe, the GRM's other flagship software, saw its 2005 version designed to be accessible to the newer computer platforms, Windows XP, and Mac OS X. Version 3 of the Acousmographe was released in 2007 with additional features and ready-to-use libraries of graphic symbols, rendering it more user-friendly.[16]

Modernization of the Acousmonium

The last major step in this multiyear renovation plan involved the Acousmonium, which had remained essentially unchanged for some fifteen years. In 2004, the Acousmonium consisted of forty speaker enclosures, each containing several loudspeakers, as well as eight sound "trees" and two analog consoles, one for handling sources, especially from live instruments, and since 1995 an EAA for diffusion, with eighteen input channels and ten output channels, plus auxiliaries. In 2005, a new Yamaha PM5D digital console (32 bits internal, 96kHz dynamic range) replaced the two Acousmonium consoles (one for sound diffusion and one for input from live sources, in the case of mixed music). This new console (48 analog + 36 digital inputs, and 48 analog + 24 digital outputs) substantially simplified setup work for concerts, with its self-contained integrated processing and effects (such as reverb and delay), twenty-four potentiometers, eight digitally controlled amplifiers (DCA), and extensive input controls for the stereo channels. Finally, renovation of the Acousmonium also included the acquisition of four new powerful L Acoustics loudspeakers (with 115 XF cones) associated with a SB 118 subwoofer.

IRCAM

In 2001 Bernard Stiegler took over the direction of the IRCAM. Taking advantage of this opportunity—a friend of the GRM being appointed to IRCAM—the GRM responded positively to IRCAM's calls to engage in joint projects, research seminars, and long-term research projects such as Écrins (Environnement de Classification et de Recherche Intelligente des Sons—Intelligent Sound Classification and Retrieval Environment). But it remains clear that, despite warm personal relationships between members of the two institutes, fundamental differences remain in terms of musical conceptualization. IRCAM remains primarily an institution dedicated to written music, while the GRM is devoted to the music of "doing." Bernard Stiegler left the IRCAM in 2005 before the end of his appointment.

International Society for Contemporary Music—SIMC

In 2002, Daniel Teruggi reactivated the International Society for Contemporary Music (ISCM), originally founded by Arnold Schoenberg in 1922. In the years since its foundation, the French chapter, the SIMC (Société internationale pour la musique contemporaine), had fallen dormant on several occasions. It had even been reestablished in 1948, under the presidency of Pierre Capdevielle, with Nadia Boulanger, Roland Manuel, Henri Martelli, and Henri Dutilleux. After a new interruption, Jean-Claude Eloy resurrected it in 1974. Since then, it has been presided over successively by Michel Decoust, François-Bernard Mâche, Michaël

Levinas, Fernand Vandenbogaerde, François Bousch, and Daniel Teruggi. The revival of the ISCM in France in 2002 was celebrated with the organization of the first Forum for Young Musical Creation in L'Archipel Hall in Paris, with the new president, Daniel Teruggi, pursuing the original objective of the society founded in 1922 in Salzburg—discovering new talent. The World New Music Days are organized each year in a different country. In 2005, the ISCM had forty-nine member countries. Among the early members of the society were Anton Webern, Bela Bartok, Arnold Schoenberg, Paul Hindemith, Arthur Honegger, Kodaly, Darius Milhaud, Francis Poulenc, and later Alban Berg, Charles Koechlin, Maurice Ravel, Ottorino Respighi, Albert Roussel, Florent Schmitt, Olivier Messiaen, Henri Dutilleux, and Iannis Xenakis. The ISCM received financial support from the Cultural Division of the SACEM, directed by Olivier Bernard, as well as from the GRM.

The ADAC

The electroacoustic music course led by Christian Eloy and Régis Renouard-Larivière at the ADAC closed definitively in 2005, following a decision by Paris city hall, which financed the ADAC artistic workshops program, later renamed Paris Ateliers.[17] Even though the course was doing well, it no longer fell within the cultural and financial objectives of the city's workshops. Christian Eloy and Régis Renouard-Larivière were forced to look for a new location, and above all new funding, perhaps from the municipal conservatories of Paris, which woefully lacked electroacoustic courses. In 2005, only the conservatory of the twentieth arrondissement offered such a course, with a half-time teaching position entrusted to the composer Gino Favotti. In 2006, the conservatory of the seventh arrondissement, directed by Edith Canat de Chizy, opened its doors to Régis Renouard-Larivière and to the former electroacoustic course of the ADAC.

The Conservatoire National Supérieur de Musique de Paris

In 2005, Luis Naon and Yann Geslin signed their new contracts with the National Conservatory of Music in Paris to teach electroacoustic music, appointed as associate professors. Training in electroacoustics remains part of the general composition curriculum, but primarily from the perspective of exposure to new technology. The purely musical aspects of the electroacoustic approach have not yet found legitimacy. Yann Geslin remains to this day the only close link between the GRM and the Paris Conservatory, because of his simultaneously holding part-time employment in each institution.

Since 1986, there has been no national examination for the Certificate of Aptitude (CA) for teaching electroacoustics, which would make it possible to standardize the status of teachers across France. Even worse, the process of obtaining diplomas in electroacoustics remains frozen. In 2024, there was still no State

Diploma (*Diplôme d'État*) in electroacoustics, yet a DE is required to apply for the CA. Adding to the confusion, there is a labyrinth of different organizations and authorities providing diverse training leading to a diploma.

The Association of Electroacoustic Professors and Composers

In 2002, the Association des Professeurs d'Electroacoustique, AECME, was created at the initiative of Christian Eloy, Fernand Vandenbogaerde, and Henry Fourès. The following year in Lyon, together with Bertrand Dubedout and Gino Favotti, the association organized Journées Électro, the "Electro Days," four concerts held October 9–10, 2003, to promote the nonprofit and to widely attract more teachers to join together in the aim of influencing decision-makers in their strategic choices in cultural policy. The members of the electroacoustic community feel more and more threatened by the mercantile objectives displayed by different actors in the cultural sector. They are seeking recognition, in particular through the establishment of a state diploma in electroacoustics in the Centres de Formation des Musiciens Intervenants (CFMIs, Centers for the Training of Intervening Musicians).

The suggested training curriculum would span over three to five years, leading to the *Diplôme d'Enseignement Musical* (DEM), preceded by the CFEM diploma, which could be awarded at the end of the second or third year. The program would be available in both national and regional conservatories as well as in municipal music schools and at some nonprofit organizations and universities. It should also be noted that these courses are more and more often titled Computer-Assisted Music (MAO—*Musique assistée par ordinateur*), terminology that is sometimes misappropriated to refer merely to technical training in the service of instrumental music, far from compositional concerns centered on sound matter.

In 2007 there were twenty-eight electroacoustic classes in France conservatories, and thirty-three in 2024. In parallel, the historical centers (Bourges, Marseille, Lyon, Albi, Reims, etc.) continue to organize training courses, while many universities also offer an introduction to electroacoustics in their musicology sections, including the University of Rennes, Saint-Étienne, Rouen, and Paris-Sorbonne.[18]

François Bayle, president of the symphonic music commission at SACEM since 2001, organized a day of reflection on Friday, February 13, 2004, among composers to evaluate the situation of the four hundred professional composers registered, and to reflect on the future of the profession. With globalization and widespread economic challenges, composers are finding it increasingly difficult to earn a living, even though in order to survive almost all of them have already been forced to take on second activities such as conservatory directors, teachers, and radio producers.

Faced with professional difficulties, musicians come together. In 2005, the first public event of the association *Futur composé* was held over two days at the Opéra-

Comique in Paris.* These concerts-encounters brought together artists from the Paris region in a joint effort to win over audiences and to denounce the neglect by public authorities.

The National Centers for Musical Creation

In 2006, in addition to the IMEB in Bourges,[19] the Ministry of Culture awarded the title of Centre National de Création to Grame in Lyon, GMEM in Marseille, CIRM in Nice, GMEA in Albi, CÉSARÉ in Reims, and La Muse en Circuit in Alfortville. These centers have received financial support from the French government to carry out their musical creation and research activities. And when this book went to press there were two new others, one in the town of Saint-Nazaire and one in Corsica.

EVENTS, PUBLICATIONS, PERFORMANCES, PEOPLE

The new institutional landscape of the 2000s created new production imperatives at the GRM. Research and development had to adapt to these new constraints. Since 2002, four major axes have emerged:

— The specifics of concert seasons have guided policies concerning hosting composers and the types of equipment necessary for composition and dissemination, particularly in correlation with Radio France's "Présences électronique" festival since 2005. "The GRM is seeking to open up electroacoustic and acousmatic thought to creators working with technology interested in exploring new domains."[20]
— Multimedia production has led to developing software and the databases.
— The activity of preserving heritage has opened a new space for exchange and research.
— Development of signal processing tools, essential to creation, has continued its course. But development has also begun focusing on the problems of pattern recognition in order to develop tools to assist in visualization of music, such as the Acousmographe. Other work has been directed towards innovative publishing tools.

Every year the GRM hosts composers in its 116A, B, and C Studios. The works commissioned from these guest composers are then premiered most often during the following season's concerts. An increasing number of young composers are

---

* Translator's note: A play on words between the French verb tense *futur composé* and musical composition for the future.

also coming forward to present finished works composed in their home studios, in the hope of their work being heard in GRM concerts or on radio programing.

In 1998–99 the concert season was renamed Multiphonies. This was the twenty-first season since the first concert of the Acousmatic Cycle initiated by Ivo Malec, who was in charge of the program. Since the beginning, the GRM has received regular support from Radio France's musical directorate. In February 2005, Présences, the annual festival of Radio France created in 1991, introduced a new concert formula, "Présences *électronique*," delegating four days to the GRM for electronic works. The festival's artistic director, René Bosc, appointed Christian Zanési artistic director for these concerts.

Radio France's confidence in the GRM to program these four days of completely electronic music originated from the success of a first concert organized by the GRM on October 11, 2003, under the title "GRM Expérience." Christian Zanési had taken up the challenge of performing in person before a live audience with two well-known musicians from the electronic scene: the Finn Mika Vaïnio of the group Pan Sonic and the Austrian Christian Fennesz.

A Selection of Works Created by the GRM

Note: passages in quotation marks are taken directly from program notes authored by the composers themselves.

1998, *Elementa* (18 minutes) by Jean-Claude Risset. Voices, Irène Jarsky and Maria Tegzes. First performance on February 23, 1998. "The elements invoked here are those of Empedocles, earth, water, air, and fire. The material comes primarily from recordings of sound manifestations of these four elements. The sound sources are not masked: I have taken advantage of their connotations and symbolic implications. The piece also includes sounds synthesized using the Music V program, mimicking the *allures* of the four states of matter—'solid' sound objects, fluid textures, aeolian and noisy murmurs, and high-pitched, mobile and dissociated 'ionized' timbres." (J.-C. R.)

1998, *Extrémités lointaines* (17 minutes) by Hans Tutschku. The sound material for this work comes from recordings made in Asia.

1998, *Step Under* (16 minutes) by Benjamin Thigpen. Composed in the GRM studios on the M.A.R.S. station, and spatialized in eight channels by the "Octopode" program developed by the composer.

1998, *Saphir, sillons, silences* (18:05) by Christian Zanési. First performance March 16, 1998, at the Salle Olivier Messiaen. Citing Helmut Lachenmann, Christian Zanési declared that "composing is constructing your own instrument," and that for this piece, "the instrument was quite simple, an electrophone, a vinyl record, and my hand, delicately or brutally positioning the sapphire needle on the grooves of the record, not at random but with precision. With a little practice one quickly becomes skilled ... On the other end, the less obvious aspect: software collecting and manipulating events snatched from their context."

1998, *Vox alia* (15 minutes) by Annette Vande Gorne. "Is it because it is the signature, the sound imprint with which each living being identifies itself and by which it locates the Other, and detects its intentions? The voice remains the primary emotional and musical vector, the ancestral support of all direct communication. *Vox alia* is a series of short vocal studies which, through electroacoustic processing, retain the trace of the melodic, rhythmic, and timbral qualities of this marvelous medium, in service of a distinctive character, a particular affect." (A. V. G.)

1998, *Manuel de résurrection* (13 minutes) by François-Bernard Mâche, for two samplers and soprano. Soprano, Françoise Kubler. Samplers, Martine Joste and Fuminori Tanada. "*The formulas for coming forth by day* are more commonly called the *Egyptian Book of the Dead*. Its original title, freely translated, corresponds to this *Resurrection Manual*, commissioned by the INA-GRM and Radio France. Recently several composers have found inspiration in this text, but this is the first time the original text is being uttered. It is delivered both as song and as (synthesized) speech, in an interplay illustrating the archetype of the responsorial litany. The ancient Egyptians believed in the agency of the word, whether written, spoken, chanted or sung, and their *Manual* repeats its refrain, 'It has been effective millions of times'. They would no doubt appreciate this attempt to resurrect, if not their dead, at least the aural substance of their words, their own energy, their *ba*, as they said ... More broadly, I wanted to revive the archetype of the litany, of late fallen from grace in Christianity." (F.-B. M.)

1998, *Terra incognita* (op. 101) (26 minutes) by Denis Dufour. A 4-track acousmatic work for two Acousmoniums, dedicated to Pierre Schaeffer and realized at the Motus studio. Sound recording and preparation of the sound elements, Agnès Poisson. "From discovery to exploration, from exploration to conquest, and from conquest to subjugation, there are stages that Schaeffer at times refused to cross, despite being the pioneer of one of the most truly novel unknown lands in 20th century music ... This *terra incognita* is here transformed into jungle, progressively populated with creatures more improbable than those of the Island of Doctor 'More' (a peninsula, in fact), from an initial crackling sound, the good old crackling sound of a wax record, like the finale experienced in reverse in Gerard de Nerval's dream in Aurélia." (D. D.)

1998, *Si l'oiseau par hasard* ... (18 minutes) by Christine Groult. The flute was recorded in the studio by Karen Fenn and the violin by Élisabeth Robert.

"*Si l'oiseau par hasard* . . . explores the labyrinth and Icarus. This musical reverie is inspired by Gaston Bachelard's imagery, which I quote freely—imagining the substance of the labyrinth in opposition to that of lightness and ethereal freedom, inconsistent and mobile materials. This labyrinthine matter lives by losing itself in its own unfolding, it is a phenomenon of viscosity, it is the conscience of a painful dough extending while murmuring; it possesses the faint movement anticipating nausea, vertigo, malaise. The associated dream instinctively recognizes this slowness." (C. G.)

1998, *L'Isle sonante* (75 minutes) by Michel Chion. Melodrama in twelve days, libretto by Michel Chion, after works by several authors. "Following *Le Prisonnier du son, Tu, Diktat, La Tentation de saint-Antoine*, and *Nuit noire*, this tribute to musique concrète and to the Schaefferian adventure, *L'Isle sonante*, is my sixth melodrama. I wanted to renew the formula, by this time involving the character of a reader, named Axelle, in reference to Jules Verne, and interpreted by Florence Chion-Mourier." (M. C.)

1998, *Lisboa, tramway 28, hommage à Fernando Pessoa* (12:30) by Elzbieta Sikora. For saxophone and tape. Saxophone, Daniel Kientzy. "Inserting a musical character into an imaginary landscape, a memory of a moment in the life of another . . . This other was a poet. He took the 28 tram, surely. He wrote, as we know, in this café named 'Brasiliera' where the city seems to pause for a moment's happiness." (E. S.)

1999, *Une Tour de Babel* (58:18) by Pierre Henry. Premiered January 7, 1999, at the Olivier Messiaen Hall. "This work is dedicated to the microphone . . . In Genesis, the story of the Tower of Babel follows the flood. The descendants of Noah, traveling towards the East, arrive in the land of Shinar in Babylonia. They speak a common language and decide to build a city and a tower whose summit penetrates the heavens, in order to make a name for themselves and remain together. We know what happened next. God, seeing men capable of erecting such a tower, decides to scatter them and to sever their linguistic unity by endowing them with languages foreign to one another. The myth of the Tower of Babel has long fascinated me. Does the Tower of Babel allow men to reach heaven, or the gods to descend from it?" (P. H.)

1999, *Bleu passe la lune* (13:40) by Jean-Loup Graton. "A triptych beginning with Japan and its myths, evoked at the beginning of the piece by a brief intrusion into a Zen ceremony. A beginning that naturally takes its source in the origin of music: the voice. A little later will come the blues—blue or black depending on the case. For me the purest and most powerful blues are those expressed by Jimmy Hendrix: sound becomes supreme reality, far beyond what rational thought can conceive. Finally, the sea and its waves that will one day stop calling us to return to our origins. Aiding in traversing these spaces without great logic is a moon—attendant to our dreams, mirror of the sounds of the chaotic world." (J.-L. G.)

- 1999, *Soundings of Angels* (15 minutes) by John Levack Drever. This piece, composed at Dartington College of Arts and at the GRM, presents and explores different readings of Alice Oswald's *Angels*, a poem that was inspired in part by Cecil Collins's painting *Angels*. The poem is read by its author as well as by others.

- 1999, *Puzzle* (20 minutes) by Michel Pascal. Created in the CIRM studios. "With *Puzzle* I have embarked on a particular path, at the boundaries between two worlds: acousmatic composition and instrumental music based on movement and the note—music for home listening as well as for public diffusion on a loud speaker system, open form, diverse aesthetics." (M. P.)

- 1999, *Le Bestiaire alchimique* (20 minutes) by Marc Favre. Sixth grimoire of *L'Illusion acoustique* in homage to François Bayle's *L'Expérience acoustique* and Francis Ponge's *Une fugue de paroles*. "This new grimoire extends the experiment into extreme regions of materials and forms, bat and insect ultrasounds transposed to lower registers . . . Behind this research, an omnipresent concern, orchestration over eight channels." (M. F.)

- 2000, *Holophonie ou la baleine rouge* (18 minutes) by Patrick Ascione. Work on the concept of space. "Unlike my previous multi-track productions, this piece is an attempt at anaglyptic spatial writing where the focus is not on different spatial planes or the legibility of the various voices, but rather on exploring the concept of sound relief. 'Holophony' as an analogy with the visual 'hologram.'" (P. A.)

- 2000, *American Triptych* (17 minutes) by Trevor Wishart. "The twentieth century was in thrall with the American dream—freedom, technological progress, and the pursuit of pleasure—represented here by the voices of Martin Luther King, Neil Armstrong, and Elvis Presley. The fall of the Berlin Wall seemed to proclaim the final triumph of this dream and the 'end of history.' Will these symbols have been ephemeral or will they persist as we enter the twenty-first century? Like other recent works, *American Triptych* explores and transforms the voices of famous people and seeks new directions for electroacoustic music." (T. W.)

- 2000, *Le contrat* (40 minutes) by Gilles Gobeil. New version with René Lussier, guitar and daxophone. "First of all, I must tell you that we are linked by friendship; 25 years ago, we were feverishly collaborating on the creation of our first 'works.' Although we have since gone our separate ways, we have called on each other regularly and our desire to make music together has remained intact. Our plan for this piece? Composing a piece inspired by Goethe's *Faust*, constructing it by respecting the very structure of the poem . . . A tough bill to fill!" (G. G).

2000, *Erradie* (40 minutes) by Kasper T. Toeplitz, for percussions, basses, and electronics. Kasper Toeplitz, basses and machines; Didier Casamitjana, percussions; Jérôme Soudan, percussions and machines.

2000, *Finnegans Tune* (14 minutes) by Jean-Yves Bosseur, for Uilleann pipes and fixed sounds. Benoît Trémollières, Uilleann pipes (Irish bagpipes). "Cross-referenced allusions to Irish musical tradition . . . , to James Joyce . . . , and to John Cage." (J.-Y. B.)

2000, *64* (64 minutes) by Andrea Liberovici. Oratorio for "memory" and actors. Sixty-four tableaux in sixty-four minutes. Sixty-four sequences, by way of the acoustic memory (the archives) of the Living Theatre of New York, notably featuring unreleased recordings of John Cage's voice, in constant dialogue with the memory of the present, embodied by the singing voice of Ottavia Fusco. With the actors Judith Malina and Hanon Reznikov.

2000, *Les joueurs de sons* (72 minutes) by Denis Dufour and Agnès Poisson. Acousmatic work in 8 tracks (primary stereo + 6 tracks) realized in the studio Préludes and in Denis Dufour's studio. Voice, Roberto Paris. Sound projection, Jonathan Prager. "*Les Joueurs de sons* tells the story of musique concrète through listening and observation of the evolution of sounds and techniques. . . . This work can be perceived as pure music as well as a teaching piece." (D. D.)

2000, *Chanson de la plus haute tour* (72 minutes) by Denis Dufour. Acousmatic work in 4 tracks, for two Acousmoniums. Texts by Thomas Brando, read by Pierre Henry, and recorded by Bernadette Mangin at the Son-Ré studio. "The title, homonym of a Rimbaud poem, expresses the visionary dimension and the necessary distance the speaker is obliged to take in order to apprehend a universe that no one had guessed anything about before him."

2001, *Archives Génétiquement Modifiées* (25 minutes) by Luc Ferrari, *Exploitation des Concepts 3*, solo for memorized sounds. Realization at Atelier Post Billig. "This composition is made with the same musical elements as *Exploitation des Concepts 1* (entitled *Archives sauvées des eaux*) for two DJs, DJ Scanner and DJ Olive, one employing CDs and the other vinyl. . . . Listening again to my work material from the 70's made me want to do something new."[21] (L. F.)

2001, *L'Étoile Absinthe* (20 minutes) by Michèle Bokanowski. "It was in a text by Ingmar Bergman about the *Seventh Seal* and the Apocalypse that I first heard of the Absinthe Star."[22] (M. B.)

2001, *Diabolus Urbanus* (12:58) by Luis Naon. For viola, Ana-Bela Chaves, and fixed sounds. Commissioned by the French government. "Fifteenth piece in the Urbana cycle and first of a quartet of solo pieces, . . . conceived as much for instruments as for instrumentalists." (L. N.)

2001, *Ciudad, chronique d'un voyage inachevé* (12 minutes) by Diego Losa. For Celtic harp and fixed sounds. Denise Mégevand, Celtic harp. "Lights, shadows, and chiaroscuro cross your future sky and are already no longer there. I crossed the city as one crosses a dream, not knowing very well any more if sleeping or awake." (D. L.)

2001, *Shin-Woon* (Le mouvement mystique) (17 minutes) by Chengbi An, for clarinets and fixed sounds; Pierre Dutrieu, clarinets. "In Korean, Shin-Woon describes the particular movement that the artistic object—painting, poem, short story, or music—creates by resonance in the human being. Here the fixed sounds express this resonance with the playing of the clarinet." (C. A.)

2002, *La forme du temps est un cercle* (59:20) by François Bayle. (1) *Concrescence*; (2) *Si loin, si proche*...; (3) *Tempi*; (4) *Allures*; (5) *Cercles*. Commissioned by the French government for the composer's seventieth birthday. The form of time is a circle. "In five stages, the listening experience will have completed a trajectory, temporal unity moving towards its finest grain, and sensory perception gradually sharpening in its appreciation of images and forms. The fleetingness of colors and the velocity of shapes will resolve itself into a spiral (this three-dimensional form of the circle), where the initial sound-image (the bells) will be prolonged into infinity in the final sound-image: that of summer crickets in the night of time, dream-filled and suspended."(F. B.)

2002, *L'ombre de la méduse* (20 minutes) by Michel Redolfi. Three movements: *Physalies, Montée de l'ombre, Cendres*. Work for glass harmonica, Ondes Martenot, and electroacoustics. Thomas Bloch, glass harmonica and Ondes Martenot; Fabrice di Falco, soprano, recorded in electroacoustic spatialization. "The title of this work was originally *Les Physalies voguaient au-dessus de nos têtes*,* Physalia designating the largest known jellyfish, whose toxic tentacles can measure up to thirty meters. . . . September 11th, 2001 brutally interrupted the work in progress. The terrifying images of New York battered instantly took on a mythical dimension, in my eyes a Dantean oceanic tableau." (M. R.)

2002, *Particules* (60 minutes) by Arnaud Rebotini. With Marc-Olivier de Nattes, violin; Jean-Paul Minali Bella, viola; Jean-Philippe Audin, cello; Pierre Roullier, flute; Thomas Savy, clarinets; Eric Faÿ, horn; Vincent Artaud, piano, electric bass; Arnaud Rebotini, electronics. Commissioned by Radio-France and INA-GRM, premiered March 23, 2002, at the Grand Auditorium of the Maison de Radio France. Arnaud Rebotini, a musician from the rock, pop, electronic, and techno scene, introduced *Particules*, "In a place where people are used to coming to listen to learned music, I try to bring elements that

---

* Translator's note: *Physalia*, also known as the Portuguese man-of-war. The original title, *The Portuguese men-of-war sailed above our heads*, would finally become *The shadow of the jellyfish*.

correspond with my career—the techno aspect and working with fundamental rhythms. It's not a matter of a discourse about rhythm as found in learned music, for example Stravinsky or certain works by Boulez, but of working with repetition and its hypnotic aspect. I am touched by American music, but its consonant harmonic approach does not satisfy me anymore, so I orient my explorations towards timbre and timbral melody. I build short rhythmic or sound cells, and by repetitions and variations, I play with them. There is also room for instrumental improvisation, spaces of freedom within a precise framework. Improvisation interests me when it is directed and controlled. With this work I want to address both an audience of specialists and an audience that reads *Inrocks*, in other words, people who are familiar with pop and electronic music." (A. R.)

2002, *Con fu sion* (30 minutes) by Michel Waisvisz. Najib Cherradi, voice. Premiered May 2, 2002, at the Grand Auditorium of the Maison de Radio France. At the crossroads of alternative musical genres that he himself contributed to inventing—Electronic Urban Sound, electroacoustic music, Rhythm & Zoom or Nara Vox—Michel Waisvisz composes music that appeals to all the senses and generates a hypnotic dimension. For his performances, he uses Mains (a prosthetic instrument with sensory sensors) that he personally developed at the STEIM Foundation in Amsterdam.

2002, *Jukurrpa—quatre rêves* (15 minutes) by Pierre Couprie. First performance June 15, 2002, at the Grand Auditorium of the Maison de Radio France. "Jukurpa, used by tribes of the central and western desert of Australia, means the Dreaming. But it should not be confused with the night dream, the little dream." (P. C.)

2003, *Labyrinthe!* (60 minutes) by Pierre Henry. First performance March 29, 2003, at the Grand Auditorium of the Maison de Radio France. Work realized from sound sequences provided by the GRM composers at the time, Philippe Dao, Évelyne Gayou, Yann Geslin, Diego Losa, Daniel Teruggi, and Christian Zanési. It was the first time that Pierre Henry did not use any of "his" own sounds, but only those of others.

2003, *GRM expérience*, with Christian Fennesz, Mika Vaïnio, and Christian Zanési. First performance October 11, 2003, at the Grand Auditorium of the Maison de Radio France. After a summer residency in the GRM studios, three artists from different backgrounds composed a collective work. This concert was the first in an international tour (Netherlands, Great Britain, Italy) as part of the European Union's Culture 2000 program.

2003, *Con brio* (12 minutes) by Francesco Giomi. First performance April 14, 2003, at the Grand Auditorium of the Maison de Radio France. This composition is inspired by different themes linked to the field of instrumental music: strings, rhythmic gestures, melodies. However, the musical "gesture" appears in an artificial manner through movement in space. The author teaches composition at the conservatory of Parma, Italy, and directs the production department of the Tempo Reale institute, founded by Luciano Berio in Florence.

2003, *Bobok* (12 minutes) by Florence Baschet. Commissioned by the French government and the 2e2m ensemble. Premiere April 15, 2003, at the Grand Auditorium of the Maison de Radio France. Pierre Roullier, flute; Jean-Marc Liet, oboe; Véronique Fèvre, clarinet; Laurent Bômont and Eric Gouillard, trumpets; Philippe Defurne, trombone; Eric Karcher, horn; Alain Huteau and Vincent Limouzin, percussions; Eric Crambes, violin; Laurent Camatte, viola; Frédéric Baldassaré, cello; Tanguy Menez, double bass and electronics. "'Bobok, Bobok, Bobok,' heard Dostoyevsky, babbled from the deepest bowels of the earth into his ear like a magnificent stammer that would have allowed the depths to rise to the surface and flow off landsliding in every direction." (F. B.)

2004, *Chronomorphoses* (23:50) by Gilles Racot. First performance on January 10, 2004, at the Grand Auditorium of the Maison de Radio France. "The object of these *Chronomorphoses*, composed in several movements, concerns in particular the effects of perception of fluid time, striated time, pulsed time, reiteration, rhythms and cycles . . . all principles of the articulation of time." (G. R.)

2004, *Radiorama*. On the occasion of "European Heritage Days," the GRM and the INA sound library presented a concert featuring their radio+electronic archives. Xavier Garcia, sampler. Alexandre Meyer, electric guitar. Lucio Recio, voice.

2004, *Zaoum* (11:23) by Évelyne Gayou. First performance May 29, 2004, at the Grand Auditorium of the Maison de Radio France. A work in homage to the Russian futurists who invented Zaoum, an imaginary language, a visible expression of their anticonformism, but also of their curiosity about origins. "Throughout the eleven minutes and twenty-three seconds of *Zaoum*, there is absolutely no silence. I wanted this continuum, symbolizing the apparent chaos of the world around us." (E. G.)

2004, *Uninterrupted Love* (18 minutes) by Kazuko Narita. First performance May 29, 2004, at the Grand Auditorium of the Maison de Radio France. "Divided into 5 sequences of 2 to 4 minutes, this piece uses only processed natural sounds and no electronic or synthesized sounds." (K. N.)

2004, *Sweet Moving* (14:20) by Jacques Stibler. First performance May 29, 2004, at the Grand Auditorium of the Maison de Radio France. "*Sweet Moving* takes on the form of a large, thick arch establishing itself, developing, and fading away, returning to reshape itself in a mirror . . . The final section, like the tail of a comet, becomes a bewitching and fluid dance, almost liquid." (J. S.)

2004, *Cronicas del tiempo* (15 minutes) by Diego Losa. First performance May 29, 2004, at the Grand Auditorium of the Maison de Radio France. "After *Ciudad*, my first commission from the GRM and an evocation of an unachieved journey, this piece is the acoustic transcription of memories in the journey of a traveler—past, present and future. The music conveys mental images of an individual as an overlay to the collective memory of a nation. The piece, composed and played in multichannel (5.1), is a work on the unconscious and a reflection on sensations, using original sounds of the city of Buenos Aires recorded at different times (1980–1996) and then transformed with GRM Tools." (D. L.)

2005, *Cercles avec "L,"* with video (19:30) by Christophe Ruetsch. First performance January 15, 2005, at the Grand Auditorium of the Maison de Radio France.

2005, *Artéfact* (12 minutes) by Pascale Criton. First performance January 16, 2005, at the Grand Auditorium of the Maison de Radio France, with Véronique Fèvre, clarinet; Patrice Petitdidier, horn; Alain Huteau, percussion; Didier Aschour, Caroline Delume, and Wim Hoogewerf, guitars; Alexis Galpérine, violin; Claire Merlet, viola; David Simpson, cello; Pierre Feyler, double bass. "In the image of a mobile world, composed of speed and distance, *Artifact* echoes the intertwining of divergent, coexisting dynamisms . . . The instrumental ensemble is organized in trios around three slightly amplified guitars, tuned in 1/12th of a tone. The tuning obtained by scordatura does not require any special fretwork. Each guitar covers a specific range, pitches extending completely from low to high, creating a continuum in 1/12th of a tone. This tuning consisting of 72 tones per octave, executed most often in evolving playing modes, allows for differences in time, pitch and timbre, an elasticity inherent to this suspended space that I have tried to render audible." (P. C.)

2005, *From Ivry* (15 minutes) by Andrea Liberovici. First performance March 11, 2005, at the Grand Auditorium of the Maison de Radio France. With the voice and violin sounds of Ivry Gitlis. "Those who know a little bit about my work, know that for the last ten years (and perhaps always) I have not composed 'pure' music (also, because I don't believe it exists), but rather music applied to languages such as poetry and theatrical settings . . . I composed *From Ivry* simultaneously, and directly, for music and images." (A. L.)

"PRÉSENCES ÉLECTRONIQUE" FESTIVAL

In February 2005, for the first time the GRM organized a special *electronic* program within Radio France's Présences festival. These four days of programming entrusted to Christian Zanési were an opportunity for the GRM to go further in reaching musicians from the international electronic scene as well as young audiences.

Invited in 2005 were Mathew Adkins (UK), Dimitri Coppe (Belgium), Arnaud Rebotini (France), Taylor Deupree / Kim Cascone (USA), Mauricio Martinucci (Italy), Christian Fennesz (Austria), Richard Chartier (USA), Sidsel Endresen / Christian Wallumrod / Helge Sten (Norway), Laurent Chanez (vidéo, France), Main (UK), Sogar / Sébastien Roux / Thomas Einfeldt (vidéo) (Germany/France/Germany), and Fabriquedecouleurs (France).

- 2005, *Nature contre nature* (13:30) by Jean-Claude Risset. Four rhythmic exercises for percussion and computer, 1996–2005. Thierry Miroglio, percussion. First performance March 12, 2005, at the Grand Auditorium of the Maison de Radio France. In this piece dedicated to Thierry Miroglio, the computer incites the percussionist to implement unnatural rhythmic protocols. These paradoxical rhythmic forms, studied by the author and also by Knowlton, Bregman, and Waren, seem to contradict the sound structure: the inner counting is not reduced to chronometric time. Such "rhythmic illusions" in fact reveal the nature of auditory perception, which sometimes contradicts the physical nature of the sound: as in the piece's title, nature against nature.
- 2005, *Voices*, for voice and electronics, by John Chowning. Maureen Chowning, coloratura soprano. First performance on March 12, 2005, at the Grand Auditorium of the Maison de Radio France. The title *Voices* alluded to the ritual of the ancient oracle, rendered by Sibyl in a sacred chamber where the walls reverberated with the words of her predictions. The technical aspect of the piece was inspired by this ritual. Precise pitches extracted from the vocal line were followed by a program written by the composer in MaxMSP. At each pitch recognized by Miller Puckette's pitch-tracking program "fiddle," accompanying sounds were generated, synthesized by a distinctive form of frequency modulation. The spectra of the synthesized sound, mostly inharmonic, were "composed" to act in the spheres of pitch, harmony, and timbre.
- 2005, *La terre ne se meut pas* (18:05) by Vincent Laubeuf. First performance May 14, 2005, at the Grand Auditorium of the Maison de Radio France. "To the old Galilean adage about the earth's movement, Vincent Laubeuf opposes, as a mischievous contradiction, *La terre ne se meut pas* (The earth does not move), echoing Husserl, while composing movements, stases, paths, and above all—as the author admits—seeking to set off our subjective perception and its trajectories." (Commentary by Sylvain Marquis.)
- 2005, *Gexin / Heart of a Pigeon* (13:25) by Marie-Hélène Bernard. First performance May 14, 2005, at the Grand Auditorium of the Maison de Radio

France. Piece composed for playback in 5+1. "While walking through the older neighborhoods of Beijing, one's attention is often drawn to strange sound trails traced across the sky in slow concentric circles, spreading like a faintly sorrowful veil over the rooftops ... The city indeed harbors a certain number of enthusiasts lovingly raising squadrons of pigeons, releasing them day after day, a tiny whistle attached to their tails, for the simple pleasure of watching and hearing them whirl." (M.-H. B.)

2005, *Marée noire* (40 minutes) by Samuel Sighicelli. Premiere October 8, 2005, at the Grand Auditorium of the Maison de Radio France. Musical film based on television archives related to the world of petroleum—derricks, tankers, pipelines, oil spills—set to a text by Tanguy Viel, with David Sighicelli, narrator.

2005, *L'Isle sonante* (2 hours 30 minutes) by Michel Chion. First performance of the definitive version December 4, 2005, at the Grand Auditorium of the Maison de Radio France. In homage to musique concrète and to Pierre Schaeffer. Dedicated to Jacques Lonchampt. Sound creation, sound shooting, editing, and mixing: Michel Chion; assistance for the editing and digital mixing: Geoffroy Montel. "*L'Isle sonante* [the resonant island], as it was known in the time of Rabelais, is a quest where the journey counts as much as the goal. It is the imaginary journey of Axelle, a passionate reader, who lives through the characters in the novels she immerses herself in." (M. C.)

2006, *Spatio intermisso (temporis)* (9 minutes) by Jacopo Baboni-Schilinghi for oboe and electronics. First performance January 7, 2006, at the Grand Auditorium of the Maison de Radio France. Christian Schmitt, oboe. Work composed in 5+1. "How long can a musical note last today? And a sound be sustained? How much time can we dedicate to listening to so-called 'new' music resulting from artistic creation? Speed and rapidity: paradigms for perception of this work.—And if we aimed at the opposite: slowness ... ?" (J. B.-S.)

2006, *Peintures Noires (le geste et la raison)* (35:10) by Jean-Marc Duchenne. First performance January 8, 2006, at the Grand Auditorium of the Maison de Radio France. Piece composed in hexadecaphony. "*Peintures Noires* is the fourth part of a series of works that began twenty years ago with *La cicatrice du geste* (1984, two final tracks), *Gaïa, Hélia, Selia* (1986, four final tracks for performance) and the *Quatre Études d'espace* (1988, octophony). They all shared the common feature of being inspired by the pictorial universe of painters (Georges Mathieu, Bernard Pomey, Yves Tanguy), and of finding inspiration in the painters' spaces of representation in order to shape and organize the sound objects in the projection space. *Peintures Noires* continues on this same path, without referring to any particular artist this time ... It is a painting in the process of being made, searching for itself, constructing and transforming itself, in a sort of reconstitution of the to-and-fro between gesture and listening, impulse and reflection." (J.-M. D.)

2006, *Dans un point infini* (20 minutes) by Beatriz Ferreyra. First performance January 8, 2006, at the Grand Auditorium of the Maison de Radio France. "Piece based on technical exercises performed by the violinist Veronica Kadlubkiewicz, to whom I dedicate this work." (B. F.)

2006, March 9–12, thirteen premieres at the Grand Auditorium of the Maison de Radio France, as part of the second annual "Présences *électronique*" festival, coproduced by the GRM (INA) and Radio France. The new works were by Pierre Henry (*Variance*, dedicated to Luc Ferrari), Ryoji Ikeda (Japan), Michel Waisvisz / Jan St. Werner (Netherlands, *in concert with very old and new electronic music instruments*), Carl Stone (US, *Attari*), Markus Popp alias OVAL (Germany), Kasper T. Toeplitz (*Lärmsmitte*, for solo "basscomputer"), Robert Hampson alias MAIN (UK, *Umbra*), Antye Greie alias AGF (Germany), Yoshihiro Hanno alias Radiq (Japan, *Fragments, Parallel timeless for it to tie*), Pan Sonic (Finland), Amon Tobin (Canada), Marc Chalosse (*Paris, New York, Tokyo, Berck-Plage*), and Jon Hassel / Lightwave / Michel Redolfi (*Shift*) (Christian Wittman & Christoph Harbonnier, synthesis. Arnaud Mercier, live processing).

2006, *Hommage à Parmerud (The Fabulous Mr. P.)* (15:24) by Natasha Barrett. First performance March 25, 2006, at the Grand Auditorium of the Maison de Radio France. "At a certain point, Åke Parmerud and I decided to embark on a project where we would leave aside basic materials and compose from our respective inspirations. To guide us, we came up with a list of rules, a sort of *10 Commandments* that we would have to follow. One of these rules was that we had to use 'new' material. Åke didn't follow this first rule at all—for me the numerous materials he gave me sounded like all the acousmatic music he had composed since the early 1990s. So I too decided to break the rules and used these same materials to plagiarize Åke's earlier work." (N. B.)

2006, *Mahoroba*, by Tomonari Higaki. First performance March 25, 2006, at the Grand Auditorium of the Maison de Radio France. A tribute to Luc Ferrari. "Mahoroba is an ancient Japanese word which means *illusion* or *utopia* and which, by derivation, designates a splendid place . . . I wanted to create a sort of 'sound garden,' both acousmatic and imaginary. This work is divided into 11 small pieces, 11 'mahoroba.' The order of the pieces is programmed in advance, while they are played according to a random mode of reproduction. The listener promenades through a sound garden in perpetual motion." (T. H.)

2006, *Glasharfe*, by Ludger Brümmer. First performance March 25, 2006, at the Grand Auditorium of the Maison de Radio France. "*Glasharfe* is a German word for a glass musical instrument, a glass harp, consisting of a series of interlocking glasses. The glasses are tuned according to a chromatic scale in order to be able to play melodies, either by rubbing their edges with the finger in a circular motion, or by hitting them with a mallet . . . Since I create sound structures using algorithmic techniques and mix them acousmatically, sounds are transposed several times from low to high during different stages of composition . . . In order not to lose too many high frequencies, I chose a sampling rate of 192kHz. Knowing that 15 minutes of recording 4-channel 24-bit at 192kHz requires 2 Gigabytes of memory, I recalled the days back in 1991 at Stanford when we only had 300 Megabyte disks to mix an entire work! This piece took 25 G-bytes!" (L. B.)

2006, *Le souffle court* (14 minutes) by Laurence Bouckaert. First performance April 23, 2006, at the Grand Auditorium of the Maison de Radio France. Composed for 5+1. The structure of the piece is based on the inexhaustible richness of the vocal material: a multitude of breaths intermingle, responding to each other and dancing with the sounds. In this journey, listeners are invited to "open the doors of their own listening, giving free rein to their imagination." (L. B.)

2006, *Le livre des foins* (20 minutes) by Philippe Mion. First performance May 20, 2006, at the Grand Auditorium of the Maison de Radio France. "For me, it is a question of rethinking a seminal experience in acousmatic music. What neophyte has not found themselves ecstatic—close up to the microphone, with headphones on—crumpling a piece of paper, a soft plastic bag, a rough cloth, a sheet of aluminum foil, a few dry twigs . . . For a long time now, my acousmatic compositions (but also to a certain extent my instrumental ones) have been more or less explicitly haunted by a kind of reference to this experience. It is the paradigm of myriads, abundances, and accumulations of micro-phenomena." (P. M.)

2006, *Le mécanisme des ombres* (16:30) by Bérangère Maximin. First performance May 21, 2006, at the Grand Auditorium of the Maison de Radio France. "In reserve, fresh fruit, dried fruit, tobacco, and water. Blue jeans baggy enough for comfort. I'd say composing begins like a long walk. Just like that I settled down once again in my studio. Folk shadow theater is what I wanted to explore . . . More than the *mechanism* of a system, in the end it is machinery I have set in motion, composed of malleable, fluid characters, bereft of any veritable scenery." (B. M.)

2006, *Schenk mir dein Ohr* (14:47) by Igor Lintz Maués. First performance May 21, 2006, at the Grand Auditorium of the Maison de Radio France.

Acousmatic study of sound morphing, for 8 channels, using sounds from seven Afro-Brazilian instruments. The title in German means "Offer me your ears" (in reference to Van Gogh's words), meaning "listen to me."

- 2006, *Désordre*, by Philippe le Goff, for voice, electric guitars, and electronics. First performance October 8, 2006, at the Grand Auditorium of the Maison de Radio France. Sylvie Deguy, mezzo-soprano; Louis Chrétiennot, electric guitars. "*Désordre* focuses on the speaking human, on the incessant flow of words uttered or suffered, sometimes to the point of being unbearable, even absurd, whether they come from our daily lives or from the incessant reservoir of the media." (P. L. G.)
- 2006, *Trois modes d'air et de lamentations* (about 16 minutes) by Françoise Barrière, for accordion and electroacoustics. Work dedicated to the victims of all wars. First performance November 12, 2006, at the Grand Auditorium of the Maison de Radio France. "I would like to thank Claudio Jacomucci for our work on the instrumental part, and Gerald Bennett and the ICST in Zurich where I did the sound processing. The final mixes were done at IMEB in April–May 2004 and in November 2005." (F. B.)
- 2006, *Rotations II* (16:16) by Roland Cahen. First performance December 10, 2006, at the Grand Auditorium of the Maison de Radio France. "*Rotations II* is—without the ambition of making a genre out of it—*kinetic electroacoustic music*. Kinetic sounds have spatial qualities, which means that if they are recorded and replayed in mono or stereo, we lose what makes them remarkable. For example, the sound of a waterfall or wind in the trees. In this piece, the distributional patterns in space are essentially rhythmic. It is a question of playing on shifting the sound samples, localized point to point, over each of the eight voices of the device, without room effects." (R. C.)
- 2007, *Lacs* (19:45) by Régis Renouard-Larivière. First performance January 13, 2007, at the Grand Auditorium of the Maison de Radio France. "I sometimes regret giving titles to my music . . . Lakes most often allow us to see the earth at their horizon. Crossing them seems easy. But their tranquility is sometimes deceiving." (R. R.-L.)
- 2007, *Musica mobile* (37:09) by Pierre-Alain Jaffrenou. Concertante and disconcerting musical environment, in three movements, for eight loudspeakers, with Cécile Daroux on flute. First performance January 14, 2007, at the Grand Auditorium of the Maison de Radio France. *Musica mobile* is a musical concept conceived from computer technologies developed for more than ten years by the research team of Grame, National Center for Musical Creation, and by the composer, implemented with an electroacoustic system using eight loudspeakers surrounding the audience. In movements 2 and 3, the author has drawn upon sounds from the world of techno.

2007, March 15–18, the third edition of the "Présences *électronique*" festival took the form of eight concerts, with seventeen premieres given at the Grand Auditorium of the Maison de Radio France, still in a spirit of bringing together varied "electro" genres, with works by composers of all nationalities and generations. The festival was a coproduction between the GRM (INA) and Radio France.

New works by Mimétic (Jérôme Soudan), *Virtual (Around me?)*; Jürgen Heckel (alias Sogar, Germany), *Mai motive*; Matmos (M. C. Schmitt and Drew Daniel, US); Hector Zazou / Bill Rieflin, Katie Jane Garside, Nils Petter Molvaer, and Lone Kent (Norway), *Corps électriques*; Atau Kanaka and Zack Settel (Japan/US), *Tibet* (duo); eRikm: *Générer—Dégénérer*; Alva Noto (Carsten Nicolai, Germany), *Xerox*; Mouse on Mars (Andi Toma and Jan St. Werner, Germany), *Interprétation—Improvisations*; Zavoloka (Ukraine), *New album*; The Books (Nick Zammuto and Paul de Jong, US/France), *Live in Paris*; Scanner (Robin Rimbaud, UK); Emilie Simon (France), *Aqua*; Toshimaru Nakamura (Japan); Jérôme Noetinger / Lionel Marchetti, *Echange cannibale*; Carl Stone, Yumiko Tanaka, and Min Xiao-Fen (US/Japan/China), *Lauburu*; Blixa Bargeld (Germany).

2007 *Kitchen-Solo* (20 minutes) by Jean-François Vrod. Jean-François Vrod, mock violin, electronics, and voice. First performance May 12, 2007, at the Grand Auditorium of the Maison de Radio France. "At the end of the 1970s, while searching for the last traditional musicians of the French Massif Central, I met Urbain Trincal, a peasant musician from the canton of Saugues in the Haute-Loire. I learned that he sometimes played the violin accompanied by his idling motorbike, hanging from the central beam of his house. Making me definitively convinced that traditional music is not just a collection of pretty melodies! If I am a musician today, it is thanks to Urbain and all his friends . . . *Kitchen-Solo* is the story of a guy 'spilling the beans.'" (J.-F. V.)

2007, *Oh là la radio* (9 minutes) by Leigh Landy. Work for 8 tracks. First performance May 12, 2007, at the Grand Auditorium of the Maison de Radio France. This work, inspired by Bernard Heidsieck's sound poem *La Semaine*, focuses, like most of my recent works, on recycling existing sounds—appropriation, 'plundering,' sampling, etc. *Oh là la radio* uses recordings made over a few days in 2006 of various French radio broadcasts. (There is only one exception: the sound at the beginning of the piece. You'll know why) . . . As far as copyrights are concerned . . . don't ask me." (L. L.)

2007, *La main vide (la fleur future, Inventions)* (10:05) by François Bayle. Octophonic version / live video, with Serge de Laubier, visual projection. "In these Inventions, five discoveries are in circulation, five gestures of the innocent hand tossing or tracing a few lines of force: inventions of volumes, signals, signal-intervals, inclusive forms, volume-intervals . . . The novelty here lies in the octophonic conception of the movement of the figures, organized according to four planes of space, as if flung to the 'four winds.'" (F. B.)

The 2007–8 season celebrated the fiftieth anniversary of the GRM, and the sixtieth anniversary of musique concrète. Pierre Henry, guest of honor at the first concert on December 9, 2007, the very day of his eightieth birthday, gave the premiere of his latest work, titled *Trajectoire*.

2008, the January 13 concert was a tribute to Jean-François Allouis, polytechnician, technical director of the GRM from 1975 to 1987, designer and inventor of Syter and much of the Studio 123 software. He had died on March 6, 2007. The works performed had all been composed with Syter: *Cummings* by Jean-François Allouis, *Diffluences* by Gilles Racot, *Seasons* by Jean Schwarz, and *E cosi via* by Daniel Teruggi.

2008, March 27, 28, 29, and 30, Festival "Présences *électronique*" in coproduction with Radio France. Four concerts in the Olivier Messiaen Hall. Works by Pierre Schaeffer, Pascal Baltazar, Christophe Ruetsch + Phonophani, Aki Onda, Chris Watson + KK.Null + Z'ev, Jean-Claude Risset, Michel Guillet, Portradium, Kaffe Matthews, Al Margolis / If, Bwana, François Bayle, Pierre Jodlowski, Colleen, Matmos, Michel Chion, Goran Vejvoda, Phil Niblock + Natalia Pschenitschnikova, and Maja Ratkje.

2008, *Variations didactiques* (22 minutes) by Yann Geslin, new version of the 1981 piece, with Alex Grillo on the vibraphone.

Radio Production

The collaboration between the GRM and the France Musique network operated perfectly. Christian Zanési, in charge of radio programming for the network, promoted the discovery of many young composers, unknown to the public. In 2005, an assistant, Alexandre Bazin, joined him to help in the preparation of programs:

Fins de mois difficiles, a two-hour musical program, devised by Christian Zanési and David Jisse. Until 2004, this program was broadcast simultaneously on France Musiques and on the satellite program Hector, in a Dolby Surround version. In 2004, the program *Electromania* took its place.

Electromania, one hour every two weeks, at 11 p.m., on terrestrial radio.

Electrain de nuit, one hour every month, Saturdays at midnight.

Tapage nocturne, 20 minutes weekly beginning in 2000 during Bruno Letort's broadcast.

On France Culture, radio creations were produced in the GRM studios, in coproduction with the Atelier de Création Radiophonique of Radio France:

Sports et divertissements d'Erik Satie, by Andrea Cohen in 2001.

Intégral, by Andrea Liberovici (major retrospective on the dictators of the twentieth century, with the voices of Mussolini, Hitler, Stalin, etc., from the archives), in 2003.

X Life, by Sheila Concari, in 2004.

Les Guyanes and *ADN Concret-mentaire*, by Hervé Birolini in 2004.

Chère-toi, by Andrea Cohen and Wiska Radkiewicz in 2004.

November 11–15, 2002, *50 ans de Création au Studio de Musique Électronique de Cologne*, five half-hour programs five mornings in a row on the history of the Cologne Studio, by Évelyne Gayou and Philippe Langlois.

Workshops

Since 1999, Denis Dufour had been organizing "workshops" in the Acousmathèque, Studio 116. Once a month, a composer or an electroacoustic professor, accompanied by his or her students, would present their work to a small audience of peers. Among the guests were Philippe Blanchard on Russolo and the futurist heritage, Thierry Besche and Jean Pallandre on musical pedagogy, Hugues Dufourt on depicting dynamics and sound morphology, Nicolas Vérin, accompanied by his students from the Chalon-sur-Saône ENM, and Gino Favotti on techno music.

In 2000, the "workshops" hosted composers in connection with their concert programming in Cycle Multiphonies, as well as researchers in musicology. Received were Roger Cochini and electroacoustic students from the Conservatory of Bourges; Roland Cahen and electroacoustic students from the Conservatory of Montbéliard; Gilles Gobeil and René Lussier; Christine Groult and electroacoustic students from the Pantin ENM; Laurent Cuniot, Luis Naon, Yann Geslin and electroacoustic students from the Paris CNSM; Gino Favotti and electroacoustic students from the Conservatory of the Twentieth Arrondissement of Paris; Anne Veitl; Bruno Bocca.

The "Imprints" sessions, similar to the "workshops" in conception, opened the door to composers elucidating their concept of electroacoustic composition in the light of their own artistic trajectories. Invited were Philippe Hurel, Guy Reibel, Edgardo Canton, Philippe Leroux, Georges Bœuf, Robin Minard, Erik Mikaël Karlsson, Javier Alvarez, and François Bayle. All of these sessions were recorded and preserved in the GRM archives.

International Colloquia

Since 2003, the GRM has co-organized international symposiums under the title EMS: Electronic Music Studies. EMS 03 took place in Paris in 2003, EMS 05 in

Montreal in 2005, EMS 06 in Beijing, EMS 07 in De Montfort, UK, EMS 08 in Paris, EMS 09 in Buenos-Aires, and EMS 10 in Shanghai. At each session, local partners from the organizing countries reinforce the team of the three founding members:

—Daniel Teruggi, director of the GRM;
—Marc Battier, professor at the Sorbonne University in Paris; and
—Leigh Landy, professor at De Montfort University in Great Britain.

The topics addressed during these colloquia concern music on media (analysis, conservation, dissemination) and their relationship with the other arts.

Joel Chadabe of the Electronic Music Foundation (EMF) in New York was the honorary president of EMS.

Books

In 1998, on the occasion of the celebrations for the fiftieth anniversary of musique concrète, two long out-of-print books by Pierre Schaeffer were reissued by Éditions Seuil: *Traité des objets musicaux* and *À la recherche d'une musique concrète*.

In 2000, the GRM volume *Ouïr, entendre, écouter, comprendre après Schaeffer* was published by Buchet-Chastel in the Bibliothèque de Recherche Musicale collection edited by François Delalande. The book gathered a series of articles from the "After Schaeffer" colloquium organized by Denis Dufour in Perpignan two years earlier on the theme of the influence of Pierre Schaeffer's thought on our way of conceiving the act of listening to music today. Denis Dufour, the editorial director of the publication, coordinated the work of the numerous authors: Jean-Christophe Thomas, Makis Solomos, Hugues Dufourt, Jean-François Augoyard, Régis Renouard-Larivière, Jean Molino, François Bayle, Jean-Claude Risset, Francis Dhomont, Denis Smalley, Lelio Camilleri, Marcel Frémiot, Pierre Schaeffer, and Sylvie Dallet.

In 2000 a book edited by Hugues Vinet and François Delalande, *Interface homme-machine et création musicale*, was published by Hermès. Stemming from a joint conference between the IRCAM and the GRM on the subject of musical creation and the interface between man and machine, it presented a series of articles by Jean-Claude Risset, Gérard Assayag, Yann Orlarey, Dominique Fober, Stéphane Letz, François Dechelle, Hugues Vinet, Michel Baudoin-Lafon, Marcello Wanderley, Philippe Depalle, Claude Cadoz, Daniel Teruggi, Gilles Racot, Emmanuel Favreau, and Philippe Manoury.

In 2001, *Le son des musiques entre technologie et esthétique* by François Delalande was published by Buchet-Chastel, based on accounts collected by Delalande from musicologists during the symposium *Le son de la musique* (The sound of music) that he had led at Studio 104 of the Maison de la Radio in 1989.

Record Releases

In 2004 the GRM, under the Radio France label Signature, produced its first audio DVD recorded exploiting the DTS standard. It was a remix of the concert *GRM expérience*, first performed in Paris on October 11, 2003, with three musicians live on stage. Christian Zanési from the GRM was joined by two musicians from the electronic scene, the Austrian Christian Fennesz and the Fin Mika Vainio (from the duo with Ilpo Väisänen, Pan Sonic), for this "experiment intertwining musical genres."

In 2005, the CD collection was approaching its one hundredth number.

LES MUSIQUES ÉLECTROACOUSTIQUES CD-ROM

The entire GRM was mobilized between 1998 and 2000 in a magnificent group effort to produce the CD-ROM *Les musiques électroacoustiques* devoted to electroacoustic music. This CD-ROM, with an original pressing of five thousand copies, was commercially released on April 26, 2000, by Éditions Hyptique in Paris, after two years of intensive work and substantial financial support from the French Ministry of Education. It presented electroacoustics in three interactive multimedia chapters: historical, with some fifty sound extracts of music and numerous photos; analytical, in the form of excerpts of works analyzed and transcribed with the help of the Acousmographe (the Acousmographe in its first version); and practical, titled the "GRM mini-studio," enabling users to try out sound transformation with the aid of a mini GRM Tools. In November 2001, the CD-ROM received numerous awards including Diapason d'Or, Grand Prix Mobius France, and a Canadian award. In 2005 it was reissued, adapted to the new Mac OS X platform, while still remaining PC-compatible.

Polychrome Portraits: Books + Multimedia

After the successful CD-ROM experiment, the 2001 launching of the *Portraits polychromes* book series, with its associated multimedia documents on the INA website,[23] inaugurated an important new editorial activity for the GRM, with support from the SACEM, the French Society of Authors, Composers and Publishers of Music. Monographs on electroacoustic music composers or composers aesthetically close to electroacoustics were published more or less twice annually: Luc Ferrari (2001), Jean-Claude Risset (2001), Gilles Racot (2002), Bernard Parmegiani (2002), Ivo Malec (2003), François Bayle (2004), John Chowning (2005), Michel Chion (2005), Jacques Lejeune (2006), Francis Dhomont (2006), and Max Mathews (2007). The 2008 issue was devoted to Pierre Schaeffer, and the next in 2011 to Denis Smalley, followed by Éliane Radigue in 2013. The *Portraits polychromes* series, coordinated and edited by Évelyne Gayou, would continue with three special thematic "Music & Technique" publications. (See the bibliography—

discography for the full list of the *Polychrome Portraits* collection, including English translations.)

The multimedia component of *Portraits polychromes* provides complementary interactive documents online on the Internet, contributing to relaunching reflection about graphic transcription and the analysis of music as well as major themes of predilection specific to research carried out at the GRM, listening as a mode of access to composition, defining an electroacoustic musical language, the search for ways to notate listening, the relationship between sound and visuals, and the challenges of multimedia in terms of music teaching.

The *Portrait polychrome* dedicated to Bernard Parmegiani was awarded the "coup de coeur" from the Académie Charles Cros in 2003, also honoring the entire *Portraits polychromes* collection. The books on John Chowning and Max Mathews were published in two versions, one in French and the other in English, thanks to coproduction with Stanford University in California, more specifically the Center for Computer Research in Music and Acoustics (CCRMA), directed by Chris Chafe.

A Former Member of the GRM

DENIS DUFOUR, 1953–

Denis Dufour joined the GRM in 1976, remaining until 2000. He began his musical studies at the CNR of Lyon in 1972, then studied composition at the CNSMP in the classes of Ivo Malec and Michel Philippot, as well as analysis with Claude Ballif. Today a composer and teacher, he is also a concert and festival organizer. He has never spared any efforts to defend the acousmatic cause since his beginnings in electroacoustics in the classes of Pierre Schaeffer and then Guy Reibel, of whom he was the assistant for a year.

His first piece of electroacoustic music, *Bocalises* (1977), contributed to his immediate admission into the circle of electroacoustic composers. In 1977, with the support of François Bayle and the GRM, he formed Trio GRM+ with Yann Geslin and Laurent Cuniot, performing the new repertoire of mixed music for tape and synthesizers. Denis Dufour left the trio in 1987, but in the meantime, he was given a teaching position in electroacoustic composition at the Conservatory of Lyon, and later at Perpignan and Paris.

Since 1993, Denis Dufour has organized the Festival d'art acousmatique et arts de support, Futura. It is also worth mentioning his investment—both financially and through his work—in extensive activity publishing book-discs concerning contemporary and electroacoustic music composers, starting with his own master, Ivo Malec. Denis Dufour amazes by his ability to carry out all these activities simultaneously. It is said that he hardly sleeps; it must certainly be true. The rigor he imposes on himself is equal to that which he demands from his collaborators and friends—the Motus association, which he founded in the Southwest of France in 1996, organizes an average of one hundred concerts a year with its rather limited means and small staff. In 2006, he also founded Synthax, active in the fields of electroacoustic and acousmatic music in Languedoc-Roussillon.

And as soon as the loudspeakers of his Acousmonium were dismantled, he immediately started composing again. In 2024, his catalog included 193 compositions. The style of his works makes us classify him as being aesthetically close to Luc Ferrari, notably in his taste for elaborately crafted anecdotal concrete sounds. Another side of Dufour's style is reminiscent of Pierre Henry, for whom he has great admiration. Pierre Henry even lent his voice in 2000 for Dufour's piece *Chanson de la plus haute tour*. Denis Dufour also composes for orchestra and voice. His works and productions are catalogued on denisdufour.fr, his bilingual website.

Postscript

This translation of the original 2007 French edition is an opportunity for me to reassure readers that, after more than fifty years of musique concrète as well as electroacoustic and acousmatic music, its founding institution, the GRM, is still evolving, following the trajectory launched decades ago. Certainly many collaborators and musicians have departed, while new ones have joined in. But they continue tirelessly, delving deeper into the multiple layers of the work of the GRM's pioneers. And as a result, composers offer us new music every day.

On the back cover of his *Traité des objets musicaux*, Pierre Schaeffer described himself as "the creator of impossible but necessary institutions," suggesting that the process of invention requires a framework. The invention of musique concrète hadn't occurred by pure chance. The serendipity of its birth was linked to the favorable environment of French radio, associated with a team of musicians and engineers endowed with not only the necessary knowledge but also the necessary curiosity and temerity. The right people at the right place at the right moment. Schaeffer also contributed a dose of interdisciplinarity and a real aptitude for leadership. The historical point in time was not inconsequential—the end of World War II, an exceptional moment of liberty, when everything seemed possible. We find the same inspiration in John Chowning's declaration about the research center he built, CCRMA: "a site of intellectual ventilation as well as coordination."[1]

Inspired by the filmmaker Jean-Luc Godard, who declared that he was not trying to show a *just image*, but *just an image*, I will simply say that electroacoustic music does not aim at a *just sound*, but at *just a sound*.

Did Pierre Schaeffer, Pierre Henry, and those who followed succeed in creating *the most universal music possible*? I believe so.

This book pays homage to this great saga by providing as many references as possible to foster understanding of the alchemy of musical invention and the path to its fulfillment.

Évelyne Gayou
Paris, 2024

NOTES

INTRODUCTION

1. Pierre Henry, *Journal de mes sons* (Actes sud, Arles, 2004), 68.
2. Pierre Schaeffer, *À la recherche d'une musique concrète* (Seuil, Paris, 1952), 15–16.
3. Michel Maffesoli, *Au creux des apparences, pour une éthique de l'esthétique* (Plon, Paris, 1990), 92–93. According to Maffesoli, baroque attitude is a relic of magical thinking, characteristic of folk wisdom.
4. Jules Michelet, *Histoire de la France* (1869; repr., Robert Laffont, Paris, 1971), 11.
5. Paul Ricœur, *La mémoire, l'histoire, l'oubli* (Seuil, Paris, 2000), 582.
6. The *TOM* and *SOS* are explored in chapter 3 and chapter 6.

CHAPTER 1. BEFORE 1948

1. In fact the exact date is ambiguous. Some documents, notably the GRM *Répertoire acousmatique*, as well as the written archives of the program's production, mention June 20, 1948. But it is the oral traces that have been immortalized, namely the October 5, 1948, broadcast.
2. Abraham Moles and Élisabeth Rohmer, *Psychologie de l'espace* (L'Harmattan, Paris, 1998), 125. Note that Abraham Moles and André Moles are the same person. Abraham changed his first name to André during the Second World War.
3. Statement by his daughter Marie-Claire, in a private interview in 2005. See also Pierre Schaeffer, *Prélude, choral et fugue* (Flammarion, Paris, 1983), where Schaeffer speaks autobiographically.
4. Schaeffer reportedly adopted the name Toto for his clandestine activities.
5. Elisabeth Schaeffer died in 1941 of illness, and not of a car accident as Antoine Goléa stated in his 1977 book.

6. Gurdjieff was a rather secretive man. His "memoirs," *Rencontres avec des hommes remarquables* (Julliard, Paris, 1960), translated as *Meetings with Remarkable Men*, should above all not be taken completely literally. See also his *Beelzebub's Tales to His Grandson* and *Life Is Real Only Then, When "I Am."*

7. Peter Demianovitch Ouspensky, *Fragments d'un enseignement inconnu* (Stock, Paris, 1950).

8. Louis Pauwels, *Monsieur Gurdjieff* (1954; repr., Albin Michel, Paris, 1974). François Bayle also attested to this attraction for Gurdjieff's thinking in *Portraits polychromes*, no. 6, *François Bayle* (2004; new and expanded edition, INA-GRM, Paris, 2007), 20–21.

9. Not to be confused with the Groupe Jeune France, founded in 1936 by Olivier Messiaen together with Yves Baudrier, Daniel-Lesur, and André Jolivet.

10. Pierre Schaeffer, *Propos sur la Coquille* (Phonurgia nova, Arles, 1990).

11. The comedian Francis Blanche had posted this name, as a joke, on the office door of his comrade.

12. Martial Robert, *Pierre Schaeffer: Des Transmissions à Orphée* (L'Harmattan, Paris, 1999), 74.

13. This text was included in the booklet of the CD set *10 ans d'essais radiophoniques, 1942–1952* (1989).

14. Booklet from the CD boxed set *10 ans d'essais radiophoniques* (1989), 75–76.

15. In 1932 the experimental Bauhaus studio was created in Dessau, Germany, by Paul Arma, Laszlo Moholy-Nagy, and Friedrich Trautwein. Moholy-Nagy had in particular experimented with manipulating records: acceleration, slow motion, reverse, etc. Schaeffer also discovered the research of Jörg Mager during this trip, later evoking it on countless occasions; Mager had begun to use the word *electro-acoustic* in 1929.

16. Martin Kaltenecker, preface to the French edition of Rudolf Arnheim's book *Radio* (Van Dieren, Paris, 2005), 34.

17. Jacques Copeau, "Écrit en 1943," in the booklet of the CDs *10 ans d'essais radiophoniques* (Phonurgia Nova, 1994), 80.

18. Louise Conte, Suzanne Flon, Nadine Alari, Daniel Ivernel, Gérard Souzay, Olivier Hussenot, Jean-Claude Léonard, Jacqueline Baudoin, Madeleine Barbulée, Catherine Toth, Marguerite Arandel.

19. Schaeffer, *Propos sur la Coquille*, 91.

20. Published in the form of four CDs with a booklet, at Phonurgia Nova.

21. Pierre Arnaud, "À la recherche de Pierre Schaeffer," in the booklet of the CDs *10 ans d'essais radiophoniques* (Phonurgia Nova, 1990), 23.

22. "Quartet of ondes" refers to the electric instrument *Ondes Martenot*, which appears often in this work.

23. Pierre Schaeffer, *Notes sur l'expression radiophonique*, 1946, in the booklet of the CDs *10 ans d'essais radiophoniques*.

24. Morphology and typology would be two of the principal axes of analysis and classification of sounds presented in Schaeffer's 1966 *Traité des objets musicaux: Essai interdisciplines (TOM)* (Seuil, Paris, 1966).

25. In Vincent Barras and Nicholas Zurbrugg, eds., *Poésies sonores* (Contrechamps, Geneva, 1992), 19, the sound poet Henri Chopin recognized that "electronics has allowed ... the voice to fly innumerable journeys, with the word, but also without it."

26. Wireless, TSF, "Télégraphie Sans Fil," "wireless telegraph," in other words, radio.

27. Published under the title "La naissance de l'univers sonore de Pierre Schaeffer: *La Coquille à planètes*" (Résonance, colloquium at the IRCAM, Paris, October 2003). See also Giordano Ferrari, "*La conchiglia* di Pierre Schaeffer" (Pierre Schaeffer's *La coquille*), *AAA-TAC: Acoustical Arts and Artifacts, Technology, Aesthetics, Communication* 1 (Istituti editoriali e poligrafici internazionali–Fondation Cini, Pisa/Rome/Venice, 2004): 11–18.

28. Schaeffer, *Notes sur l'expression radiophonique*.

29. Andrea Cohen, *Les compositeurs et l'art radiophonique* (L'Harmattan, Paris, 2015).

30. Gaston Bachelard, *La poétique de l'espace* (PUF, Paris, 2001), 111–12.

31. See the researches by the Canadian André Gaudreault and the Swiss François Albera. They report that the *bonimenteur* or *bonisseur*—"pitchman" or "barker"—often improvised sound effects to enhance the film. See also the colloquium "Le muet a la parole," Rick Altman, "Silent Film Sounds," Le Louvre, Paris, 2004; Alain Boillat, "Colloque 'Le muet a la parole' (10–13 juin 2004, auditorium du Louvre)," http://journals.openedition.org/1895/2112, https://doi.org/10.4000/1895.2112.

32. *Le silence est d'or* magnificently reconstructs the early days of the film industry, specifically the period of transition from silent film to talkies, from 1910 to 1930.

33. *La Revue musicale*: "Le film sonore, L'Écran et la musique en 1935," special issue under the direction of Henry Prunières (Paris, December 1934), 72.

34. The tape recorder began to emerge in 1945.

35. Philippe Langlois, *Les procédés électroacoustiques dans les différents genres cinématographiques* (thesis, University of Paris IV–Sorbonne, 2004).

36. Langlois, 173.

37. *La Revue musicale*, special issue December 1934, 72–73.

38. *Tönende Handschrift*, 1932 (www.youtube.com/watch?v=bDwEpPFqQCo).

39. *Ornament Sound Experiments*, 1932 (https://vimeo.com/ondemand/26951).

40. Thomas Y. Levin, *Tones from out of Nowhere* (Grey Room, New York, 2003).

41. Jean Guignebert, a member of the Studio d'Essai, was appointed minister of information following the Liberation of France.

42. Booklet from the CD boxed set *10 ans d'essais radiophoniques* (1989), 77.

43. Pierre Schaeffer, "L'élément non visuel au cinéma, I: Analyse de la bande-son," in *La revue du cinéma*, no. 1, Paris (October 1946): 45.

44. Schaeffer, 47.

45. Schaeffer, 51.

46. Schaeffer, 52.

47. Pierre Schaeffer, "L'élément non visuel au cinéma, III: Psychologie du rapport vision audition." In *La revue du cinéma*, no. 3, Paris (December 1946): 52.

48. Schaeffer, 54.

49. One encounters the same types of statements of enthusiasm and the feeling of venturing into terra incognita at the other centers of research of the time, especially in Germany.

50. Jean Tardieu had met Francis Ponge in 1934 and worked with him at Hachette (Ponge became a socialist in 1919, a member of the CGT in 1936, and a member of the Communist Party from 1937 to 1947).

51. Bernard Parmegiani would adopt this Claudel title for a 1970 composition.

52. André Jolivet (1905–74) was a former student of Paul Le Flem and Edgar Varèse.

53. Composers Arthur Honegger and Arthur Hoérée had already used this technique for the soundtrack of Dimitri Kirsanoff's film *Rapt* (1931), to illustrate the mental confusion in which the hero found himself at the moment of making an important life decision.

54. *À propos d'un essai acoustique*, Cahiers du club d'Essai, 1947.

55. It became the GRMC—Groupe de Recherches de Musique Concrète—in 1951. See chapter 6.

56. The Scott de Martinville's cylinder has been inscribed on the Memory of the World Register by UNESCO since 2015.

57. The administration numbered the studios in chronological order of their construction, which remained the pattern until 1963, when the Maison de la Radio was inaugurated.

58. Pierre Arnaud, in Schaeffer, *Propos sur la Coquille*, 19.

59. For fans of technical detail, I'll add that the Club d'Essai was equipped with Charlin loudspeakers.

60. Paul Hindemith composed *Der Lindberghflug* in 1929, in collaboration with Kurt Weill, to a libretto by Bertold Brecht.

61. Ottorino Respighi had met Busoni in 1909 in Berlin.

62. Jean-Yves Bosseur, *Musique et arts plastiques: Interactions au XXe siècle* (Minerve, Paris, 1998).

63. Pierre Schaeffer recalls this experience and others in his personal diary: *In Search of a Concrete Music*, trans. Christine North and John Dack (Berkeley: University of California Press, 2012).

64. An early form of additive synthesis in the Hammond organ, Trautonium, and Telharmonium consisted of a set of pure tone generators, tuned in harmonic series, producing basic timbres. The intensity of each harmonic could then be adjusted.

65. The Theremin, originally called an Aetherophone, was manufactured in the United States by Robert Moog beginning in 1948. Jörg Mager had founded an "electro-acoustic music" organization in Germany in 1929.

66. Martenot invented the Ondes Martenot after an encounter with Theremin, and after a voyage to India where he had noticed that twelfth-tone intervals were used.

67. Giovanni Lista, *Futuristie: Manifestes, Documents, Proclamations* (L'Age d'homme, Paris, 1973), 42.

68. Luigi Russolo, *L'art des bruits* (1954; repr., Éditions Allia, Paris, 2003), 25–26. Also see Pierre Schaeffer's "Tableau de la typo-morphologie des sons" in *TOM*, 584–85; English version: *Treatise on Musical Objects: An Essay across Disciplines* (*TMO*), trans. Christine North and John Dack (University of California Press, Oakland, 2017), 464–65.

69. Russolo, *L'art de bruits*, unnumbered pages.

70. Claude Debussy, *Monsieur Croche* (1921; repr., Gallimard, Paris, 1987), 240 (translated into English as *Monsieur Croche, The Dilitantte Hater*).

71. Before his passion for music and noise, Luigi Russolo (1885–1947) had first painted. Two of his paintings can be seen in France: *Synthèse plastique des mouvements d'une femme* at the Grenoble Museum of Modern Art and *Dynamisme d'une automobile* at the Palais de Tokyo in Paris.

72. The first *Manifeste du futurisme*, by Marinetti, had been published in the French newspaper *Le Figaro*, February 20, 1909.

73. Maurice Lemaître was a Letterist author, close to Isidore Isou. Lemaître quoted page 31 of *À la recherche d'une musique concrète*, where Schaeffer wrote in 1952, "In fact the Italians with Marinetti were precursors 20 years ago. But it consisted of concerts of straightforward noises, leading, as we have seen, to a dead end."

74. In his introduction to the 1954 reprint of *L'Art des bruits*, published by Richard-Masse, Maurice Lemaître spoke of "historical murder" by Schaeffer concerning futurism.

75. Marinetti had gone to meet Maïakovski in Russia in 1913 to internationalize his futurist movement.

76. The Koulechov effect has become a classic in film editing, playing on contrasting two identical shots (a face with a neutral expression) with two other very different shots (a child in rags and a plate of steaming soup). Juxtaposing the blank face and the unhappy child inspires sadness; juxtaposing the same face and soup inspires contentedness.

77. Georges Sadoul, *Dziga Vertov* (Champ libre, Paris, 1971), 160.

78. Sadoul, 19.

79. Sadoul, 50.

80. Conference: *Dziga Vertov et l'invention sonore* (Dziga Vertov and invention in sound), June 19, 2004, at the Louvre auditorium in Paris, as part of the *Du muet au parlant* festival.

81. Using optical sound. The sound effects were recorded on the soundtrack directly on the film, on the celluloid side. The waveform was visible.

82. Schaeffer had certainly heard of the Russian futurists from Gurdjieff.

83. Giovanni Lista, *Futurisme, manifests, documents, proclamations* (l'Age d'homme, Paris, 1973), 142.

84. *Mots en liberté* (words in liberty) is a term echoed by Marinetti in his futurist manifesto.

85. Sadoul, *Dziga Vertov*, 35–36 (based on documents borrowed from Tristan Tzara).

86. Ornella Volta, *Satie et la danse* (Plume, Paris, 1992), 72.

87. Jean Cocteau, *Le Coq et l'Arlequin* (1918; repr., Stock, Paris, 1993), 101.

88. Cocteau, 105.

89. Cocteau, 106.

90. Picabia and Satie had even published a small *Manifeste Instantanéiste* parodying the just published *Surrealist Manifesto*.

91. Francis Picabia, "Sur René Clair," *La Danse*, November 1924, reprinted in Picabia, *Écrits* (Belfond, Paris, 1978), 2:158.

92. André Breton, *Manifestes du surréalisme* (1924; repr., Gallimard, Paris, 1985).

93. Interviews by the author with Christian Rosset, François-Bernard Mâche, and Jean-Yves Bosseur for the radio network France Culture in 2002.

94. Le Groupe des six (The Group of Six) consisted of Georges Auric, Louis Durey, Arthur Honegger, Darius Milhaud, Francis Poulenc, and Germaine Taillefer.

95. Jean Cocteau, *Opium* (1930; repr., Stock, Paris, 2005), 212–13.

96. André Breton, *Premier manifeste du surréalisme* (1924; repr., Gallimard, Paris, 2001), 73.

97. Marius Constant was one of the few French composers to take an open interest in surrealism. In the 1960s, he explored chance techniques in music in *Les chants de Maldoror* (1962) and *Trait* (1969).

98. Robert Wangermée devoted an entire book to André Souris in 2002, *André Souris et le complexe d'Orphée: Entre surréalisme et musique sérielle* (Mardaga, Sprimont, 2002).

99. Wangermée, 84, quoting *Musique 2*.

100. Even though he had been Schaeffer's greatest advocate for years. Wangermée, 314.

101. Wangermée, 314.

102. Jean-Etienne Marie was designated as "Musicien Metteur en Ondes (MMO)" at the RTF, and as such he collaborated with the GRM until the late 1960s. In 1968, he founded the CIRM in Nice. See translator's note on *Metteur en Ondes* in this chapter.

103. Jean-Étienne Marie, *Musique vivante* (Privat/PUF, Paris, 1953), 79, also reminds us of the influence of the 1885 World Fair, where Claude Debussy discovered exotic music.

104. Olivier Messiaen published *Technique de mon langage musical* in 1944. The title of Messiaen's Conservatory "Class of Harmony" in 1941 was changed to "Class of Analysis, Aesthetics and Rhythm" in 1947, a designation created especially for him.

105. Marie, *Musique vivante*, 99. "Dense sound" here refers to "complex sound." This was in 1953, long before the defined vocabulary of *TOM* in 1966.

106. See Jesus Aguila, *Le Domaine musical* (Fayard, Paris, 1992), 177, or Claude Mollard, *Le 5e pouvoir* (Armand Colin, Paris, 1999).

107. To be exact, December 29, 1915.

108. Ferruccio Busoni was born in 1866 in Empoli, near Florence, and died in 1924 in Germany. He propagated many revolutionary musical ideas for his time—quarter-tone and third-tone scales. In 1907 he predicted the future of music: "In my opinion, it will move towards abstract timbre, towards a technique free from all constraint, towards absolute tonal freedom. Let's bring music back to its original essence, let's free it from architectural, acoustic and aesthetic dogmas, let it become pure invention, pure feeling, in its harmonies, forms and sound colors."

109. It should be noted that a portrait of Varèse has always adorned the office of the successive directors of the GRM, a sort of sacred relic they have always preserved, one director after another, throughout each relocation.

110. See Riccardo Chailly's 1998 CD recording, DECCA 460 208-2.

111. Barras and Zurbrugg, *Poésies sonores*, 130.

112. Bernard Heidsieck evolved from "sound poetry" to "action poetry" in 1955.

113. Released on CD as *Early Modulations*, Vintage volts CAI 2027-2, and Hat Hut Records Ltd.

114. Bergson had refined his thinking on the subject, notably through contact with Einstein.

115. In reference to cinema, see the works of Michel Chion, cited in the bibliography. Michel Chion, spiritual child of the GRM, has considerably contributed to reflection concerning sound in cinema since 1982 when he published his first book on the subject, *La voix au cinéma*.

116. For example Walther Ruttmann with his 1930 sound film *Weekend*.

117. Schaeffer, *Notes sur l'expression radiophonique*, 73.

118. The Studio, later renamed the Club d'Essai, had no official connection with the Conservatory, nor with music societies.

119. Pierre Schaeffer, *À la recherche d'une musique concrète* (Seuil, Paris, 1952), 30.

120. François Bayle, *Musique acousmatique, propositions . . . . . . positions* (INA/Buchet-Chastel, Paris, 1993), 29.

CHAPTER 2. A NAME—A SCHOOL—A STYLE OF MUSIC

1. Pierre Schaeffer, *À la recherche d'une musique concrète* (1952; repr., Edition Seuil, Paris, 1998). The first published written evocation of musique concrète, by Pierre Schaeffer, appears in the review *Polyphonie* of 1950, devoted to mechanized music. The first written trace goes back to a memorandum by Pierre Schaeffer, dated June 20, 1948, sent on the 28th, addressed to the director general of Radio France.

2. Introductory text for the events of the twenty-fifth anniversary of Radio-Geneva, brochure from 1950. Internal document, GRM, not catalogued.

3. The term *musiques expérimentales* had first been used in 1953 by Pierre Schaeffer and Abraham Moles.

4. Pierre Schaeffer, *De la musique concrète à la musique même* (1977; repr., Mémoire du livre, Paris, 2002), 259.

5. Pierre Schaeffer, *Traité des objets musicaux: Essai interdisciplines* (*TOM*) (Seuil, 1966), 24; Schaeffer, *Treatise on Musical Objects, An Essay across Disciplines* (*TMO*), trans. Christine North and John Dack (University of California Press, Oakland, 2017), 8.

6. Pierre Henry, article from 1952, "La musique concrète et le XXè siècle," in Pierre Schaeffer, *De la musique concrète à la musique même* (Mémoire du livre, Paris, 2002), 146.

7. First oral use of the phrase during the presentation of the *Concert de bruits*, on the radio October 5, 1948. David Vaughn, the translator of the present book, chose a literal translation for the phrase, seeing the situation similarly to the translators of the *TOM*, Christine North and John Dack, who commented:

> One expression, "the most general instrument possible," caused some debate; we considered translating it as "the most universal instrument possible" but finally for the sake of accuracy stayed with the literal translation, as indeed we have done with the whole vocabulary of generalization and laws. This aspect of the treatise is important to the understanding of Schaeffer's thought, and to modify it would be to fail in our aim to be as faithful as possible to the text. It also gives a fascinating insight into a different mode of thought.

> In Pierre Schaeffer, *Treatise on Musical Objects: An Essay across Disciplines* (University of California Press, Oakland, 2017), introduction, xxiii–xxiv.

8. Michel Chion, "20 ans de musique électro-acoustique ou une quête d'identité," *Musique en jeu*, no. 8 (1972): 26–27.
9. Michel Chion, in *Portraits polychromes*, no. 8, *Michel Chion* (INA-GRM, Paris, 2005), 35.
10. Pierre Schaeffer, *Machines à communiquer, 1: Genèse des simulacres* (Seuil, Paris, 1970), 152.
11. Celestin Deliège, *Cinquante ans de modernité musicale: De Darmstadt à l'IRCAM* (Mardaga, Sprimont, 2003), 149.
12. Jean-Étienne Marie, *Musique vivante* (Privat/Puf, Paris, 1953). This text can be found on the final page of his book, a little like a declaration of faith.
13. Patent, Pierre Schaeffer/Jacques Poullin no. 605.467, February 26, 1951, in France.
14. See next chapter, concerning tools.

CHAPTER 3. CONCEPTS—PEDAGOGY—TOOLS

Epigraph: Igor Stravinsky, *Chroniques de ma vie* (1962; repr., Denoël, Paris, 2000), 14.

1. Jean Cocteau, *Le Coq et l'Arlequin* (1918; repr., Stock, Paris, 1993), 45.
2. According to Louise Hirbour, *Edgar Varèse, écrits* (Christian Bourgois, Paris, 1983), 146.
3. Robert Francès, *La perception de la musique* (Vrin, Paris, 1958). Schaeffer often referred to this work in *TOM*.
4. Pierre Schaeffer, *Traité des objets musicaux (TOM)* (Seuil, Paris, 1966), reedited in 1998, for the fiftieth anniversary of musique concrète.
5. At the time of writing, this version is still only in manuscript form.
6. *Solfège de l'objet sonore (SOS)*, boxed set of three CDs and a book, copublished by INA–Publications and Musidisc, no. 292582.
7. Michel Chion, *Guide des objets sonores* (INA/Buchet-Chastel, Paris, 1983).
8. Chion, 13.
9. Chion, 9.
10. A bit like Leonardo da Vinci, who saw ten possible properties to an object: lightness and darkness, color and substance, form and position, distance and approach, movement and immobility.
11. The table of musical objects in *TOM* is a beautiful example, on 584–87.
12. Italicized in the original. Pierre Schaeffer, *Treatise on Musical Objects: An Essay across Disciplines (TMO)*, trans. Christine North and John Dack (University of California Press, Oakland, 2017), 213.
13. Schaeffer, 229.
14. Edmund Husserl, *Méditations cartésiennes* (J. Vrin, Paris, 1947); Husserl, *Logique formelle et logique transcendantale* (PUF, Paris, 1957).
15. Schaeffer, *TOM*, 267; Schaeffer, *TMO*, 210. Schaeffer's concept of the object is derived from his reading of Merleau-Ponty. The concept of objects appropriate, or not, to music is probably linked to his reading of Lucretius.
16. Schaeffer, *TOM*, 153; Schaeffer, *TMO*, 113.
17. Roman Jakobson, *Essais de linguistique générale* (Minuit, Paris, 1963).
18. Schaeffer, *TOM*, 35; Schaeffer, *TMO*, 16.

19. Schaeffer, *TOM*, 350; Schaeffer, *TMO*, 277.
20. Schaeffer, *TOM*, 662; Schaeffer, *TMO*, 529.
21. See the examples presented in chapter 1, from the recordings *10 ans d'essais radiophoniques, 1942–1952*.
22. Roland Barthes, *La chambre claire: Note sur la photographie* (Cahiers du cinéma–Gallimard–Seuil, Paris, 1980), 123–24.
23. Schaeffer, *TOM*, 23; Schaeffer, *TMO*, 7.
24. Schaeffer, *TOM*, 584; Schaeffer, *TMO*, 464–67.
25. Chion, *Guide des objets sonores*, 101.
26. Schaeffer, *TOM*, 361; Schaeffer, *TMO*, 286.
27. Brigitte Van Wymeersch, *Descartes et l'évolution de l'esthétique musicale* (Mardaga, Sprimont, 1999), 124.
28. Schaeffer, *TOM*, 23; Schaeffer, *TMO*, 179.
29. Chion, *Guide des objets sonores*, 154.
30. Schaeffer, *TOM*, 533–34 Schaeffer, *TMO*, 425–26.
31. *Musique animée*, fifteen-minute broadcast by Philippe Arthuys and Jérôme Peignot, Chaîne Nationale, no. LUR 301, 1955.
32. Also worth mentioning is the series *Akousma*, broadcast every summer since 1994 by the GRM on France Musiques, offering works from the GRM repertoire.
33. Hans-Robert Jauss, *Pour une esthétique de la réception* (Gallimard, Paris, 1978). Hans-Robert Jauss (inspired by Husserl) is a leading voice in reception theory and the Constance school, in Switzerland.
34. Pierre Schaeffer, in *L'oreille oubliée*, collective exhibition catalog, under the direction of Georges Pigeat (Centre Georges Pompidou/CCI, Paris, 1982), 15. Also see translator's note on *jeu* in this chapter.
35. Jauss, *Pour une esthétique*, 153.
36. The earliest teaching of electroacoustics in France dates from 1953, at the Schola Cantorum under the guidance of Jean-Étienne Marie.
37. Guy Reibel, *L'homme musicien: Musique fondamentale et création musicale* (Edisud, Aix-en-Provence, 2000), 337.
38. Reibel, 338–39.
39. Reibel, 339.
40. This article, "Trois idées clés pour une pédagogie musicale d'éveil," was first published in the *Programme-Bulletin* no. 4, February 6, 1974 (distributed at the beginning of the concert).
41. Claire Renard, composer and pianist, joined the GRM in the composition class of the Conservatoire, class of 1971–73. Impassioned by pedagogy and critical of traditional teaching (especially concerning the piano), she was at the origin of the first GRM experiments in musical pedagogy in schools in 1973. Her collaboration ended in 1978.
42. Until Wiska Radkiewicz left for the United States to work on her doctorate in music at Princeton.
43. Until 1979, Wiska Radkiewicz and Claire Renard's primary activity was to lead training courses for teachers.
44. And by 2016, twenty-three issues. See the bibliography for the complete list.

45. In France, this teaching was introduced in the 1930s and has since spread internationally (United States, Great Britain), continuing to this day.

46. For the Beaune Workshop, see chapter 1.

47. This fee, a *cachet*, corresponded to a certain number of "work sessions" devoted to a specific task. A "session" lasted four hours. The word *session* should be understood here in the same sense as a "service" in the context of orchestra rehearsals.

48. Student interns generally engage for periods of two to six months. Some work-study programs run for two years, with the trainee present at the GRM every other week.

49. Pierre Schaeffer, "La Musique Concrète et la recherche scientifique," in *7 ans de Musique Concrète 1948–1955* (Centre d'Études Radiophoniques, RTF, Paris, 1955), 35.

50. Schaeffer, 36.

51. See also chapter 8.

52. Yann Geslin, article from 1998, "Environnements de transformations sonores et musicales: Une expérience de vingt années au Groupe de Recherches Musicales," published in English as "Digital Sound and Music Transformation Environments: A Twenty-Year Experiment at the 'Groupe de Recherches Musicales,'" trans. Riccardo Walker, *Journal of New Music Research* 31, no. 2 (June 2002): 99–107.

53. François Bayle, *Komposition und Musikwissenschaft im Dialog IV (2000–2003): L'image de son; Technique de mon écoute / Klangbilde; Technik meines Hörens* (LIT, Cologne, 2003), 192.

54. Bayle, 50.

55. Bayle, 192.

56. Schaeffer, *TOM*, 43; Schaeffer, *TMO*, 25.

57. André Leroi-Gourhan, *L'homme et la matière* (Albin Michel, Paris, 1971), 10.

58. Schaeffer, *TOM*, 357; Schaeffer, *TMO*, 283.

59. Pierre Schaeffer, *Faber et Sapiens* (Belfond, Paris, 1985), 79.

60. See chapter 4, concerning the attempts at spatialization.

61. Schaeffer, *TOM*, 34; Schaeffer, *TMO*, 16.

62. See chapter 6 for the description of sound projection of Messiaen's *Timbres-Durées* by Pierre Henry.

63. Radio France began recording concerts in stereo in 1959, but stereo broadcasting did not begin until 1964.

64. 1990 interview with the German radio producer Rudolf Frisius, in Pierre Schaeffer, *Propos sur la coquille* (Phonurgia nova, Arles, 1990), 116.

65. Jacques Poullin, "L'apport des techniques d'enregistrement dans la fabrication de matières et de formes musicales nouvelles: Application à la musique concrète," *Ars sonora* no. 9 (1999): 25 (article originally published in the journal *L'Onde électrique* in 1954).

66. 1960 biographical note on Schaeffer, GRM document.

67. Enrico Chiarucci, physicist by training and radar specialist, left the GRM in the early 1970s.

68. The national sociopolitical unrest of "May '68" included strikes at the ORTF and the participation of numerous collaborators in the "events."

69. LF, low frequencies, and VLF, very low frequencies.

70. Excerpt from the leaflet for *Conférences des journées d'études du Festival international du son de Paris*, GRM internal document not numbered, 1972, p. 6.

71. Pierre Schaeffer (assisted by Bernard Dürr) used the Moog synthesizer to compose *Le trièdre fertile*. A voltage control amplifier (VCA) is a means for electronic control of sound modules.

72. Then in 2001 studio 116B and in 2003 studio 116C.

73. Bénédict Mailliard, "À la recherche du studio musical," 1981 conference, at Festival du Son, in *La revue musicale*, "Recherche musicale au GRM," quadruple issue no. 394, 395, 396, 397 (Richard-Masse, Paris, 1986), 51.

74. Francis Régnier, "La nouvelle situation de la recherche musicale, face à l'ordinateur, instrument de synthèse," *Les cahiers de l'atelier d'informatique 1971*, no. 8, text from a conference at Journées d'Études du Festival International du Son 1971. Document at GRM, unpublished, uncatalogued, pp. 6–7.

75. Régnier, 7.

76. DEC, a company specializing in digital computing, was a smaller competitor of IBM. Today, DEC has been bought by Compacq. Engineers Bénédict Mailliard and Jean-François Allouis took over responsibility for computer projects from Francis Régnier in 1975.

77. Mailliard, "À la recherche du studio musical," 54–56.

78. Mailliard, 56.

79. Daniel Teruggi, *Le système Syter* (thesis, University of Paris VIII, 1998), 204.

80. See the detailed list of these works in chapter 9, under the year 1985.

81. The Apple II was introduced in 1977. Macintosh is a registered trademark, as is Pro Tools.

82. After graduating from École Nationale Supérieure des Télécom (ENST) in 1985, Hugues Vinet took his first job at the GRM, becoming scientific director of IRCAM in 1994.

83. ICMC, International Computer Music Conference.

84. Jean-François Allouis's strikingly original Syter ball interpolator would only start being implemented in personal computers in the 1990s, due to their earlier limited computational power.

85. Max was launched in 1971 by Miller Puckette. The name Max was chosen as a tribute to Max Mathews.

86. Schaeffer, *TOM*, 631; Schaeffer, *TMO*, 504.

87. Iannis Xenakis used the UPIC (Unité Polyagogique Informatique du Cemamu) to compose his piece *Mycenae Alpha* in 1978. Also realized with the aid of this tool were works by François-Bernard Mâche, Jean-Claude Eloy, Patrice Bernard, and Yuasa Yuasa.

88. Terminology of musical semiology—*esthesic*, *poietic*, and *neutral level*—is presented in the following chapter.

89. See the horseshoe-shaped 1943 studio of the Club d'Essai in figure 2.

90. Patents registered by Mathews, 1980, and Mathews and Boie, 1987, in the United States.

91. Association pour la Création et la Recherche sur les Outils d'Expression (Association for Creation and Research on Tools of Expression) specialized in artificial intelligence and applied mathematics.

92. François Delalande authored a book dedicated to the subject, *Le son des musiques entre technologie et esthétique* (INA/Buchet-Chastel, Paris, 2001).
93. Geneviève Mâche, cousin of the composer François-Bernard Mâche, and wife of François Bayle.
94. The late Philippe Carson had come from the Musée de l'Homme, where he had become convinced of the importance of conservation and indexing of archives in ethnology.
95. Interview with Laurent Cuniot in 1988. Personal document, held by Geneviève Mâche.
96. Statement gathered by the author in 2005.
97. K for a recording speed of 19 cm/sec., L for 38 cm/sec., and M. for 76 cm/sec. (7.5, 15, & 30 ips).
98. These letters corresponded to the initials of the recording studios: P for Pistor, E for Erard, B for Bourdet, R for Rodin, U for University, LY for Lyon, and LLL for Lille.
99. DAT is the abbreviation for *digital audio tape* (cassette format), and CD signifies *compact disc*.
100. Schaeffer, *TOM*, 64–65; Schaeffer, *TMO*, 42.
101. The first phase of development in 2001 cost €38,000.
102. See the next chapter.

CHAPTER 4. SPACE—CONCERT—AUDIENCE

Epigraph: Interview with the author in 2004. GRM archives.

1. *Faire sonner* the space, literally *to make the space sound*, i.e., to best exploit the available technical resources in the context of the specific site and its acoustic characteristics.
2. Évelyne Gayou, "Intentions de création en musique électroacoustique, chez les compositeurs invités au GRM" (Creative intentions in electoacoustic music among guest composers at the GRM), in *Intentions et création dans la musique d'aujourd'hui*, in "Actes de la journée du 24 mars 2004, 1res Rencontres Interartistiques de l'OMF," series "Conférences et Séminaires" no. 19, University of Paris IV–Sorbonne, 2005, 21–27.
3. "L'espace du son," *Lien, revue d'esthétique musicale*, ed. Francis Dhomont (Musiques & Recherches, Ohain, Belgium), 1988, 21, 24.
4. Henri Bergson, *La pensée et le mouvant* (1938; repr., PUF, Paris, 2003), 11.
5. Michel Rigoni, *Stockhausen . . . un vaisseau lancé vers le ciel* (Millénaire III, Rouen, 1998), 337.
6. In particular his *Licht* cycle, begun in 1977.
7. Rigoni, *Stockhausen*, 337.
8. Interview from May 2001.
9. Abraham Moles and Élisabeth Rohmer, *Psychologie de l'espace* (1972; repr., L'Harmattan, Paris, 1998).
10. Gilles Deleuze and Felix Guattari, *Mille plateaux* (Minuit, Paris, 1980), 598.
11. Peter Szendy, "Actes d'une dislocation," *Les cahiers de l'IRCAM, Recherche et musique*, no. 5 (Espaces, IRCAM/Centre Pompidou, Paris, 1994), 53–64.
12. Several photos of this installation were taken that same day in 1951, with either Pierre Henry or Pierre Schaeffer at the controls. See figure 12 in chapter 6.

13. *Orphée 51 ou Toute la lyre*, a forty-five-minute work by P. Schaeffer and P. Henry, premiered July 6, 1951, at Théâtre de l'Empire in Paris.

14. Michel Chion, *L'art des sons fixés, ou La Musique Concrètement* (Metamkine/Nota-Bene/Sono-Concept, Fontaine, 1991); Elsa Justel, *Les structures formelles de la musique de production électronique* (thesis, University of Paris VIII, 2000).

15. Peter Szendy, "Propos impromptus sur l'aura et l'espace du son," *Les cahiers de l'IRCAM, Recherche et musique*, no. 5 (Espaces, IRCAM/Centre Pompidou, Paris, 1994), 184–85.

16. *Portraits polychromes*, no. 7, *John Chowning* (2005; new and expanded edition, INA-GRM, Paris, 2007), 52. In this volume, Jean-Claude Risset cites the work of physicist Daniel Kastler and mathematician Alain Connes. English translation by David Vaughn: *John Chowning, Polychrome Portraits* no. 18 (GRM/INA, Paris, 2013), 45.

17. François Bayle was influenced by his readings of Bachelard, notably *La Poésie de l'espace*.

18. François Bayle, *Musique acousmatique, propositions . . . . . . positions* (INA-GRM/Buchet-Chastel, Paris, 1993), 42–43.

19. June 17, 1953.

20. Bernhardt and Garrett believed they had invented stereophony, but discovered afterwards that the Philips firm had registered a patent before the war.

21. This device had strongly impressed the architect Le Corbusier, who would employ it, with Iannis Xenakis, for the inauguration of the Cité Radieuse in Marseille.

22. Internal document, GRM, uncatalogued.

23. *La revue musicale*, no. 236 (1953): 110.

24. See chapter 6, at the year 1954, for greater detail on the work *Déserts* and the memorable scandal at its premiere.

25. Bayle, *Musique acousmatique*, 44–46.

26. The two GMEB directors, Christian Clozier and Françoise Barrière, were former GRM collaborators who had left in 1968.

27. Christian Clozier, "Un instrument de diffusion: Le gmebaphone," in *Lien, revue d'esthétique musicale*, ed. Francis Dhomont (Musiques et Recherches, Ohain, Belgium, 1988), 56.

28. Notably Jonathan Prager, from the Motus team, directed by Denis Dufour since 1996.

29. François Bayle, interview in *Portraits polychromes* no. 6 (Michel de Maule / INA, 2003), 43.

30. The France Musique radio network did not broadcast in stereo until 1964, but stereo recording of concerts had begun in 1959. Radio remains the principal diffuser of electroacoustic musical works, in terms of volume, although since the early 2000s this has decreased slightly, particularly with the emergence of the Internet.

31. Moles and Rohmer, *Psychologie de l'espace*.

32. René Thom, *Paraboles et catastrophes* (Flammarion, Paris, 1989).

33. Christian Zanési informed me that this was still the case during the GRM Experience concert that he presented in Rome in June 2005. However, this audience reaction has become rare.

34. Bernard Lortat-Jacob, *Musiques en fête* (Société d'ethnologie, Nanterre, 1994), 130.

35. Pierre Schaeffer, *Traité des objets musicaux* (*TOM*) (Seuil, Paris, 1966), 116–28; Schaeffer, *Treatise on Musical Objects, An Essay across Disciplines* (*TMO*), trans. Christine North and John Dack (University of California Press, Oakland, 2017), 80–93.

36. Lortat-Jacob, *Musiques en fête*, 129. For forty years, Bernard Lortat-Jacob was the director of the ethnomusicology laboratory of the Musée de l'Homme, in Paris.

37. *Musiques en fête*, 134.

38. Bruce Chatwin, *The Songlines* (Penguin Books, London, 1987), 2; French version: Chatwin, *Le chant des pistes* (Grasset, Paris, 1988), 10.

39. Michel Maffesoli, *Au creux des apparences, pour une éthique de l'esthétique* (Plon, Paris, 1990), 219.

40. Maffesoli, 220. Also see Maffesoli in Michel Gaillot, *La Techno, un laboratoire artistique et politique du présent* (Dis Voir, Paris, 2003).

41. Program leaflet for Festival de la Recherche 1960, GRM archives, p. 1.

42. See chapters 7 and 8, where some of these works are described.

43. "Le concert pourquoi? Comment?," in *Cahiers Recherche/Musique*, no. 5, ed. Michel Chion (INA-GRM, Paris), 93.

44. In 1998–99, France witnessed the arrival of high-speed internet and the explosion of personal production systems (home studios).

45. Seminar led by François Delalande, on the theme of "experimental creation in the 'interstices' of the system," with the participation of Bernard Delage (Bureau d'Etudes Acoustiques Delage & Delage) and Bruno Letort from Radio France.

46. Maffesoli, *Au creux des apparences*.

47. Antonio R. Damasio, *Spinoza avait raison: Joie et tristesse, le cerveau des émotions* (Odile Jacob, Paris, 2003), 32.

48. See the research by François Delalande carried out in Italy on the initial sensory-motor experiences of children from nursery school age, which would be at the origin of their capacity to perceive music.

49. Study titled *Composer sur son ordinateur: Les pratiques musicales en amateur liées à l'informatique*, Ministère de la culture et de la communication, Direction de la musique, de la danse, du théâtre et des spectacles vivants, Département des études et de la prospective, Serge Pouts-Lajus, Sophie Tiévant, Jérôme Joy, and Jean-Christophe Sevin, 2002.

50. One hour per week in a middle school class of approximately thirty students aged about eleven to fourteen

51. Reminiscent of the not-so-old debate about the relationship between jazz and classical music.

52. Gilbert Rouget, *La musique et la transe* (Gallimard, Paris, 1990), 53. In the same work, Rouget uses the term *sujets musiqués* (musicked subjects) in certain types of communal trances induced by music.

53. Friedrich Nietzsche, *La naissance de la tragédie* (Le livre de Poche, Paris, 1994).

54. For example in the frenzy of the techno free party.

55. Michel Maffesoli, "interview," in *La Techno, un laboratoire artistique et politique du présent* (Dis Voir, Paris, 2003), 111.

56. Bastien Gallet, *Le Boucher du prince Wen-Houei: Enquêtes sur les musiques électroniques* (Musica Falsa, Paris, 2000).
57. Commissioned by the GRM for the Multiphonie season at the Grand Auditorium of the Maison de Radio France, in Paris.
58. The Australian sound design project, webarchive.nla.gov.au.
59. Aleatory and multichannel works, in 1955.
60. Nicolas Schöffer, *Perturbation et chronocratie* (Denoël/Gonthier, Paris, 1978), 225.
61. François Donato left the GRM in September 2005.

CHAPTER 5. IN SEARCH OF MUSIC WRITING

Epigraph: Henri Bergson, excerpt from the 1911 conference "L'intuition philosophique," in *La pensée et le mouvant* (1938; repr., PUF, Paris, 2003), 130.

1. Stéphane Roy, *L'analyse des musiques électroacoustiques: Modèles et propositions* (L'Harmattan, Paris, 2003); Pierre Couprie, *La musique électroacoustique: Analyse morphologique et représentation analytique* (thesis, University of Paris IV–Sorbonne, 2003).
2. Pierre Schaeffer, *Traité des objets musicaux: Essai interdisciplines* (*TOM*) (Seuil, Paris, 1966), 19; Schaeffer, *Treatise on Musical Objects, An Essay across Disciplines* (*TMO*), trans. Christine North and John Dack (University of California Press, Oakland, 2017), 4.
3. Schaeffer, *TOM*, 19; Schaeffer, *TMO*, 4.
4. Schaeffer, *TOM*, 699–700, 686; Schaeffer, *TMO*, 559, 548.
5. Pierre Schaeffer, *Machines à communiquer*, vol. 1: *Genèse des simulacres*, vol. 2: *Pouvoir et communication* (Seuil, Paris, 1970).
6. Schaeffer, 1:311.
7. Schaeffer, 1:313.
8. Schaeffer, 1:312.
9. Schaeffer, 1:163. This 1953 text was published as "Vers une musique expérimentale," *La revue musicale* no. 256, ed. Pierre Schaeffer (Paris, 1957), 11–27.
10. Schaeffer, 1:169.
11. François Delalande, "L'analyse des musiques électroacoustiques," *Musique en jeu*, no. 8 (1972): 51.
12. Schaeffer, *TOM*, 114–28; Schaeffer, *TMO*, 82–93.
13. Schaeffer, *Machines à communiquer*, 1:167.
14. Schaeffer, 1:314.
15. Nattiez explained:

> By "poietic" I understand describing the *link* among the composer's intentions, his creative procedures, his mental schemas, and the *result* of this collection of strategies; that is, the components that go into the work's material embodiment. Poietic description thus also deals with a quite special form of hearing (Varèse called it "the interior ear"): what the composer hears while imagining the work's sonorous results, or while experimenting at the piano, or with tape.... By "esthesic" I understand not merely the artificially attentive hearing of a musicologist, but the description of perceptive behaviors within a given population of

listeners; that is how this or that aspect of sonorous reality is captured by their perceptive strategies

Jean-Jacques Nattiez, *Music and Discourse: Toward a Semiology of Music* (Princeton University Press, Princeton, 1990). Translation by Carolyn Abbate of *La musique et le Discours: Apologie de la musicologie* (Éditions Fides, Anjou, Québec, 2011).

16. François Delalande, "L'articulation interne/externe et la détermination des pertinences en analyse," *Observation, analyse, modèles: Peut-on parler de l'art avec les outils de la science?* Proceedings of the 2nd International Colloquium on Musical Epistemology (IRCAM/L'Harmattan, Paris, 2002), 3.

17. Delalande, p. 17.

18. Jean Molino, "Fait musical et sémiologie de la musique," *Musique en jeu*, no. 17 (1975), also printed in Jean Molino, *Le singe musicien: Sémiologie et anthropologie de la musique* (Actes Sud/INA, 2009), 73–115; Jean-Jacques Nattiez, *Fondements d'une sémiologie de la musique* (Union Générale d'Édition, coll. 10–18, Paris, 1975).

19. Roy, *L'analyse des musiques électroacoustiques*.

20. Schaeffer, *TOM*, 701; Schaeffer, *TMO*, 560.

21. Jean Molino, "La musique et l'objet," in *Ouïr, entendre, écouter, comprendre après Schaeffer* (Buchet-Chastel, Paris, 1999), 121.

22. Roy, *L'analyse des musiques électroacoustiques*, 21–31.

23. Philippe Mion, Jean-Jacques Nattiez, and Jean-Christophe Thomas, *L'envers d'une œuvre: De Natura sonorum de Bernard Parmegiani* (INA-GRM/Buchet-Chastel, Paris, 1983).

24. Jean-Pierre Richard, *Littérature et sensation* (Seuil, Paris, 1954).

25. *Portraits polychromes*, no. 6, *François Bayle* (Michel de Maule / INA-GRM, Paris, 2004; new and expanded edition, INA-GRM, Paris, 2007).

26. Marie-Noëlle Moyal, "Vagues figures ou les promesses du flou: L'art acousmatique comme rapport 'flou' au réel," *Proceedings of the seventh CICADA colloquium, 5, 6, 7 December 1996* (Publications of the University of Pau PUP, Pau, 1997), 282.

27. Moyal, 281.

28. François Delalande, *Le son des musiques* (Buchet-Chastel, Paris, 2001).

29. François Delalande, "Il gesto musicale, Dal senso motorio al simbolico—Aspetti ontogenettici" [Musical gesture, from sensory-motor to symbolic—Ontogenetic aspects], in *Dall'atto motorio alla interpretazione musicale*, Actes du 2e colloque international de psychologie de la musique (Proceedings of the Second International Conference on the Psychology of Music) (Ravello, October 1990), Editions 10/17, Salerne, Italy, 1992.

30. Denis Dufour, ed., *Ouïr, entendre, écouter, comprendre après Schaeffer* (Buchet-Chastel, Paris, 1999), 231.

31. The initial presentation of this research to other members of the GRM took place during an in-house seminar January 13, 1993.

32. CD-ROM, *Les Unités sémiotiques temporelles: Éléments nouveaux d'analyse musicale* (Laboratoire Musique et Informatique de Marseille, Marseille, 1996), distributed by ESKA, 5 av. de l'Opéra 75001 Paris, Documents Musurgia collection.

33. CD-ROM, *Les Unités sémiotiques temporelles*.

34. CD-ROM, *Les musicographies* (INA-GRM, 1995, not sold commercially).
35. CD-ROM, *Les musiques électroacoustiques* (GRM, Hyptique, Paris, 2000).
36. See the full list of *Portraits polychromes* monographs, including translations, in the bibliography—discography.
37. CD-ROM, *Les musicographies*.
38. Terminology developed by Roy in his book *L'analyse des musiques électroacoustiques*, 26.
39. However, in the case of Jean-Christophe Thomas, it should be noted that he attended the two-year electroacoustic composition course at the CNSMP.
40. Sonagramme is derived from the brand name of the device that generates it, the Sonagraphe. Technically, Sonagraphe is a misnomer, but is now in common use. Spectrogram would be more accurate.
41. In François Bayle, *Musique acousmatique, propositions . . . . . . positions* (INA-GRM/Buchet-Chastel, Paris, 1993), 165.
42. See Samuel Rousselier and Benjamin Levaux's transcriptons in "Portrait Parmegiani" on the GRM site, www.ina-grm.com.
43. Those specifications were drafted by Jean-Baptiste Thiébault.
44. Michel Chion, *Sur la transcription graphique des œuvres de musique acousmatique/concrète*, GRM seminar, April 3, 1994, unpublished document.
45. Henri Bergson, *La pensée et le mouvant* (1938; repr., PUF, Paris, 2003), 13.
46. Quote borrowed by Schaeffer from Hoffmann. Schaeffer, *TOM*, 9; Schaeffer, *TMO*, xxxvii.
47. Olivier Cullin, *L'image musique* (Fayard, Paris, 2006), 12.
48. Rudolf Arnheim, *Radio*, 1936; repr., Van Dieren, 2005), 144; English translation: Arnheim, *Radio*, trans. Margaret Ludwig and Herbert Read (Faber & Faber, London, 1936), 137.
49. Rudolf Arnheim, *La pensée visuelle* (1969; repr., Flammarion, Paris, 1976), 107.
50. "L'espace du son," in *Lien, revue d'esthétique musicale*, ed. Francis Dhomont (Ohain, Belgium, 1988), 37.
51. François Bayle, *Komposition und Musikwissenschaft im Dialog IV (2000–2003): L'image de son; Technique de mon écoute* (LIT, Cologne, 2003), 52.
52. Claude Bailblé, professor in the cinema department at the University of Paris VIII, at La Fémis, and at the Louis Lumière School, is frequently invited to lecture in the professional training programs for sound technicians at Radio France.
53. Claude Bailblé, *La perception et l'attention modifiées par le dispositif cinéma* (thesis, University of Paris VIII, 1999), 245.
54. Claude Bailblé also tackled these themes during a radio program produced by the GRM on France Culture, *Le son*, 10:30–11 a.m., September 20, 2001.
55. Bailblé, *La perception et l'attention modifiées*, 246.
56. Conference by Bernard Stiegler, at the GRM seminar *Analyses croisées, Les objets temporels*, March 18, 1997.
57. Jean-Yves Bosseur, *Musique et arts plastiques: Interactions au XXe siècle* (Minerve, Paris, 1998).

58. The partially graphic score of André Boucourechliev's *Archipelago I* (1967) resembles the map of an archipelago suggesting various routes from one island to another.

59. Jean-Yves Bosseur, *Musique et arts plastiques*, 217–19.

60. Cécile Regnault-Bousquet, *Les représentations visuelles du sonore, Application à l'urbanisme* (thesis, Université de Nantes, École d'Architecture de Grenoble and École Polytechnique, 2001).

61. Roy, *L'analyse des musiques électroacoustiques*, 25.

62. Schaeffer, *TOM*, 83–84; Schaeffer, *TMO*, viii, 57.

63. Jacques Siron, *La partition intérieure: Jazz, musiques improvisées* (1992; repr., Outre Mesure, Paris, 2004), 17.

64. Roy, *L'analyse des musiques électroacoustiques*, 95.

65. Gilles Deleuze, *Cinéma 1: L'Image mouvement, Cinéma 2: L'Image-temps* (Minuit, Paris, 1983).

66. Jean-Yves Bosseur, *Du son au signe: Histoire de la notation musicale* (Alternatives, Paris, 2005), 133.

CHAPTER 6. 1948–1958

Epigraph: Cocteau, *Le Coq et l'Arlequin* (1918; repr., Stock, Paris, 1993), 47; Cocteau, *Cock and Harlequin: Notes Concerning Music* (Egoist Press, London, 1921), 9.

1. Interview with Luc Ferrari by François Delalande and Évelyne Gayou, for radio, 1998. accessible at ina.fr.

2. Danièle Cohen-Levinas, "Les icônes de l'écoute, Schoenberg, les oh! et les ah! de l'histoire," in *De l'écoute à l'oeuvre: Études interdisciplinaires; Actes du colloque tenu en Sorbonne les 19 et 20 février 1999*, ed Michel Imberty (Editions L'Harmattan, Paris, 2001), 37.

3. Pierre Schaeffer, *À la recherche d'une musique concrète* (1952; repr., Seuil, Paris, 1998). This booklet, from the four-record boxed set *Pierre Schaeffer, L'œuvre musicale*, published by the INA/GRM and the Séguier bookstore in 1990, also included the different *Études*.

4. Schaeffer, 24.

5. Jean-Étienne Marie, *Musique vivante* (Privat/PUF, Paris, 1953), 203–4.

6. Presented by Serge Moreux.

7. President, Raymond Bayer, professor of general philosophy.

8. The work was based on Théophile Gautier's play *Une larme du diable*, performed by Gérard Philippe, Marcelle Derrien, and Danièle Delorme. The Italia award, organized by Italian Radio (RAI), is an annual award for a creative radio work, open to productions from all over the world. In 1952, a boxed set of six vinyl discs was released: *10 ans d'essais radiophoniques 1942–1952*, where the principal radio experiments from the Studio d'Essai and the beginnings of the Club d'Essai discussed in chapter 1 can be heard, notably an extract from *Une larme du diable*.

9. The stereo effect involved only the sound levels. The material on the two tracks was identical, except that changes in volume levels gave the illusion of movement.

10. *10 ans d'essais radiophoniques*, disc B, tracks 21, 22, 23.

11. See a description of the installation in chapter 4, including Schaeffer's comments on sound projection and relief.

12. See the detailed description of phonogènes in chapter 3, as well as photographs of the phonogènes.

13. Orpheus, poet and musician, descended into the underworld of Hades to recover his beloved Eurydice, only to lose her forever after defying the imposed rule not to look back. Schaeffer would often identify himself with Orpheus.

14. In 1950, Pierre Henry had composed *Aube*, the first piece of musique concrète for a film. Directed by Jean-Claude Sée.

15. In "Le concert pourquoi? Comment?," *Cahiers recherche/Musique*, no. 5 (1977): 102.

16. Jean-Étienne Marie, *Musique vivante* (Privat/PUF, Paris, 1953), 207.

17. Marie, 205.

18. See chapter 3, the biographical paragraph about Michel Philippot.

19. This étude was revised by the author for the release of the compilation disc of his early works under the title *Le Microphone bien tempéré*.

20. Interview with the author 2005.

21. Evoked in these terms by Pierre Schaeffer in the boxed set: *Pierre Schaeffer, L'œuvre musicale* (INA-GRM Librairie Séguier, Paris, 1990), 49.

22. Presentation notes for the concert, archives of the GRM.

23. Presentation notes.

24. Varèse coined the term *organized sounds*.

25. Note the futurist influence.

26. Fernand Ouelette, *Edgard Varèse* (Christian Bourgois, Paris, 1989).

27. Tape conserved in the GRM archives, no. 32 LUR B. Released as a CD by the INA in 2007, along with interviews with Georges Charbonnier.

28. Presentation leaflet for *Festival de la recherche*, 1960, p. 9.

29. Jacques Bureau, electronics engineer and jazz enthusiast, founded the Hot Club in 1932.

30. In fact, Béjart had already staged a first short ballet to musique concrète at Fontaine des Quatre Saisons in 1954, with music by Philippe Arthuys, based on an idea by Prévert.

31. Pierre Henry, *Journal de mes sons* (Actes sud, Arles, 2004), 49–50.

32. And reciprocally Meyer-Eppler kept himself informed of the work at the Paris Studio.

33. Pierre Schaeffer, *Traité des objets musicaux: Essai interdisciplines* (TOM) (Seuil, Paris, 1966), 65; *Treatise on Musical Objects, An Essay across Disciplines* (TMO), trans. Christine North and John Dack (University of California Press, Oakland, 2017), 42.

34. These questions concern every musical genre. On July 5, 1954, the American Elvis Presley recorded the song "It's All Right," which became "The Sound" of rock.

35. Jean-Jacques Nattiez, in *Pierre Boulez/John Cage, Correspondance* (Christian Bourgois, Paris, 1991), 237.

36. See *Portraits polychromes, Eliane Radigue*, no. 17 (INA-GRM, Paris, 2013).

37. See Jesus Aguila's book, *Le Domaine musical* (Fayard, Paris, 1992).

38. Nattiez in *Correspondance*. For reference, remember that Pierre Boulez also composed *Le Marteau sans maître*, for contralto, five instruments, and percussion in 1954.

39. Excerpt from Pierre Boulez's text in *Encyclopédie Fasquelle* (Paris, 1958), 577.

40. Pierre Henry and Philippe Arthuys had also experimented with prepared piano in 1949.

41. Joël Chadabe, *Electric Sound: The Past and Promise of Electronic Music* (Prentice Hall, Upper Saddle River, NJ, 1997), 55; account of Earle Brown, for the 1998 radio broadcast: *Opus, Earle Brown*, France Culture, June 19, 1999, by Jean-Yves Bosseur and Évelyne Gayou. The *I Ching* was the foundational text of Chinese civilization for thirty-five centuries.

42. Pierre Boulez, "Alea," *La Nouvelle Revue Française*, no. 59, November 1, 1957, also in Boulez, *Relevés d'apprenti* (Seuil, Paris, 1966), 41.

43. Series of five radio broadcasts for France Culture, *Les 50 ans de la musique électronique au studio de Cologne*, November 11–15, 2001, by Évelyne Gayou and Philippe Langlois.

44. H. Eimert had composed a dodecaphonic string quartet in 1924, as well as a booklet on the subject. Collaborative works by H. Eimert and R. Beyer, with the assistance of the sound engineer Wilfried Bierhals, included *Klangstudie I* (1952) with sounds from the Melochord, *Klang im unbegrenzten Raum* (1952), *Ostinate Figuren und Rhythmen* (1952), *Klangstudie II* (1952), and *Klangstudie III* (1952).

45. Werner Meyer-Eppler coined the term *electronic music* in 1949, in his book *Elektrische Klangerzeugung*.

46. The Bode Melochord, the keyboard instrument invented by Harald Bode, produces electronic sounds at fixed intervals, while Friedrich Trautwein's Trautonium is capable of producing a wide gamut of scales.

47. Film composers, notably for Fred L. Wilcox's *Forbidden Planet* in 1956.

48. In France, Pierre Henry would be the first to use electronic sounds in musique concrète in his 1956 work *Haut-Voltage*, two years after *Gesang der Jünglinge*.

49. On the same model, Francisco Kropfl founded the Estudio de Fonologia de Buenos Aires, in Argentina.

50. Paolo Donati and Ettore Pacetti, eds., *C'erano una volta nove oscillatori..., Lo studio di fonologia della Rai di Milano nello sviluppo della nuova musica in italia* (RAI Teche, RAI Eri, Rome, 2002), 7.

51. Excerpt from the program of the 1950 lecture. GRM document.

52. Schaeffer evoked this collaboration with André Moles (aka Abraham Moles) in the brochure of *2 concerts de musique concrète*, May 21 and 25, 1952, at Salle de l'ancien Conservatoire, in Paris.

53. Admission to these events was 200 francs for each event, or 1,000 old francs for the entire series [or approximately $5 and $25, respectively, in inflation-adjusted US dollars in 2023—Trans.].

54. *Musique animée*, 15-minute radio broadcast by Philippe Arthuys and Jérôme Peignot (Chaîne Nationale, no. LUR 301, 1955).

55. See Michel Chion's book *Pierre Henry* (Fayard/Sacem, Paris, 2003).

56. Schaeffer, *TOM*, 24.

CHAPTER 7. 1958–1968

1. *Traité des objets musicaux* (*TOM*) was signed by Pierre Schaeffer but was the product of collective work with his collaborators.

2. Since 1957, the RTF had the status of EPIC, *établissement public à caractère industriel et commercial*.
3. *La Revue musicale*, "Vers une musique expérimentale," no. 236 (Richard-Masse, Paris, 1957), iv–v.
4. *La Revue musicale*, no. 236, v.
5. Internal document of the Service de la Recherche, 1961.
6. See below in this chapter *TOM* and *SOS*, and note 32.
7. Biographical note by Pierre Schaeffer, internal document of the GRM, approximately 1970. Not catalogued.
8. Interview in *Portraits polychromes*, no. 1, *Luc Ferrari* (INA-GRM/CDMC, Paris, 2001; new and expanded edition, INA-GRM, Paris, 2007), 30.
9. Asger Jorn, "Nouvelles de l'Internationale," in *Internationale Situationiste 1958–69* (Van Gennep, Amsterdam, 1972), 27, referring to the recent publication of his book *Pour la forme* (1958; repr., Allia, Paris, 2001).
10. When the GRM moved from the Bourdan Center to the Maison de la Radio in 1975, Studio 116 C of the Maison de la Radio was equipped with the material from the former Studio 52 of the Bourdan Center.
11. *Programme-Bulletin* no. 8 (ORTF, Paris, 1974), 6.
12. Jesus Aguila, *Le Domaine musical* (Fayard, Paris, 1992), 324–25.
13. Interview with Leopold von Knobelsdorff by Évelyne Gayou and Philippe Langlois, for France Culture, November 2001.
14. Jean-Yves Bosseur, *Le sonore et le visuel* (Dis Voir, Paris, 1992), 140.
15. Pierre Schaeffer, *Machines à communiquer, tome 1* (Editions Seuil, Paris, 1970), 168.
16. Schaeffer, 169.
17. Joël Chadabe, *Electric Sound* (Prentice Hall, Upper Saddle River, NJ, 1997), 15.
18. *La Revue musicale*, "Expériences musicales, musiques concrète électronique exotique," ed. Pierre Schaeffer and François-Bernard Mâche, no. 244 (Richard-Masse, Paris, 1959), 63.
19. *Le Service de la recherche* (internal presentation leaflet of the RTF, 1961), 33.
20. *Portraits polychromes*, *François Bayle* (2004; new and expanded edition, INA-GRM, Paris, 2007), 22.
21. *Portraits polychromes*, *François Bayle*, 26.
22. André Boucourechliev had just spent three years collaborating at the RAI studio in Milan.
23. The diffusion installation consisted of two tape recorders, the first with two tracks and the second with one.
24. *Le Service de la recherche*, 33.
25. ORTF stands for Office de Radio Télévision Française. French national radio—and later television—has changed names several times over its long history.
26. This "group" consisted of only one person, Sophie Brunet. She joined the GRM in 1957 and spent most of her career in the shadow of Pierre Schaeffer as a specialist and advisor on theoretical questions. She focused, among other things, on questions of language and semiology. She was also gifted as a writer.

27. This commentary accompanying the concert *Retour aux sources* would also appear in Pierre Schaeffer's *Machines à communiquer I* (Seuil, Paris, 1970), 164–82.

28. Spelling varies; one encounters Constantin or Konstantin, Simonovic or Simonovitch. We have retained the spellings used in the source documents.

29. Interview with the author in 1998.

30. Excerpt from François Delalande and Évelyne Gayou, "Xenakis et le GRM," in *Présences de Iannis Xenakis*, ed. Makis Solomos (CDMC, Paris, 2001), 32–33.

31. Pierre Schaeffer, *Traité des objets musicaux* (Seuil, Paris, 1966), referred to as the *TOM*, reedited in 1998 for the fiftieth anniversary of musique concrète, translated into English only recently by Christine North and John Dack, as *Treatise on Musical Objects, An Essay across Disciplines* (*TMO*) (University of California Press, Oakland, 2017). For discussion of the *TOM*, see especially chapter 3.

32. Pierre Schaeffer, *Solfège de l'objet sonore*, boxed set of three vinyl records and a booklet (Seuil, Paris, 1967), referred to as the *SOS*. When it was reissued as a CD in 1998, the English text was completely revised, and a Spanish text replaced the German text.

33. Cited in Sylvie Dallet and Anne Veitl, *Du sonore au musical* (L'Harmattan, Paris, 2001), 155–56.

34. Guy Reibel, *L'Homme musicien: Musique fondamentale et création musicale* (Edisud, Aix-en Provence, 2000), 335.

35. The spelling of Xenakis's first name took a long time before stabilizing as Iannis. The GMAP was composed of Pierre Barbaud (specialist in set theory), Roger Blanchard, Jeannine Charbonnier, Jean Germain, and Brian de Martinoir. During the first Festival de la Recherche in 1960, they presented a joint work, calculated by a machine, *7! (factorial 7)*, exploring the formula $7! = 1 \times 2 \times 3 \times ... \times 7 = 5{,}040$ combinations of the series 6, 5, Z, 7, X, 9, 1, 3, 0, 8, 4, 2 (F sharp, F, B, G, B-flat, A, C-sharp, D-sharp, C, G-sharp, E, D) chosen randomly.

36. After leaving the GRM, in 1966 Iannis Xenakis founded the Équipe de Mathématiques Musicales (EMAMU) with Théodore Guilbaud and Marc Barbut, which became the Centre de Mathématique Musicale (CEMAMU) in 1972.

37. Most of the works cited have been released on CD. Please refer to the GRM website for further information: https://inagrm.com/fr.

38. *Portraits polychromes, François Bayle*, 27–28.

39. Delalande and Gayou, "Xenakis et le GRM."

40. The New York school, revolving around John Cage: Morton Feldman, Christian Wolff, and Earle Brown.

41. The Maison de la Radio had been inaugurated by General Charles de Gaulle, president of the French Republic, in 1963.

42. *Kontakte*, the first entirely electronic piece, by Karlheinz Stockhausen, dated from 1960.

43. In *Portraits polychromes*, no. 5, *Ivo Malec* (Michel de Maule / INA, Paris, 2003), 77.

44. Frémiot in *Ouïr, entendre, écouter, comprendre après Schaeffer* (Buchet-Chastel, Paris, 1999), 232.

45. In *Portraits polychromes, François Bayle*, 92.

46. In *Portraits polychromes*, no. 4, *Bernard Parmegiani* (INA/CDMC, Paris, 2002; new and expanded edition, INA-GRM, Paris, 2011), 96.

47. Interview with Luc Ferrari by Évelyne Gayou and François Delalande, January 20, 1998, unpublished.

48. Following the Sorafom, founded in 1956, the Office de Coopération Radiophonique (OCORA), founded in 1962, was intended to provide assistance to African radio and television stations.

49. CD, *Xenakis, Electronic Music*, EMF CD 003.

50. François Delalande, *Il faut être constamment un immigré*, interviews with Iannis Xenakis (Buchet-Chastel, Paris, 1997).

CHAPTER 8. 1968-1978

Epigraph: Friedrich Nietzsche, *Crépuscule et idoles, maximes et pointes* §33 (1888), in *Œuvres complètes de Frédéric Nietzsche*, vol. 12, trans. Henri Albert (Mercure de France, Paris, 1908), 113.

1. In November 1971, François Bayle became a member of the SACEM's commission for classical music, the Commission de la Musique Symphonique, later becoming president of the commission from 2001 through 2005. His inclusion was an event of considerable symbolic importance for electroacoustic music.

2. Tape recording, no. KUR 762, 2 boxes, GRM archives.

3. GRI: Groupe de Recherche Image; GRT: Groupe de Recherche Technique; GEC: Groupe d'Étude Critique; GER: Groupe d'Enseignement Recherche. See also chapter 3, note 51.

4. Tape recording, no. LUR 2328, 7 boxes, GRM archives.

5. François Delalande, "L'analyse des musiques électroacoustiques," *Musique en jeu*, no. 8 (1972): 50–56.

6. Jean-Jacques Nattiez, "Situation de la sémiologie musicale" *Musique en jeu*, no. 5 (1971): 3–18.

7. *Portraits polychromes*, no. 6, *François Bayle* (Michel de Maule / INA-GRM, Paris, 2004; new and expanded edition, INA-GRM, Paris, 2007), 33.

8. Interview with Karlheinz Stockhausen for a radio broadcast on France Culture, November 15, 2002.

9. Jesus Aguila, *Le Domaine musical* (Fayard, Paris, 1992), 135.

10. Pierre-Albert Castanet, *Hugues Dufourt, 25 ans de musique contemporaine* (TUM/ Michel de Maule, Paris, 1995).

11. Célestin Deliège, *Cinquante ans de modernité musicale: De Darmstadt à l'IRCAM* (Mardaga, Sprimont, 2003), 873–79.

12. Anne Veitl and Noémi Duchemin, *Maurice Fleuret, une politique démocratique de la musique: 1981–1986* (Comité d'histoire du ministère de la Culture, Paris, 2000), 189.

13. Unpublished internal document, mimeographed, uncatalogued.

14. Pierre Schaeffer, "La musique et les ordinateurs," reprinted from an article in *Musique et technologie*, proceedings of the international conference organized by UNESCO in Stockholm, June 1970, published in *La Revue musicale*, no. 268–69 (Paris, 1971): 57–88.

15. With the permission of Max Mathews. At the time, the scientific community readily exchanged information.

16. "Vers un projet 'ordinateur' au Groupe de Recherches Musicales" [Towards a 'computer' project at the GRM], unpublished internal document of the Research Department, by Francis Régnier, reference GEC/486/10T, July 9, 1970.

17. Pierre-Alain Jaffrenou, also a composer, collaborated with the GRM from 1971 to 1977. Together with James Giroudon in Lyon in 1981, he founded GRAME, Générateur de Ressources et d'Activités Musicales Exploratoires.

18. Schaeffer, "La musique et les ordinateurs," 70.

19. In *Portraits polychromes, François Bayle*, 42.

20. École nationale supérieure des Telecom (ENST), the prestigious French national engineering school for telecommunications, which offered an audiovisual option in its curriculum as early as 1974.

21. Afterwards, Xavier Nouaille was recruited by Radio France, later becoming its technical director, while Jean-François Allouis remained at the GRM.

22. CD-ROM created with Dominique Besson, in 1995, not commercialized.

23. Polydor, 1992; *News for Lulu*, Sonny Clark, Hat Art Records 188.

24. Interview with Xavier Nouaille, September 2003.

25. *Cahiers recherche/musique*, no. 3 (INA-GRM, Paris, 1975), 1–3.

26. VCO, Voltage Control Oscillator; VCA, Voltage Control Amplifier.

27. *Krisie*, his intern study at the GRM, was dated 1974.

28. Robert Cohen-Solal (born in 1943) collaborated in the "applied music" sector of the GRM from 1965 to 1973. He then continued his career as a composer and teacher, mainly in the fields of theater and cinema.

29. Jacques Rouxel, a great admirer of Great Britain, chose the initials GB to designate the Gibi, the hero of this cartoon. As for the word Shadok, it means nothing, it simply "sounded" Anglo-Saxon, that's all. Interview by Évelyne Gayou with Jacques Rouxel, broadcast in five parts on France Culture, from February 26 to March 1, 1996.

30. *Portraits polychromes*, no. 4, *Bernard Parmegiani* (2002; new and expanded edition, INA-GRM, Paris, 2011), 35.

31. Jean-Yves and Dominique Bosseur, *Révolutions musicales, La musique contemporaine depuis 1945* (Minerve, Paris, 1993). In calling his project "art bruit," Bayle was citing the fine arts term that Jean Dubuffet had coined in 1955.

32. *Portraits polychromes, François Bayle* (INA-GRM, Paris, 2007), 103.

33. Commentary by Alain Savouret, for the program notes for his composition.

34. *Portraits polychromes, Bernard Parmegiani*, 33.

35. Gilles Deleuze and Felix Guattari, *Mille plateaux* (Minuit, Paris, 1980), 592–625, concerning the opposition between smooth and striated space, which Boulez had been the first to describe in these terms.

36. Philippe Langlois offered further comments on the audiovisual achievements of Parmegiani and the composers of the ORTF Research Department in his thesis *Les Procédés électroacoustiques dans les différents genres cinématographiques: Une étude transversale au XXe siècle* (University of Paris IV–Sorbonne, 2004).

37. André Boucourechliev, 1925–97, left his native Bulgaria, settling in France in 1949. Pianist, then composer and author of essays on music, he also taught at the École normale de musique de Paris until 1960.

38. Before moving into the Maison de la Radio in 1975, the GRM was housed in the Centre Bourdan, a private mansion in the sixteenth arrondissement of Paris, just a stone's throw from the future "round Maison."

39. Pierre Schaeffer in program notes for *Trièdre fertile*, in the catalog for the concert series "Pierre Henry 25 ans d'œuvres," December 17, 18, and 19, 1976, p. 21.

40. Following the election of François Mitterand as president of the Republic, and the Socialist Left coming to power.

41. ARC (Animation-Recherche-Confrontation), created in 1967 at the Museum of Modern Art of the City of Paris by Pierre Gaudibert, curator, had become one of the most active venues for contemporary art. The first issue of the ARC review had been published in 1958.

42. *Portraits polychromes, François Bayle*, 41.

43. Around Alain Trutat, the initial ACR team in 1969 was composed of René Farabet, Janine Antoine, Viviane van den Broek, and Marcel Grenier.

44. Adaptation of Rabelais.

45. René Farabet left the ACR at retirement age.

46. See Andrea Cohen, *Les compositeurs et l'art radiophonique* (L'Harmattan, Paris, 2015).

47. Olivier Corroenne, *L'atelier de création radiophonique de France culture*, dissertation for a degree in journalism and communication, Free University of Brussels, 1985.

48. Guy Reibel, *L'homme musicien: Musique fondamentale et création musicale* (Edisud, 2000), 333–45.

49. Reibel, 336.

50. At the Maison de la Radio, studio numbers always began with the numeral 1. The GRM operated Studio 123 (until December 2004) and Studio 116, subdivided into three sections: 116A, 116B, and 116C. The numbers of editing booths ("cellules") began with a 2. The GRM had three on the second floor of the building, next to 123: no. 209, 210, and 211 studios, which would be transferred to the sixth floor in 2005, when Radio France evacuated the previous premises to comply with fire safety regulations. The GRM also had three other soundproofed booths on the third floor, near its offices and the administrative offices.

51. They all continued their careers in electroacoustics. Notably, Dufour became an eminent member of the French electroacoustic scene, Caesar set up a center in Brazil, and Teruggi became the director of the GRM.

52. See the specific references in the bibliography.

53. Pierre Schaeffer, *Machines à communiquer 1* (Seuil, Paris, 1970), 7.

54. *La Revue musicale*, no. 274–75 (1971).

55. A5 format (5.8 × 8.3 inches), the pages attached with a single staple.

56. Michel Chion, *Guide des objets sonores* (INA/Buchet-Chastel, Paris, 1983).

57. Michel Chion and Guy Reibel, *Les Musiques électroacoustiques* (Édisud/INA-GRM, Aix-en-Provence, 1976.

58. Under the authority of the French Ministry of National Education.

59. Reibel, *L'homme musicien*, 333.

60. Reibel, 130.

61. Reibel, 303.

62. Guy Reibel, *Jeux vocaux* (Salabert, Paris, 1985).
63. Reibel, *L'homme musicien*, 188.
64. See chapter 9.
65. Reibel, *L'homme musicien*, 347.
66. Reibel, 308, 318–19.
67. Michel Chion, *L'art des sons fixés ou La Musique Concrètement* (Métamkine/Nota-Bene/Sono-Concept, Fontaine, 1991), 97.
68. Lionel Marchetti, *La musique concrète de Michel Chion* (Métamkine, Rives, 1998), 253.
69. It is worth noting that until 1972, the examination (in fact a miniature selection competition) was not held until the end of the first year. After 1972, the examination took place before the entry into the first year, based on much more traditional criteria of musical knowledge.
70. Marchetti, *La musique concrète de Michel Chion*, 248.
71. *Portraits polychromes*, no. 8, *Michel Chion* (INA-GRM, Paris, 2005).
72. The GRM even maintained tape recorders in functioning order in Studio 116C, almost exclusively for the use of Michel Chion.

CHAPTER 9. 1978–1988

1. See the article by Jean-Baptiste Barrière, "La fin de la recherche et le triomphe de la production?," in *Avidi Lumi*, no. 12 (Palerme, 2001).
2. See, on this subject, Anne Veitl and Noémi Duchemin's book *Maurice Fleuret, une politique démocratique de la musique, 1981–1986* (Comité d'histoire du ministère de la culture, Paris, 2000).
3. Veitl and Duchemin, 273.
4. Max Mathews had entrusted his program to Jean-Claude Risset, who distributed it as widely as possible in Europe, notably at the GRM and IRCAM.
5. Internal document of the GRM: a series of interviews on the history of sound synthesis.
6. See *Portraits polychromes*, no. 7, *John Chowning* (Michel de Maule / INA-GRM, Paris, 2005; new and expanded edition, INA-GRM, Paris, 2007); English translation: no. 18 (2013).
7. Andrew J. Nelson, *The Sound of Innovation: Stanford and the Computer Music Revolution* (MIT Press, Cambridge, MA, 2015).
8. Six years for David Wessel, five years for John Chowning, and three years for Jon Appleton.
9. As explained in the preceding chapter.
10. *Portraits polychromes*, no. 6, *François Bayle* (2007), 38.
11. Bénédict Mailliard passed away June 27, 1993.
12. Bernard Dürr, a self-taught electronics engineer, was passionate about studio technique and live instrumental playing. He had contributed to the development of Studio 54 at the Bourdan Center in the early 1970s.
13. At DEC, the PDP-11/23 and then the PDP-11/73 were successors to the 11/60. IRCAM also had a PDP-11, followed by a 32-bit VAC.

14. Interview by the author with Jean-François Allouis, in July 2004.

15. Guy Reibel, *L'homme musicien: Musique fondamentale et création musicale* (Edisud, Aix-en-Provence, 2000), 23.

16. Jean-François Allouis, "Syter et le temps réel," *La revue musicale*, "Recherche musicale au GRM," quadruple issue no. 394, 395, 396, 397 (Richard-Masse, Paris, 1986), 69.

17. Jean-Claude Risset, interviewed by Olivier Meston in *Portraits polychromes*, no. 2, *Jean-Claude Risset* (INA-GRM/CDMC, Paris, 2001), 40; new and expanded version, 2008; translated into English in *Jean-Claude Risset: Polychrome portraits*, no. 19 (INA, Paris, 2013), 38.

18. Giuseppe Di Giugno also went by Pepino Di Giugno.

19. In 1985, Pierre Boulez composed *Dialogue de l'ombre double*, for clarinet, tape, and spatialization system.

20. In 2007, Denis Lorain acted as coordinator for its five digital studios.

21. Veitl and Duchemin, *Maurice Fleuret*, 181. The Conservatoire de Musique de Paris had been established during the French Revolution by the law of 16 Thermidor year III (August 3, 1795).

22. In 2004, Michel Larquié took over as director of ADAC, which was holding 450 workshops for the 250 disciplines taught.

23. At the site of today's Cité de la Musique de la Villette, in Paris.

24. Since 2004, the workshop has been handled by ADAC.

25. Pierre-Alain Jaffrenou was a member of the GRM from 1971 to 1977.

26. See the Grame internet site.

27. François Bayle's collaborator and wife.

28. See "le catalogue commenté des œuvres de François Bayle," by Régis Renouard-Larivière, in *Portraits polychromes, François Bayle*.

29. An open letter to Pierre Schaeffer, serving as preface, concerning *Pierres réfléchies*, April 1982, GRM Archives.

30. *UPIC du Cemamu* stands for Unité Polyagogique Informatique du Centre d'Études de Mathématiques et Automatique Musicale (Polyagogical Computer Unit of the Center for Studies of Mathematics and Musical Automation). The UPIC is a computer system enabling music making through drawing. Program note from the concert, GRM Archives.

31. *Portraits polychromes*, no. 4, *Bernard Parmegiani* (INA-GRM, Paris, 2002), 106. Catalog of Bernard Parmegiani's works with commentary by Régis Renouard-Larivière; new and expanded edition (INA-GRM, Paris, 2011), 125.

32. "Recherche musicale au GRM," *La revue musicale*, quadruple issue no. 394, 395, 396, 397 (Richard Masse, Paris, 1986), 239.

33. This work has been analyzed by Daniel Teruggi in *Portraits polychromes*, no. 2, *Jean-Claude Risset* (INA-GRM/CDMC, Paris, 2001), 61–68.

34. *Portraits polychromes*, no. 3, *Gilles Racot* (INA-GRM, Paris, 2002), 101.

35. A member of the French Resistance during the war, Agnès Tanguy was part of the circle of Pierre Schaeffer's most faithful collaborators.

36. Named after the piece by François Bayle that premiered that day.

37. According to Jean-Claude Risset, Joël Chadabe's 1978 piece *Solo*, was most likely the first work to be composed interactively. With the aid of two Theremin antennas patched to

the Synclavier digital synthesizer, Chadabe controlled certain compositional parameters by gesture.

38. Bernard Dürr assisted Berio for his *Chants parallèles* in 1975.

CHAPTER 10. 1988-1998

Epigraph: Inscription by Man Ray under a reproduction of his work *Le violon d'Ingres*.

1. The Grand Auditorium of the Maison de Radio France was renamed Salle Olivier Messiaen on October 22, 1992, in homage to the composer, who had died a few months earlier.

2. To give the reader some guidance: the year 1973 saw the birth of the Ethernet network at Xerox PARC; 1989 saw the release of the first Nintendo Game Boy and 1997 the first MSN Instant Messaging.

3. Bernard Stiegler (1952–2020) was also the author of several important books on technology and time, including *La technique et le temps* (2 vols., Éditions Galilée, Paris, 1994 and 1996).

4. The *s* at the end of the word *musique* appeared with the arrival of Pierre Bouteiller as director of the network in 2000. [In French, using the plural *musiques* stresses aesthetic and cultural diversity.—Trans.]

5. The first two workshops were led by Jean-François Allouis, and then all the others by Daniel Teruggi.

6. This project was launched under the name Genesis in 1997, independently of the GRM, which had chosen to focus on lightweight computer systems.

7. Interview with Denis Dufour, by Jonathan Prager, *Ars Sonora Revue*, no. 5 (Paris, 1997): 12.

8. *Les musicographies*, CD-ROM (INA-GRM, 1995, not sold commercially).

9. After leaving IRCAM, which he had directed since 1975, Pierre Boulez decided to devote himself exclusively to conducting and composing. Between 1991 and 1995 he reworked *Explosante fixe* for flutes, ensemble, and computer. He signed an exclusive contract with Deutsche Grammophon.

10. Marco Stroppa was one of the last composers to work at the Cologne Studio, composing his work *Zwielicht* from 1994 to 1999.

11. Philippe Mion collaborated with the GRM from 1979 to 1989, notably as a specialist in executing concert works on the Acousmonium. From 1989 to 1991 he was in residence at the Villa Medicis in Rome. Today, in parallel to his career as a composer, he teaches at the Conservatoire de Vitry and the CFMI in Lyon, as well as in Belgium.

12. The MIDI norm had existed since 1983.

13. The ADAC workshop closed in 2005, as a result of budgetary restrictions imposed by the City of Paris on ADAC workshops.

14. These analyses of *La terrasse des audiences* from Debussy's *Clair de lune* were published before the Congress, in a special edition of the journal *Analyse Musicale* (no. 16 in French, and no. 16 bis in English). Compiling such a large number of analyses of one single work by researchers from such different horizons was a first.

15. By the journalist Pierre Gervasoni, *Le Monde* (newspaper), January 30, 2004.

16. See François Bayle, *Musique acousmatique, propositions ... ... positions* (INA-GRM/Buchet-Chastel, Paris, 1993), 190. Bayle's concept of "Son-Mu" evokes the territory "between sound and music."

17. Presentation leaflet of the Seminar, document from the GRM archives.

18. Presentation leaflet of the Seminar.

19. "Empreintes" presentation leaflet, document from the GRM archives.

20. Bayle, *Musique acousmatique*.

21. Brochure realized by the ADDA of the Department of Tarn, the GMEA, and Ouïe-dire Production.

22. Notable alumni include Samuel Sighicelli and Benjamin de la Fuente.

23. The Syter system, entrusted to Yann Geslin, had already been present since 1986.

24. It should also be mentioned that an attempt to create a "computer" class under the auspices of IRCAM had been made at the conservatory in 1985, the conservatory acquiring its very first computer, a Macintosh Mac Plus. David Wessel, Tristan Murail, and Arnaud Petit directed the workshop in turn until 1990, when IRCAM created its own computer music course.

CHAPTER 11. 1998 AND BEYOND

1. Inamedia.com, an extranet-type site, with access restricted only to authorized users.

2. Cavada's successor in 2005, François Cluzel, would continue the same approach.

3. Since 1975, the GRM had produced between twenty and twenty-five hours of radio programming per year for France Culture.

4. In radio the archive began in 1933, but it was only in 1975 that serious documentation and indexing began. In 2003, it was estimated that 60 percent of the collection lacked all documentation.

5. In fact, the project would end "on schedule," in 2022, following a new timeline established in 2018.

6. In particular the "general" sound library, i.e., all the tapes other than musical ones, essentially conferences and radio programs dating from before 1977, including the very first programs of the Studio d'Essai.

7. Suzanne Bordenave, the administrator who had followed Schaeffer from French Overseas Radio—Sorafom—in the 1950s, retired in 1987.

8. Each year the Italia award, organized by RAI, Italy, honors a creative work of radio broadcasting.

9. The EMTec factory is located in the vicinity of Strasbourg.

10. At the time, the franc was still in use. The considerable sum of 250,000 French francs was the equivalent of €38,000, almost two year's salary for a young composer/researcher at the GRM at the time.

11. Solange Barrachina was involved in this activity from 2002 to 2007.

12. The French projects Épicure and then Écoute—Environnement de Publications à Caractère Industriel (Industrial Publication Environment). This development project brought the GRM and IRCAM together with the Université Technologique de Compiègne and the French region of Picardy, which financed it from 2003 to 2005. The aim was to use

378   NOTES TO CHAPTER 11

the GRM web radio as a model, developing the GRM site as a prototype, and then rendering it accessible to various partners interested in distributing all sorts of documents, such as tourist trailers, safety advice, and company presentations with commentary.

13. In 2005, 4 percent of these one hundred million hours had been conserved, and less than 1 percent had been digitized. After the end of the project, INA continued to digitize its archives. And when this book went to press, INA was no longer behind in digitizing.

14. Nadia Nadah, a student at the University of Technology of Compiègne (UTC), carried out a four-month research project in ontology at the GRM, for her DEA (masters) in science and technology, on the topic of the *Réalisation d'une ontologie des musiques électroacoustiques*, in 2003.

15. See chapter 3 describing the origins of GRM Tools.

16. See chapter 3 describing the origins of the Acousmographe.

17. Christian Eloy was also a professor of electroacoustic composition at the Bordeaux Conservatory from 1989 to 2013. He cofounded and, until 2021, served as artistic director of the SCRIME (Studio de Création et de Recherche en Informatique et Musique Électroacoustique), a tripartite organization involving the Conservatory of Bordeaux, the Bordeaux I Science University, and the ENSFIRB engineering school.

18. The 1998–99 academic year saw the launching of the first master's seminar in collaboration with the GRM, at the Sorbonne's Music and Musicology of the 20th Century Department, titled "Theories, Analyses and Technological Environment."

19. The Institut International de Musique Electroacoustique de Bourges (IMEB) closed definitively in 2011.

20. Interview by the author with Daniel Teruggi, December 2005.

21. The title *Archives sauvées des eaux* (Archives rescued from the water) evokes Ferrari's experience of flooding in his home, obliging him to salvage damaged sound archives. In the process, he rediscovered his own work, leading to his recycling excerpts.

22. The Absinthe Star is mentioned in the King James Bible, Revelation 8: "And when he had opened the seventh seal, there was silence in heaven about the space of half an hour. . . . The name of the star is Wormwood."

23. www.ina-grm.com.

POSTSCRIPT

1. Andrew J. Nelson, *The Sound of Innovation* (MIT Press, Cambridge, MA, 2015), 11.

BIBLIOGRAPHY—DISCOGRAPHY

BOOKS AND ARTICLES

Adorno, Theodor W. *Philosophie de la nouvelle musique*. Gallimard, Paris, 1962.
Aguila, Jesus. *Le Domaine musical*. Fayard, Paris, 1992.
Allouis, Jean-François. "Syter et le temps réel." *La revue musicale*, "Recherche musicale au GRM," quadruple issue no. 394, 395, 396, 397 (Richard-Masse, Paris, 1986), 64–69.
Altman, Rick. *Silent Film Sound*. Columbia University Press, New York, 2004.
Arnaud, Jean-Pierre. *Freud, Wittgenstein et la musique: La parole et le chant dans la communication*. PUF, Paris, 1990.
Arnheim, Rudolf. *La pensée visuelle*. Flammarion, Paris, 1976.
———. *Radio*. 1936; repr., Van Dieren, Paris, 2005.
———. *Radio*. Translated by Margaret Ludwig and Herbert Read. Faber & Faber, London, 1936.
Augoyard, Jean-François, and Henry Torgue. *À l'écoute de l'environnement: Répertoire des effets sonores*. Parenthèses, Marseille, 1995.
Bachelard, Gaston. *L'air et les songes*. Le livre de poche, Paris, 1999.
———. *L'eau et les rêves*. Le livre de poche, Paris, 1999.
———. *La poétique de l'espace*. PUF, Paris, 2001.
Bailblé, Claude. *La perception et l'attention modifiées par le dispositif cinéma*. Thesis, University of Paris VIII, 1999.
Baltrusaïtis, Jurgis. *Anamorphoses*. Flammarion, Paris, 1996.
Bara, Guillaume. *La Techno*. Librio, Paris, 1999.
Barras, Vincent, and Nicholas Zurbrugg, eds. *Poésies sonores*. Contrechamps, Genève, 1992.
Barrière, Jean-Baptiste. "La fin de la recherche et le triomphe de la production?," In *Avidi Lumi*, no. 12. Palerme, 2001.

Barthes, Roland. *La Chambre claire: Note sur la photographie*. Cahiers du cinéma–Gallimard–Seuil, Paris, 1980.

———. *Œuvres complètes*. Seuil, Paris, 2002.

Battier, Marc. "L'approche gestuelle dans l'histoire de la lutherie électronique. Etude de cas: le theremin." In *Les nouveaux gestes de la musique*, edited by Hugues Genevoix and Raphaël de Vivo. Parenthèses, Marseille, 1999, 139–150.

———. "Electroacoustic Music Studies and the Danger of Loss." *Organised Sound* 9, no. 1 (2003): 47–53.

Battier, Marc, and Pierre Couprie. "L'Acousmographe: Un outil pour l'analyse informatique de documents sonores." In *Les cahiers de l'OMF*, no. 4. University of Paris IV–Sorbonne, Paris, 1999, 59–63.

Bayle, François. *Komposition und Musikwissenschaft im Dialog IV (2000–2003): L'image de son; Technique de mon écoute*. Klangbilde, LIT, Münster, 2003.

———. *Musique acousmatique, propositions ... ... positions*. INA-GRM/Buchet-Chastel, Paris, 1993.

Bergson, Henri. *L'Évolution créatrice*. PUF, Paris, 2003.

———. *La pensée et le mouvant*. 1938; repr., PUF, Paris, 2003.

Bobillot, Jean-Pierre. *Bernard Heidsieck, poésie action*. Jean-Michel Place, Paris, 1996.

Bosseur, Dominique, and Jean-Yves Bosseur. *Révolutions musicales: La musique contemporaine depuis 1945*. Minerve, Paris, 1993.

Bosseur, Jean-Yves. *Du son au signe, Histoire de la notation musicale*. Alternatives, Paris, 2005.

———. *John Cage*. Minerve, Paris, 1993.

———. *Musique et arts plastiques: Interactions au XXe siècle*. Minerve, Paris, 1998.

———. *Le sonore et le visuel*. Dis voir, Paris, 1992.

Boucourechliev, André. *Igor Stravinsky*. Fayard, Paris, 1982.

———. *Le Langage musical*. Fayard, Paris, 1993.

Boulez, Pierre. *Penser la musique aujourd'hui*. Denoël/Gonthier, Paris, 1963.

———. *Relevés d'apprenti*. Seuil, Paris, 1966.

Breton, André. *Manifestes du surréalisme.* 1924; repr., Gallimard, Paris, 1985.

Brunet, Sophie. *Pierre Schaeffer*. Richard-Masse, Paris, 1969.

Cage, John. *Silence*. Wesleyan University Press, Middletown, 1961.

Carlson, Carolyn. *Le soi et le rien*. Actes Sud, Arles, 2001.

Cassirer, Ernst. *Langage et mythe*. Minuit, Paris, 1973.

Castanet, Pierre-Albert. "Grands et petits moyens musicaux au service du futurisme," *Les cahiers du CIREM*, no. 8–9, Rouen, September 1988.

———. *Hugues Dufourt, 25 ans de musique contemporaine*. TUM/Michel de Maule, Paris, 1995.

———. "La Théorie des catastrophes de René Thom et l'avènement du temporel dans les œuvres musicales de François Bayle et d'Hugues Dufourt," *Analyse musicale et création*, L'Harmattan, Paris, 2001.

———. *Tout est bruit pour qui a peur: Pour une histoire sociale du son sale*. Michel de Maule, Paris, 1999.

Caux, Jacqueline. *Presque rien avec Luc Ferrari*. Main d'œuvre, Nice, 2002.

Cazaban, Costin. *Temps musical/espace musical comme fonctions logiques.* L'Harmattan, Paris, 2000.
Céleste, Bernadette, François Delalande, and Elisabeth Dumaurier. *L'enfant du sonore au musical.* INA-GRM/Buchet-Chastel, Paris, 1982.
Chadabe, Joël. *Electric Sound: The Past and Promise of Electronic Music.* Prentice Hall, Upper Saddle River, NJ, 1997.
Challet-Haas, Jacqueline. *Grammaire de la notation Laban,* vol. 1 and 2. Centre national de la danse, Pantin, 1999.
Chatwin, Bruce. *Le chant des pistes.* Grasset, Paris, 1988.
———. *The Songlines.* Penguin Books, London, 1987.
Chion, Michel. "20 ans de musique électro-acoustique ou une quête d'identité." *Musique en jeu,* no. 8 (1972): 26–27.
———. *L'art des sons fixés ou La Musique Concrètement.* Metamkine/Nota-Bene/Sono-Concept, Fontaine, 1991.
———. *Un art sonore, le cinéma.* Cahiers du cinéma, Paris, 2003.
———. *L'audio-vision.* Nathan, Paris, 1990.
———. *La Comédie musicale.* Cahiers du cinéma / Scérén-CNDP, Paris, 2002.
———, ed. *François Bayle, parcours d'un compositeur.* Lien, revue d'esthétique musicale. Musiques et Recherches, Ohain (Belgium), 1994.
———. *Guide des objets sonores.* INA/Buchet-Chastel, Paris, 1983.
———. *La musique au cinéma.* Fayard, Paris, 1995.
———. *Musiques médias et technologies.* Flammarion, Paris, 1994.
———. *Pierre Henry.* Fayard/Sacem, Paris, 2003.
———. *Le poème symphonique et la musique à programme.* Fayard, Paris, 1993.
———. *Le promeneur écoutant, Essais d'acoulogie.* Plume, Paris, 1993.
———. *Le Son.* Nathan, Paris, 1998.
———. *La symphonie à l'époque romantique, de Beethoven à Mahler.* Fayard, Paris, 1994.
———. *La voix au cinéma.* Cahiers du cinéma, Paris, 1993.
Chion, Michel, and Guy Reibel. *Les musiques électroacoustiques.* Edisud, Aix-en-Provence, 1976.
Chopin, Henri. *Poésies sonores.* Contrechamps, Geneva, 1992.
Claudel, Paul. *Œuvres en prose.* Gallimard, La Pléiade, Paris, 1965.
Clozier, Christian. "Un instrument de diffusion: Le gmebaphone." In *Lien, revue d'esthétique musicale,* edited by Francis Dhomont. Musiques et Recherches, Ohain (Belgium), 1988, 56–57.
Cocteau, Jean. *Le Coq et l'Arlequin.* 1918; repr., Stock, Paris, 1978.
———. *Opium.* 1930; repr., Stock, Paris, 2005.
Cohen, Andrea. *Les compositeurs et l'art radiophonique.* L'Harmattan, Paris, 2015.
Cohen-Levinas, Danièle. "Les icônes de l'écoute, Schoenberg, les oh! et les ah! de l'histoire." In *De l'écoute à l'oeuvre: Études interdisciplinaires; Actes du colloque tenu en Sorbonne les 19 et 20 février 199,* edited by Michel Imberty. Editions L'Harmattan, Paris, 2001, 31–41.
Colette, Marie-Noëlle, Marielle Popin, and Philippe Vendrix. *Histoire de la notation du Moyen Âge à la Renaissance.* Minerve, Paris, 2003.

Corroenne, Olivier. *L'atelier de création radiophonique de France Culture*. Dissertation, Free University of Brussels, 1985.
Couprie, Pierre. *La musique électroacoustique: Analyse morphologique et représentation analytique*. Thesis, University of Paris IV–Sorbonne, 2003.
Cullin, Olivier. *L'image musique*. Fayard, Paris, 2006.
Dallet, Sylvie. *Pierre Schaeffer, itinéraire d'un chercheur*. Centre d'études et de recherche Pierre Schaeffer, Montreuil, 1998.
Dallet, Sylvie, and Anne Veitl. *Du sonore au musical*. L'Harmattan, Paris, 2001.
Damasio, Antonio R. *L'Erreur de Descartes*. Odile Jacob, Paris, 1995.
———. *Le Sentiment même de soi: Corps émotion, conscience*. Odile Jacob, Paris, 1999.
———. *Spinoza avait raison: Joie et tristesse, le cerveau des émotions*. Odile Jacob, Paris, 2003.
Damisch, Hubert. *L'origine de la perspective*. Flammarion, Paris, 1987.
Daniélou, Alain. *Origines et pouvoirs de la musique*. Les cahiers du mleccha, Kailash, Paris, 2003.
Debord, Guy. *La société du spectacle*. Gallimard, Paris, 1992.
Debussy, Claude. *Monsieur Croche*. 1921; repr., Gallimard, Paris, 1987.
Delalande, François. "L'analyse des musiques électroacoustiques," *Musique en jeu*, no. 8 (1972): 50–56.
———. "L'articulation interne/externe et la détermination des pertinences en analyse." In *Observation, analyse, modèles: Peut-on parler d'art avec les outils de la science?* Proceedings of the 2nd International Colloquium on Musical Epistemology. IRCAM/L'Harmattan, edited by Jean-Marc Chouvel and Fabien Lévy (Paris, 2002), 3.
———. "D'une rhétorique de la forme à une déontologie de la composition." With the participation of Ivo Malec, Luc Ferrari, and Jean-Yves Hameline. *Analyse Musicale* 20, no. 3, Paris, 1990, 41–51.
———. "Il gesto musicale, Dal senso motorio al simbolico—Aspetti ontogenettici" [Musical gesture, from sensory-motor to symbolic—Ontogenetic aspects]. In *Dall'atto motorio alla interpretazione musicale*, Actes du 2e colloque international de psychologie de la musique (Proceedings of the Second International Conference on the Psychology of Music) (Ravello, October 1990). Editions 10/17, Salerne, Italy, 1992.
———. *Il faut être constamment un immigré: Entretiens avec Xenakis*. INA/Buchet-Chastel, Paris, 1997.
———. *La musique est un jeu d'enfant*. INA/Buchet-Chastel, Paris, 1984.
———. "Perception des sons et perception des œuvres." In *Quoi, quand, comment: La recherche musicale*, edited by Tod Machover. Christian Bourgois/IRCAM, Paris, 1985, 197–209.
———. *Le son des musiques entre technologie et esthétique*. INA/Buchet-Chastel, Paris, 2001.
———. "La terrasse des audiences au clair de lune de Debussy: Essai d'analyse esthésique." *Analyse musicale*, no. 16, Paris, 1989.
———. "Trois idées-clés pour une pédagogie musicale d'éveil." *Cahiers recherche/musique* no. 1. INA-GRM, Paris, 1976, 10–30.
Delalande, François, and Évelyne Gayou. "Xenakis et le GRM." In *Présences de Iannis Xenakis*, edited by Makis Solomos. CDMC, Paris, 2001, 29–36.

Delalande, François, and Pascal Gobin. "Les Unités Sémiotiques Temporelles: Un niveau d'analyse de l'organisation musicale du temps." In *Les universaux en musique*, proceedings of the 4th International Congress on Musical Meaning, edited by Costin Miereanu and Xavier Hascher. Publications de la Sorbonne, Paris, 1998, 573–87.
Deleuze, Gilles. *Cinéma 1: L'Image-mouvement, Cinéma 2: L'Image-temps.* Minuit, Paris, 1983.
Deleuze, Gilles, and Félix Guattari. *Mille plateaux.* Minuit, Paris, 1980.
Deliège, Célestin. *Cinquante ans de modernité musicale: De Darmstadt à l'IRCAM.* Mardaga, Sprimont, 2003.
Delorme, Pierre. *Le Kendo à Rudrâ, l'école de danse de Maurice Béjart.* Guy Trédaniel, Paris, 2003.
Derrida, Jacques. *La Voix et le phénomène.* Quadrige/PUF, Paris, 2003.
Deshays, Daniel. *De l'écriture sonore.* Entre/vues, Marseille, 1999.
Donati, Paolo, and Ettore Pacetti, eds. *C'erano una volta nove oscillatori . . . , Lo studio di fonologia della Rai di Milano nello sviluppo della nuova musica italiana.* RAI Teche, RAI Eri, Rome, 2002.
Duchesneau, Michel. *L'Avant-garde musicale à Paris de 1871 à 1939.* Mardaga, Sprimont, 1997.
Dufour, Denis, ed. *Ouïr, entendre, écouter, comprendre après Schaeffer.* Buchet-Chastel, Paris, 1999.
Dufourt, Hugues. *Musique, pouvoir, écriture.* Christian Bourgois, Paris, 1991.
Durand, Gilbert. *Figures mythiques et visages de l'œuvre.* Dunod, Paris, 1992.
———. *Les structures anthropologiques de l'imaginaire.* Dunod, Paris, 1992.
Duteurtre, Benoît. *Requiem pour une avant-garde.* Pocket-Agora, Paris, 2002.
Eco, Umberto. *L'œuvre ouverte.* Seuil, Paris, 1965.
Farabet, René. *Bref éloge du coup de tonnerre et du bruit d'ailes.* Phonurgia nova, Arles, 1994.
Ferrari, Giordano. "La conchiglia di Pierre Schaeffer" (Pierre Schaeffer's *La coquille*). *AAA-TAC: Acoustical Arts and Artifacts, Technology, Aesthetics, Communication* 1. Istituti editoriali e poligrafici internazionali–Fondation Cini, Pisa/Rome/Venice (2004): 11–18.
———. *La naissance de l'univers sonore de Pierre Schaeffer: La Coquille à planètes.* Résonance, colloquium at the IRCAM, Paris, October 2003.
Fort, Bernard, and Philippe Gonin. *Du son à l'œuvre: Un chemin vers les nouvelles musiques.* Lugdivine, 2002.
Foucault, Michel. *Dits et écrits.* Gallimard, Paris, 1994.
Francès, Robert. *La perception de la musique.* Vrin, Paris, 1958.
Gaillot, Michel. *La Techno, un laboratoire artistique et politique du présent.* Dis Voir, Paris, 2003.
Gallet, Bastien. *Le Boucher du prince Wen-Houei: Enquêtes sur les musiques électroniques.* Musica Falsa, Paris, 2000.
Gaudibert, Pierre. *Du culturel au sacré.* Casterman, Paris, 1981.
Gavoty, Bernard. *Les Français sont-ils musiciens?* Conquistador, Paris, 1950.
Gayou, Évelyne. "Intentions de création en musique électroacoustique, chez les compositeurs invités au GRM" (Creative intentions in electoacoustic music among guest composers at the GRM). In *Intentions et création dans la musique d'aujourd'hui*, in "Actes de

la journée du 24 mars 2004, 1res Rencontres Interartistiques de l'OMF," series "Conférences et Séminaires" no. 19. University of Paris IV–Sorbonne, 2005, 21–27.

Geslin, Yann. "Environnements de transformations sonores et musicales Une expérience de vingt années au Groupe de Recherches Musicales." Published in English as "Digital Sound and Music Transformation Environments: A Twenty-Year Experiment at the 'Groupe de Recherches Musicales.'" Translated by Riccardo Walker. *Journal of New Music Research* 31, no. 2 (June 2002): 99–107.

Gilly, Cécile. *Pierre Boulez, l'écriture du geste*. Christian Bourgois, Paris, 2002.

Giner, Bruno. *Weimar 1933, La musique aussi brûle en exil*. Le temps des cerises, Pantin, 2001.

Goléa, Antoine. *La musique de la nuit des temps aux aurores nouvelles*, vol. 1 and 2. Alphonse Leduc et Cie, Paris, 1977.

Gurdjieff, Georges Ivanovitch. *Rencontres avec des hommes remarquables*. Julliard, Paris, 1960.

Hennebelle, Guy, with Alain and Odette Virmaux. *Les grandes "écoles" esthétiques*. CinémAction. Corlet/Télérama, Paris, 1989.

Henry, Pierre. *Journal de mes sons*. Actes sud, Arles, 2004.

Hirbour, Louise. *Edgar Varèse, écrits*. Christian Bourgois, Paris, 1983.

Husserl, Edmund. *Logique formelle et logique transcendantale*. PUF, Paris, 1957.

———. *Méditations cartésiennes*. J. Vrin, Paris, 1947.

Jakobson, Roman. *Six leçons sur le son et le sens*. Minuit, Paris, 1976.

Jankélévitch, Vladimir. *Le Je-ne-sais-quoi et le Presque-rien: 1. La Manière et l'occasion; 2. La Méconnaissance, Le Malentendu; 3. La Volonté de vouloir*. Seuil, Paris, 1980.

Jauss, Hans-Robert. *Pour une esthétique de la réception*. Gallimard, Paris, 1978.

Jorn, Asger. "Nouvelles de l'Internationale." In *Internationale Situationiste 1958–69*. Van Gennep, Amsterdam, 1972, 3–30.

———. *Pour la forme*. 1958; repr., Allia, Paris, 2001.

Justel, Elsa. *Les structures formelles dans la musique de production électronique*. Thesis, University of Paris VIII, 2000.

Kandinsky, Wassily. *Du spirituel dans l'art et dans la peinture en particulier*. Gallimard, Paris, 1989.

Landy, Leigh. *Experimental Music Notebooks*. Harwood Academic Publishers, Chur, Switzerland, 1994.

———. *Understanding the Art of Sound Organization*. MIT Press, London, 2007.

———. *What's the Matter with Today's Experimental Music? Organized Sound Too Rarely Heard*. Harwood Academic Publishers, Chur, Switzerland, 1991.

Langlois, Philippe. *Les Procédés électroacoustiques dans les différents genres cinématographiques: Une étude transversale au XXe siècle*. Thesis, University of Paris IV–Sorbonne, 2004.

Laubier (de), Serge. "MIDI Formers." *Computer Music Journal* 21, no. 1 (1997): 39–40.

Lemoine, Jean-Jacques. *La Société des Auteurs et Éditeurs de Musique, SACEM 1850–1950*. SACEM, Paris, 1950.

Leroi-Gourhan, André. *L'homme et la matière*. Albin Michel, Paris, 1971.

Leroy, Jean-Luc. *Vers une épistémologie des savoirs musicaux*. L'Harmattan, Paris, 2003.

Levin, Thomas Y. *Tones from out of Nowhere: Rudolph Pfenninger and the Archaeology of Synthetic Sound*. Grey Room, New York, 2003.

Lista, Giovanni. *Futurisme, manifestes, documents, proclamations.* L'Age d'homme, Paris, 1973.
Lortat-Jacob, Bernard. *Musiques en fête.* Société d'ethnologie, Nanterre, 1994.
Lucretius. *De rerum natura.* Flammarion, Paris, 1998.
Lyotard, Jean-François. *La condition postmoderne.* Minuit, Paris, 1979.
Mâche, François-Bernard. *Entre l'observatoire et l'atelier.* Kimé, Paris, 1998.
———. *Musique, mythe, nature ou les dauphins d'Arion.* Klincksieck, Paris, 1983.
———. *Un demi-siècle de musique... et toujours contemporaine.* L'Harmattan, Paris, 2000.
Maffesoli, Michel. *Au creux des apparences, pour une éthique de l'esthétique.* Plon, Paris, 1990.
———. *La contemplation du monde, Figures du style communautaire.* Grasset, Paris, 1993.
———. *Du nomadisme.* Le livre de poche, Paris, 1997.
———. *L'instant éternel: Le retour du tragique dans les sociétés postmodernes.* Denoël, Paris, 2000.
———. *L'ombre de Dionysos.* Klincksieck, Paris, 1985.
Mailliard, Bénédict. "À la recherche du studio musical." In *La revue musicale,* "Recherche musicale au GRM," quadruple issue no. 394, 395, 396, 397 (Richard-Masse, Paris, 1986), 51–63.
———. "Simulation par ordinateur du studio électroacoustique et applications à la composition musicale." In *Conférences des journées d'étude, Festival international du son Haute-fidélité.* Éditions Radio France, Paris, 1981.
Marchetti, Lionel. *La musique concrète de Michel Chion.* Métamkine, Rives, 1998.
Marie, Jean-Etienne. *Musique vivante.* Privat/PUF, Paris 1953.
Marin, Louis. *De la représentation.* Seuil/Gallimard, Paris, 1994.
———. *De l'entretien.* Minuit, Paris, 1997.
Mathews, Max. "The Digital Computer as a Musical Instrument," *Sciences,* no. 142 (November 1963): 553–57.
Menger, Pierre-Michel. *Le paradoxe du musicien.* Flammarion, Paris, 1983.
Merleau-Ponty, Maurice. *L'Œil et l'Esprit.* Gallimard, Paris, 1964.
———. *Phénoménologie de la perception.* Gallimard, Paris, 1945.
———. *Le visible et l'invisible.* Gallimard, Paris, 1964.
Michelet, Jules. *Histoire de la France.* 1869; repr., Robert Laffont, Paris, 1971.
Mion, Philippe, Jean-Jacques Nattiez, and Jean-Christophe Thomas. *L'envers d'une œuvre: De natura sonorum de Bernard Parmegiani.* INA-GRM/Buchet-Chastel, Paris, 1983.
Moindrot, Gérard. *Approches symboliques de la musique d'André Jolivet.* L'Harmattan, Paris, 1999.
Moles, Abraham. *Art et ordinateur.* Blusson, Paris, 1990.
———. *Théorie de l'information et perception esthétique.* Flammarion, Paris, 1958.
———. *Théorie structurale de la communication et société.* Masson Cnet ENST, Paris, 1995.
Moles, Abraham, and Élisabeth Rohmer. *Labyrinthes du vécu.* Librairie des Méridiens, Paris, 1982.
———. *Psychologie de l'espace.* L'Harmattan, Paris, 1998.
Moles, Abraham, and Wladimir Ussachevsky. "L'emploi du spectrographe acoustique et le problème de la partition en musique expérimentale." *Annales des télécommunications* 12, no. 9 (September 1957): 299–304.

Molino, Jean. "Fait musical et sémiologie de la musique." *Musique en jeu*, no. 17 (1975).
———. "La musique et l'objet." In *Ouïr, entendre, écouter, comprendre après Schaeffer.* Buchet-Chastel, Paris, 1999, 119–36.
———. *Le singe musicien: Sémiologie et anthropologie de la musique.* Actes Sud / INA, 2009.
Mollard, Claude. *Le 5e pouvoir.* Armand Colin, Paris, 1999.
More, Thomas. *L'Utopie.* GF Flammarion, Paris, 1987.
Murray-Schafer, Raymond. *Le paysage sonore.* Jean-Claude Lattès, Paris, 1979.
Nattiez, Jean-Jacques. *Correspondance, Pierre Boulez / John Cage.* Christian Bourgois, Paris, 1991.
———. *Fondements d'une sémiologie de la musique.* Union Générale d'Édition, coll. 10–18, Paris, 1975.
———. *Music and Discourse: Toward a Semiology of Music.* Translated by Carolyn Abbate. Princeton University Press, Princeton, 1990.
———. "Situation de la sémiologie musicale." *Musique en jeu*, no. 5 (1971): 3–18
Naumann, Francis M. *Marcel Duchamp L'art à l'ère de la reproduction mécanisée.* Hazan, Paris, 1999.
Nelson, Andrew J. *The Sound of Innovation: Stanford and the Computer Music Revolution.* MIT Press, Cambridge, MA, 2015.
Nietzsche, Friedrich. *La naissance de la tragédie.* Le livre de Poche, Paris, 1994.
Ouelette, Fernand. *Edgard Varèse.* Christian Bourgois, Paris, 1989.
Ouspensky, Peter Demianovitch. *Fragments d'un enseignement inconnu.* Stock, Paris, 1950.
Pauwels, Louis. *Monsieur Gurdjieff.* 1954; repr., Albin Michel, Paris, 1974.
Peirce, Charles Sanders. *Le raisonnement et la logique des choses.* Cerf, Paris, 1995.
Picabia, Francis. "Sur René Clair." *La Danse,* November 1924. Reprinted in Picabia, *Écrits,* Belfond, Paris, 1978, 2:158.
Pineau, Marion, and Barbara Tillmann. *Percevoir la musique: Une activité cognitive.* L'Harmattan, Paris, 2001.
Ponge, Francis. *Pratiques d'écriture, où l'inachèvement perpétuel.* Hermann, Paris, 1984.
Porcile, François. *Les conflits de la musique française 1940–1965.* Fayard, Paris, 2001.
Poullin, Jacques. "L'apport des techniques d'enregistrement dans la fabrication de matières et de formes musicales nouvelles: Application à la musique concrète." *Ars sonora*, no. 9 (1999): 25–34. (Article originally published in the journal *L'Onde électrique* in 1954.)
Quignard, Pascal. *La haine de la musique.* Gallimard, Paris, 1996.
Ramault-Chevassus, Béatrice. *Musique et postmodernité.* PUF/Que sais-je?, Paris, 1998.
Regnault-Bousquet, Cécile. *Les représentations visuelles des phénomènes sonores: Application à l'urbanisme.* Thesis, Université de Nantes, École d'architecture de Grenoble, 2001.
Reibel, Guy. *L'homme musicien: Musique fondamentale et création musicale.* Edisud, Aix-en-Provence, 2000.
———. *Jeux vocaux.* Salabert, Paris, 1985.
Ricœur, Paul. *La mémoire, l'histoire, l'oubli.* Seuil, Paris, 2000.
———. *Temps et récit 3, Le Temps raconté.* Éd du Seuil, Paris, 1985.
Richard, Jean-Pierre. *Littérature et sensation.* Seuil, Paris, 1954.
Rigoni, Michel. *Stockhausen . . . un vaisseau lancé vers le ciel.* Millénaire III, Rouen, 1998.
Risset, Jean-Claude. "Évolution des outils de création sonore." In *Interfaces homme-machine et création musicale*, edited by Hugues Vinet and François Delalande. Hermès, Paris, 17–36.

———. "Musique, électronique et théorie de l'information." *L'Onde électrique*, no. 451 (1964): 1055–63.

———. "Musique et perception." In *Quoi, quand, comment / La recherche musicale*. Christian Bourgois/IRCAM, Paris, 1985, 211–19.

———. "Musique, recherche, théorie, espace, chaos." In *Inharmoniques* no 8, Christian Bourgois / IRCAM, Paris, 1991, 273–316.

———. "Paradoxes de hauteur." *Rapport Ircam no. 10*, with sound examples on cassette, 1978.

———. "Rubrique sons," *Encyclopedia Universalis*, 1973, 168–71.

———. "Synthèse et matériau musical." *Les cahiers de l'IRCAM*, no. 2: *La synthèse sonore*, 1993, 43–65.

———. "Timbre et synthèse des sons." In *Le timbre*, coordinated by Jean-Baptiste Barrière. Christian Bourgois/IRCAM, Paris, 1991, 239–60.

Risset, Jean-Claude and Wessel, David. "Les illusions auditives," *Universalia—Encyclopedia Universalis*, 1979, 167–91

Roads, Curtis. *L'audionumérique*. French version: Jean de Reydellet, Dunod, Paris, 1998.

———, ed. *Composers and the computer*. Kaufmann, Los Altos, CA, 1985.

Robert, Martial. *Pierre Schaeffer, Communication et Musique en France de 1936 à 1986*. L'Harmattan, Paris, 1999, 2000, 2002. Volume 1: *Des Transmissions à Orphée*; Volume 2: *d'Orphée à Mac Luhan*; Volume 3: *De Mac Luhan au fantôme de Gutenberg*.

Rosset, Clément. *L'anti-nature*. PUF, Paris, 1973.

———. *L'objet singulier*. Minuit, Paris, 1979.

Rouget, Gilbert. *La musique et la transe*. Gallimard, Paris, 1990.

Roy, Stéphane. *L'analyse des musiques électroacoustiques: Modèles et propositions*. L'Harmattan, Paris, 2003.

Russolo, Luigi. *L'art des bruits*. Richard-Masse, Paris, 1954; repr., Éditions Allia, Paris, 2003.

Sadoul, Georges. *Dictionnaire des cinéastes*. Seuil, Paris, 1965.

———. *Dictionnaire des films*. Seuil, Paris, 1965.

———. *Dziga Vertov*. Champ libre, Paris, 1971.

Schaeffer, Pierre. *À la recherche d'une musique concrète*. 1952; repr., Seuil, Paris, 1998.

———. *De la musique concrète à la musique même*. 1977; repr., Mémoire du livre, Paris, 2002.

———. "L'élément non visuel au cinéma, I: Analyse de la bande-son." In *La revue du cinéma*, no. 1, Paris (October 1946): 45–48.

———. "L'élément non visuel au cinéma, II: Conception de la musique." In *La revue du cinéma*, no. 2, Paris (November 1946): 62–65.

———. "L'élément non visuel au cinéma, III: Psychologie du rapport vision audition." In *La revue du cinéma*, no. 3, Paris (December 1946): 51–54.

———. *Faber et Sapiens*. Belfond, Paris, 1985.

———. *In Search of a Concrete Music*. Translated by Christine North and John Dack. Berkeley: University of California Press, 2012.

———. *Machines à communiquer, 1: Genèse des simulacres; 2: Pouvoir et communication*. Seuil, Paris, 1970.

———. *La Musique concrète*. PUF/Que sais-je?, Paris, 1967.

———. "La Musique Concrète et la recherche scientifique." In *7 ans de Musique Concrète, 1948–1955*. Centre d'Études Radiophoniques, RTF, Paris, 1955.

———. "La Musique et les ordinateurs," in *La revue musicale*, no. 268–69 (Paris, 1971): 57–88.

———. *Prélude, choral et fugue*. Flammarion, Paris, 1983.

———. *Propos sur la Coquille*. Phonurgia nova, Arles, 1990.

———. *Traité des objets musicaux: Essai interdisciplines (TOM)*. Seuil, Paris, 1966.

———. *Treatise on Musical Objects: An Essay across Disciplines (TMO)*. Translated by Christine North and John Dack. University of California Press, Oakland, 2017.

Schaeffer, Pierre, and Michel Chion. *Guide des objets sonores*. Buchet-Chastel, Paris, 1983.

Schöffer, Nicolas. *Perturbation et chronocratie*. Denoël/Gonthier, Paris, 1978.

Siron, Jacques. *La partition intérieure: Jazz, musiques improvisées*. 1992; repr., Outre Mesure, Paris, 2004.

Solomos, Makis. *Iannis Xenakis*. P. O. Éditions, Paris, 1999.

———, ed. *Présences de Iannis Xenakis*. CDMC, Paris, 2001.

Stiegler, Bernard. *La technique et le temps*. 2 vols. Galilée, Paris, 1994 and 1996.

Stockhausen, Karlheinz. *Internationalen Ferienkursen für Neue Musik in Darmstadt 1951–1996*. Stockhausen verlag, Kuerten, 2001.

Stravinsky, Igor. *Chroniques de ma vie*. Denoël, Paris, 1962.

Szendy, Peter. "Actes d'une dislocation." *Les cahiers de l'IRCAM, Recherche et musique*, no. 5. Espaces, IRCAM/Centre Pompidou, Paris, 1994, 53–64.

———. *Écoute, une histoire de nos oreilles*. Minuit, Paris, 2001.

———. "Propos impromptus sur l'aura et l'espace du son." *Les cahiers de l'IRCAM, Recherche et musique*, no. 5. Espaces, IRCAM/Centre Pompidou, Paris, 1994, 184–85.

Teruggi, Daniel. *Le système SYTER: Son histoire, ses développements, sa production musicale, ses implications dans le langage électroacoustique d'aujourd'hui*. Thesis, Université Paris VIII, 1998.

Thom, René. *Paraboles et catastrophes*. Flammarion, Paris, 1989.

Vallier, Dora. *L'art abstrait*. Hachette, Paris, 1980.

Van Wymeersch, Brigitte. *Descartes et l'évolution de l'esthétique musicale*. Mardaga, Sprimont, 1999.

Vaudrin, Marie-Claude. *La musique techno ou le retour de Dionysos*. L'Harmattan, Paris, 2004.

Veitl, Anne, and Noémi Duchemin. *Maurice Fleuret, une politique démocratique de la musique: 1981–1986*. Comité d'histoire du ministère de la culture, Paris, 2000.

Vidal de La Blache, Paul. *Principes de géographie humaine*, posthumous work. 1921; repr., Utz, Paris, 1995.

Vinay, Gianfranco. *Charles Ives*. Michel de Maule, Paris, 2001.

Vinet, Hugues. "DSP Station, HyperCard Environment for DSP Sound Processing Algorithms." *Proceedings of the ICMC*, Montréal, 1991.

Vinet, Hugues, and François Delalande, eds. *Interfaces homme-machine et création musicale*. HERMES Science Publications, Paris, 1999.

Volta, Ornella. *Satie et la danse*. Plume, Paris, 1992.

Virilio, Paul. *L'art du moteur*. Galilée, Paris, 1993.

Wangermée, Robert. *André Souris et le complexe d'Orphée: Entre surréalisme et musique sérielle*. Mardaga, Sprimont, 2002.

Weid (von der), Jean-Noël. *La musique du XXe siècle*. Hachette, Paris, 1997.
Winckel, Fritz. *Vues nouvelles sur le monde des sons*. 1959; translation by A. Moles, Dunot, Paris, 1960.
Xenakis, Iannis. *Musiques formelles*. Stock, Paris, 1981.
Zemp, Hugo. *Écoute le bambou qui pleure*. Gallimard, Paris, 1995.

OTHER PRINTED SOURCES

Actes musicaux 1993. Texts compiled by Jérôme Dorival. Medicis, Lyon, 1994.
À propos d'un essai acoustique. Cahiers du club d'Essai, 1947.
Ars Sonora. "Revue électronique," edited by Jean-Yves Bosseur. Paris, nine issues from 1995 to 1999.
Les cahiers de l'IRCAM: Recherche et musique. Edited by Peter Szendy. Two issues a year, from 1992 to 2001.
Cahiers de médiologie/IRCAM, no. 18. "Révolutions industrielles de la musique." Coordinated by Nicolas Donin and Bernard Stiegler. Fayard, Paris, 2004.
Cahiers recherche/musique. Six issues. INA-GRM, Paris.

  No. 1: "Pédagogie musicale d'éveil," edited by François Delalande, 1975.
  No. 2: "Le Traité des objets musicaux, 10 ans après," edited by Michel Chion, 1976.
  No. 3: "Synthétiseur/ordinateur," edited by Bénédict Mailliard, 1976.
  No. 4: "La musique du futur a-t-elle un avenir?," edited by Michel Chion, 1977.
  No. 5: "Le Concert pourquoi? Comment?," edited by Michel Chion, 1977.
  No. 6: "Le Pouvoir des sons," edited by Élisabeth Dumaurier, 1978.

La chambre d'écho. Cahiers du Club d'Essai de la radiodiffusion française. Paris, 1947.
Claude Ballif. Les cahiers du CIREM. Collective work, edited by Alain Poirier, no. 20–21, Rouen, July–September 1991.
Computer Music Journal. MIT Press, Cambridge, MA.
De l'écoute à l'œuvre: Études interdisciplinaires. Proceedings of the conference held at the Sorbonne on February 19 and 20, 1999, edited by Michel Imberty. L'Harmattan, Paris, 2001.
L'écoute. Texts compiled by Peter Szendy. Les cahiers de l'IRCAM. L'Harmattan, IRCAM–Centre Pompidou, Paris, 2000.
Encyclopédie de la musique. Fasquelle, Paris, 1959.
"L'espace du son." In *Lien, revue d'esthétique musicale*, edited by Francis Dhomont. Musiques et Recherches, Ohain (Belgium), 1988.
Espaces. Les cahiers de l'IRCAM, Recherche et musique, no. 5. IRCAM–Centre Pompidou, Paris, 1994.
Les Femmes et la création musicale. Proceedings of the symposium of the International Council for Music and of the CDMC, UNESCO, March 7 and 8, 1996. CDMC/CIM, Paris, 2002.
"Il gesto musicale, Dal senso motorio al simbolico—Aspetti ontogenetici." In *Dall'atto motorio alla interpretazione musicale*, edited by François Delalande, proceedings of the 2nd international symposium on the psychology of music (Ravello, October 1990), Ed. 10/17, Salerno, Italy, 1992.

Musique en jeu, no. 8, "Musiques électroacoustiques," quarterly review. Seuil, Paris, 1972.

Observation, analyse, modèles: Peut-on parler de l'art avec les outils de la science? Proceedings of the 2nd International Colloquium on Musical Epistemology, edited by Jean-Marc Chouvel and Fabien Levy. L'Harmattan/IRCAM, Paris, 2002.

L'oreille oubliée. Collective work, edited by Jean-Paul Pigeat. Centre Georges Pompidou/CCI, Paris, 1982.

Portraits polychromes, no. 1, *Luc Ferrari*, INA-GRM/CDMC, Paris, 2001; new and expanded edition, INA-GRM, Paris, 2007.

Portraits polychromes.

> No. 2: *Jean-Claude Risset*. INA-GRM/CDMC, Paris, 2001; new and expanded edition, INA-GRM, Paris, 2008; English translation, no. 19, INA, Paris, 2013.
>
> No. 3: *Gilles Racot*. INA-GRM, Paris, 2002.
>
> No. 4: *Bernard Parmegiani*. INA-GRM, Paris, 2002; new and expanded edition, INA-GRM, Paris, 2011.
>
> No. 5: *Ivo Malec*. Michel de Maule / INA-GRM, Paris, 2003; new and expanded edition, INA-GRM, Paris, 2007.
>
> No. 6: *François Bayle*. Michel de Maule / INA-GRM, Paris, 2004; new and expanded edition, INA-GRM, Paris, 2007.
>
> No. 7: *John Chowning*. Michel de Maule / INA-GRM, Paris, 2005; new and expanded edition, INA-GRM, Paris, 2007; English translation, no. 18, 2013.
>
> No. 8: *Michel Chion*. INA-GRM, Paris, 2005.
>
> No. 9: *Jacques Lejeune*. INA-GRM, Paris, 2006.
>
> No. 10: *Francis Dhomont*. INA-GRM, Paris, 2006.
>
> No. 11: *Max Mathews*. INA-GRM, Paris, 2007; English translation, no. 12, 2007.
>
> No. 13: *Pierre Schaeffer*. INA-GRM, Paris, 2008, with 4 CDs; English translation (without CDs), no. 14, 2009.
>
> No. 15: *Denis Smalley*. INA-GRM, Paris, 2010; English translation, no. 16, 2011.
>
> No. 17: *Éliane Radigue*. INA-GRM, Paris, 2013.
>
> No. 21: Special thematic issue, Musique et technologie: Préserver, Archiver, Re-Produire. INA-GRM, Paris, 2013.
>
> No. 22: Special thematic issue, Musique et technologie: Éveiller, Enseigner, Créer. INA-GRM, Paris, 2015.
>
> No. 23: Special thematic issue, Musique et technologie: Regards sur les musiques mixtes. INA-GRM, Paris, 2016.

Répertoire international des musiques électroacoustiques, Hugh Davies, MIT Press, Cambridge, MA, GRM Service de la Recherche, Electronic Music Review no. 2/3, April–July, 1967.

REVUE et CORRIGÉE, quarterly review, edited by Olivier Masson, Nota Bene, since 1989.

La Revue musicale.

> Special issue: "Le film sonore, L'Écran et la musique en 1935," edited by Henry Prunières. Paris, December 1934.

No. 236: "Vers une musique expérimentale," edited by Pierre Schaeffer. Richard-Masse, Paris, 1957.

No. 244: "Expériences musicales, musiques concrète électronique exotique," edited by Pierre Schaeffer and François-Bernard Mâche. Richard-Masse, Paris, 1959.

No. 274 and 275: "De l'expérience musicale à l'expérience humaine," edited by Pierre Schaeffer. Richard-Masse, Paris, 1971.

No. 394, 395, 396, and 397: "Recherche musicale au GRM," edited by Michel Chion and François Delalande. Richard-Masse, Paris, 1986.

Sons & lumières: Une histoire du son dans l'art du XXè siècle, catalog of the exhibition presented at the Centre Pompidou, from September 22, 2004, to January 3, 2005. Centre Pompidou, Paris, 2004.

La Techno, un laboratoire artistique et politique du présent. Michel Gaillot, interviews with Jean-Luc Nancy and Michel Maffesoli. Dis Voir, Paris, 2003.

DISCOGRAPHY

Given the vast quantity of existing electroacoustic music recordings, we refer the reader to a few "relatively stable" and well-documented Internet sites.

www.ina-grm.com
   Site of the Groupe de Recherches Musicales, France.
http://ears.huma-num.fr
   ElectroAcoustic Resource Site, Music, Technology and innovation Research Center at De Montfort University, Leicester, Grande-Bretagne.
http://on1.zkm.de/zkm/
   Zentrum für Kunst und Medientechnologie, Karlsruhe, Germany.
www.electrocd.com
   Empreintes Digitales, record producer and publisher, Canada.
www.fondation-langlois.org
   Ricardo Dal Farra, Latin American electroacoustic music collection.
www.furious.com/perfect/ohm
   OHM- The Early Gurus of Electronic Music 1948–80.
http://www.discogs.com/label/Metamkine
   Metamkine, record publisher and distributor, France.

INDEX

4A; 4X (digital systems), 225, 234, 260, 265, 268
10 ans d'essais radiophoniques 1942-1952 (1955 vinyl disc collection) (RTF), 24–25, 27
"27 games—exercises" (Renard), 69

acetate discs, 28, 30, 34
Acousmaline database, 64, 100, 101–2
Acousma-Raves, 124
Acousmathéque, 64, 100–101, 148
acousmatics, 48–49, 64–65, 148
Acousmographe, 64, 93–94, 123, 132, 141–49, 233
Acousmonium, 49, 54, 64, 96–97, 103, 104, 112–14, 116, 121–22, 148, 229, 230–31, 236–37
ACR (Atelier de Création Radiophonique de France Culture), 197
ACROE, 96
Adac program, 84, 294
Adami, Valerio, 195
Adkins, Mathew, 335
Adrienne Mesurat (1949 production) (Green), 26
Aguila, Jesus, 198, 226
AKS, 83
À la recherche d'une musique concrète (Schaeffer), 24
Alari, Nadine, 350n18
Aléa (Brown), 179
Allouis, Jean-François, 87, 88, 89, 141–42, 233–34, 235
Altman, Rick, 19

Alvarez, Javier, 307, 342
L'Amen de verre (Haubenstock-Ramati), 175
Amériques (1921 piece) (Varèse), 41
Amy, Gilbert, 198, 205, 226
An, Chengbi, 331
"L'analyse des musiques électroacoustiques" (Delalande), 135–36
Anderson, Barry, 293
Andrews, John, 214
Andrieux, Dominique, 299
Animographe, 240
Antiphonie (1951) (Henry), 170
Antoine, Janine, 373n43
Antunes, Jorge, 239
Apocalypse de Jean (Henry), 178, 238
Apollinaire, Guillaume, 36, 37, 60
Appleton, Jon, 96
Apsome Studio, 188, 197–98, 238
Aragon, Louis, 21
Arandel, Marguerite, 350n18
L'arbre et coetera (1972 work) (Savouret), 243
Arcana (1927 piece) (Varèse), 41
Arcane (1955 ballet) (Béjart), 166, 176
Arezzo, Guy d', 132
Aristotle, 56
Aristoxenus of Taranto, 56
Arman, 195, 197
Arnaud, Pierre de Chassy-Poulay, 13, 15–16, 28
Arnheim, Rudolf, 149, 151

393

INDEX

Arrieu, Claude, 15, 16, 25
Ars Nova (ensemble), 238, 240
Artaud, Antonin, 214
Artaud, Pierre-Yves, 198
Artaud, Vincent, 331
Arthuys, Philippe, 48, 52, 70, 171–72, 173, 174, 176, 186, 187, 193, 204
Artikulation (Ligeti), 198, 203
L'art précolombien (1953 short) (Fulchignoni), 171
The Art of Noises (Russolo), 32, 33, 35
Aschour, Didier, 334
Ascione, Patrick, 279, 297, 329
Aspect sentimental (1957 piece) (Sauguet), 174–75
Assayag, Gérard, 343
Astrologie (1952 film) (Grémillon), 166
Atelier de Création Radiophonique de France Culture (ACR), 197
atonality, 40
De natura sonorum (1975 piece) (Parmegiani), 83
Auber, Dominique, 304
Audin, Jean-Philippe, 331
Audiosculpt software, 132, 292
Augoyard, Jean-François, 343
Auric, Georges, 354
avant-gardes, 3, 4–5, 9, 34, 41, 44–45, 47, 52, 61, 113, 121, 161, 194–97. See also musique concrète

Babbitt, Milton, 41, 181–82, 200
Baboni-Schilinghi, Jacopo, 336
Bach, Johann Sebastian, 79, 272
Bachelard, Gaston, 18, 138, 213, 298–99, 308, 328, 361n17
Bai, Li, 305
Bailblé, Claude, 150, 365nn52,54
Balagna, Francis, 131
Baldassaré, Frédéric, 333
Ballif, Claude, 205, 206, 207
Baltazar, Pascal, 341
Bandt, Ros, 126
Barbaud, Pierre, 210
Barbizet, Pierre, 236
Barbulée, Madeleine, 350n18
Barbut, Marc, 370n36
Bargeld, Blixa, 340
Baronnet, Jean, 188
Barrachina, Solange, 316, 318, 377n11
Barraqué, Jean, 50, 53, 70, 131, 161, 167, 174, 186
Barras, Robert, 185
Barras, Vincent, 185, 351n25
Barrat, Pierre, 245

Barrett, Natasha, 337
Barrière, Françoise, 236, 239, 339, 361n26
Barrière, Jean-Baptiste, 374n1
Barron, Louis and Bebe, 179, 180, 182, 185
Barthes, Roland, 60, 138, 233
Bartok, Bela, 40, 185
Baschet, Bernard, 97, 187, 209
Baschet, Florence, 333
Bathygraphe, 147
Battier, Marc, 262, 272, 343
Baudoin, Jacqueline, 350n18
Baudoin-Lafon, Michel, 343
Baudrier, Yves, 170, 185
Baudrillart, Alain, 134
Bayer, Raymond, 366n7
Bayle, François, ii, ix, 1, 11, 39, 44, 48–49, 50, 51, 56, 58, 64, 66, 67, 70, 71, 74–75, 75, 80, 83, 85, 88, 96, 99–100, 105, 108, 112–16, 121, 138–39, 146, 148, 150, 163, 174, 186, 187, 197, 202–3, 205, 206, 207, 208, 211, 213, 214, 219, 224, 228–29, 230, 231, 233, 234, 241–42, 292, 345
Bayle, Laurent, 292
Bazaine, Jean, 195
Bazin, Alexandre, 317, 341
Beaune workshop, 13–15, 23, 50, 70
Becce, Giuseppe, 23
Beck, Francis, 312
Becker, Jacques, 188
Béjart, Maurice, 164, 166, 172, 176, 188, 193, 197, 238, 367n30
Belar, Herbert, 200
Bell, Alexander Graham, 28
Bell Laboratories, 200, 201, 225, 231, 232, 234, 241
Benglia, Habib, 25, 166
Benjamin, George, 293
Benjamin, Walter, 108
Bennett, Gerald, 225
Benoît, Denise, 166
Benoît, Jean-Christophe, 166
Berberian, Cathy, 198
Berg, Alban, 40, 161
Bergeron, Arthur, 275
Bergson, Henri, 43, 105, 129, 149
Berio, Luciano, ix, 51, 182, 183, 198, 203, 213, 225, 239
Bernard, Marie-Hélène, 304, 335
Bernard, Olivier, 323
Bernard, Patrice, 359n87
Bernard, Pierre, 103, 239
Bernardini, Nicola, 307
Bernhart, José, 109, 165

Bernier, Jean-Yves, 87, 235
Bertolina, Lucien, 236
Besche, Thierry, 237
Besson, Dominique, 143, 148, 156
Bettelheim, Bruno, 303
Beyer, Robert, 180, 181, 182
Biagi, Vittorio, 214, 242
Bidule en ut (Schaeffer and Henry), 48, 163, 164, 165, 170
Bierhals, Wilfried, 368n44
Bir, Catherine, 215
Birolini, Hervé, 342
Blackburn, Maurice, 270
Blanchard, Philippe, 342
Blanchard, Roger, 370n35
Blanche, Francis, 350n11
Bleuse, Marc, 269
Blin, Bernard, 165
Bloch, Thomas, 331
Bocca, Bruno, 342
Bode, Harald, 31, 177
Bœhmer, Konrad, 181
Boesch, Rainer, 310
Boeswillwald, Pierre, 114, 236, 239
Bœuf, Georges, 236, 342
Bohor (1958 composition) (Xenakis), 77, 211, 218
Bokanowski, Michèle, 239, 243, 276, 330
Bollery, Jean, 305
Bômont, Laurent, 333
Bonfils, Muriel, 143
Bonnet, François, 128, 187, 317
The Books, 340
Bordenave, Suzanne, 377n7
Borg, René, 239
Borgeaud, Nelly, 272
Borowzyck, Valerian, 241
Bosc, René, 326
Bosseur, Jean-Yves, 29, 151, 156, 199
Bouckaert, Laurence, 338
Boucourechliev, André, 151, 183, 203
Bouffaïd, Omar, 283
Bouhalassa, Ned, 271
Boulanger, Nadia, 10, 40
Boulanger, Richard, 96
Boulez, Pierre, ix, 50, 53, 70, 131, 161, 167, 170, 173, 177, 178–80, 185, 186, 198, 224, 225, 226, 238
Bouquet, Michel, 25
Bour, Ernest, 215
Bourdan Center, 98, 99, 210, 229, 230, 245, 258, 369n10, 373n38, 374n12
Bourget, Jean-Paul, 315

Bourgoin, Françoise, 239, 250
Bousch, François, 323
Brady, Tim, 304
Brahms, Johannes, 40
Brando, Thomas, 279
Brau, Jean-Louis, 42
Braun, Peter-Michael, 226
Breton, André, 38, 39, 215
Breton, Jean-Serge, 241
Brière, Valérie, 245
Brissot, Jacques, 206
Brown, Earle, ix, 5, 51, 151, 179–80, 212
Bruck, Charles, 208
Bruges-Renard, Bernard, 315
bruitism, 32, 34, 35, 40–41, 44
Brümmer, Ludger, 338
Brun, Arno-Charles, 206
Brün, Herbert, 231
Brunet, Sophie, 252, 369n26
Bruzdowicz, Joanna, 239
Bucchla, Don, ix
Buffet, Gabrielle, 41
Bulgakowa, Oksana, 35
Bulski, Richard, 235
Bultiauw, Didier, 142, 144
Bureau, Jacques, 175
Burgos, Gilles, 302
Burliuk, David, 17, 34
Busnel, Louis, 206
Busoni, Ferruccio, 41

Cadoz, Claude, 96
Caesar, Rodolfo, 239
Cage, John, ix, 5, 42, 49, 51, 151, 163, 177, 178–80, 183, 185, 196, 198, 199
Cahen, Robert, 239
Cahen, Roland, 221
Cahiers du cinéma, 196, 197
Cahill, Thaddeus, 31, 41
Caillat, Stéphane, 243
Calder (Cage), 185
Calon, Christian, 271, 301
Camatte, Laurent, 333
Camilleri, Lelio, 343
Camus, Albert, 21
Canat de Chizy, Edith, 323
Cantate pour elle (Malec), 213
Canton, Edgardo, 205, 206, 207, 208, 213, 215, 239
Capdevielle, Pierre, 322
Capparos, Olivier, 249
Capture éphémère (Parmegiani), 214

carbon microphones, 28
Cardew, Cornelius, 151
Caron, Elise, 303
Carson, Philippe, 99, 205, 206, 207, 208, 211
Carter, Elliott, 40
Cartridge Music (Cage), 198
Casamitjana, Didier, 330
Cascone, Kim, 335
Castanet, Pierre-Albert, 213
Castellana, Marcello, 295
Castiglioni, Niccolo, 183
Cavada, Jean-Marie, 313
Cazeneuve, Maurice, 25–26
CCRMA (Center for Computer Research in Music and Acoustics), 263, 345, 347
CD (compact disc), 49, 54, 97, 99, 101, 117, 140, 153, 268, 282, 314
CDMC (Center for Documentation of Contemporary Music), 228
Céleste, Bernadette, 282, 283
Cendras, Blaise, 195, 202
Center of Radiophonic Studies, 72
Centre International de Recherche Musicale (CIRM), 237
centripetal/centrifugal concept, 107, 116
CER (Centre d'Études Radiophoniques), 23, 24–25, 46, 109, 185
Le cercle (1956 ballet) (Henry), 176
Ceremony (1969 album) (Henry and Spooky Tooth), 121, 238
CERT (Centre d'Études de la Radio et de la Télévision; Centre d'Études de Radio-Télévision), 50, 203
César, 195
CÉSARÉ, 325
Chabrol, Claude, 196
Chadabe, Joël, 200, 343, 375–76n37
Chafe, Chris, 96
Chagas, Paulo, 181
Chalosse, Marc, 337
Chamass-Kyrou, Mireille, 97, 202, 204, 206, 210
Chamberlain, John, 195
La Chambre claire (Barthes), 60
Chambure, Alain de, 175, 210, 217–18
Chaminé, Jorge, 275
Chanez, Christian, 335
Charbonnier, Georges, 232, 367n27
Charbonnier, Jeannine, 370n35
Charette, Marin de, 276
Charles, Daniel, 66

Chartier, Richard, 335
Chatwin, Bruce, 119
Chautemps, Jean-Louis, 213
Chaves, Ana-Bela, 330
Cherradi, Najib, 332
Chiarucci, Enrico, 82, 86, 196, 205, 209, 210, 229, 241
Chion, Michel, 6, 48, 49, 51, 56–57, 62, 63–64, 95, 107–8, 116, 149, 209, 221, 239, 244
Chion-Mourier, Florence, 328
Chojnacka, Elisabeth, 279
Chomsky, Noam, 182
Chopin, Henri, 42
Chowning, John, 40, 108, 234, 262, 263, 335, 344, 345, 347
Chowning, Maureen, 335
Chrétiennot, Louis, 339
Christo, 195
Christophe Colomb (1947 production) (Claudel), 25
cine-eye concept, 35
cinéma vérité, 197
CIRM (Centre International de Recherche Musicale), 237
Clair, René, 19, 37, 109, 165
Clapaud, Charles, 239
Clarke, Kenny, 243
Claudel, Paul, 15, 16, 25, 33
Clementi, Aldo, 183
closed groove, 30–31, 43
Clozier, Christian, 114, 236, 239
Club d'Essai, 9, 22–27, 42, 44, 50, 65, 109, 162, 164, 165, 173, 174, 177, 203, 216
Cluzel, Jean-Paul, 314
CNSML (Conservatoire National Supérieur de Musique de Lyon), 268–69
CNSMP (Conservatoire National Supérieur de Musique de Paris; National Conservatory of Paris), 47, 50, 51, 53, 66, 68, 220, 246, 249–51, 255, 268–69, 310, 345, 365n39
Cochereau, Pierre, 269
Cochini, Roger, 221, 236, 239
Cocteau, Jean, 25, 37, 38, 56, 162
Cohen, Andrea, 18, 342
Cohen-Levinas, Danièle, 161
Cohen-Solal, Robert, 221, 239, 240, 372n28
Coissac, Nina, 258
Collaer, Paul, 39
Collard, Jeannine, 171
Collardey, Dominique, 239
Colleen (Cécile Schott), 341

Cologne Studio, 5, 51, 99, 180, 181, 182–83, 198–200, 210, 225–26, 244. *See also* WDR (Westdeutscher Rundfunk) (Cologne)
Colon, Claude, 236
Coltrane, John, 243
Columbia Princeton Electronic Music Center, 41, 181
computer-assisted music (CAM), 50, 54, 200, 231–32
Computer Music Workshop, 231
Concari, Sheila, 342
Concert Collectif, 206–8, 216, 217, 239
Concert de bruits (*Concert of Noises*) (Schaeffer), 9, 42, 44, 120, 162, 184, 212
Concertino Diapason (Schaeffer, Grünenwald), 39, 163
Concerto des ambiguïtés (1950) (Henry), 176
Concerto for Prepared Piano (Cage), 178
concrete method, 27
Concret PH (1958) (Xenakis), 112, 174, 201, 218
Connes, Alain, 361n16
Constant, Marius, 40, 238
Conte, Louise, 350n18
Continuo (Schaeffer and Ferrari), 201
Copeau, Jacques, 14, 15, 17, 23, 70
Coppe, Dimitri, 335
Le Coq et l'Arlequin (Cocteau), 37, 38, 56
La Coquille à planètes (1943–44) (Schaeffer and Arrieu), 15–18, 25, 27
Le Corbusier, 174
Corregia, Enrico, 307
Correspondance (group), 39
Corroenne, Olivier, 249
Coulombe Saint-Marcoux, Micheline, 239
Coupigny, Francis, 79, 82, 169, 196, 210, 229, 232, 241
Coupigny Synthesizer, 82, 83–84, 242
Couprie, Pierre, 132
Couraud, Michel, 243
Cowell, Henry, 163
Le crabe qui jouait avec la mer (Arthuys), 174, 186
Crambes, Eric, 333
Credo (Cage), 42
Crénesse, Pierre, 21
Cristal (Bayle), 234
Critical Studies Group (GEC), 73
Criton, Pascale, 308, 334
Croix, Philip de la, 307
Cros, Charles, 28, 345
C Sound, 201
Cullin, Olivier, 149

Cuniot, Laurent, 238, 239
Cunningham, Merce, 176
Cuny, Alain, 214
Curjol, Jocelyne, 315

Dadaism, 6, 9, 36–37, 44, 176
Dallet, Sylvie, 295, 343
Damasio, Antonio, 122–23
Dao, Philippe, 96, 116, 128
Daoust, Yves, 271
Darmstadt, 180, 183, 198, 215
Darnis, Jacques, 96
Daroux, Cécile, 305, 339
Daumal, René, 11, 272
Davies, Hugh, 390
Debade, Nicolas, 317
Debord, Guy, 42, 134, 195
Debussy, Claude, 33, 40, 44
DEC (Digital Equipment Corporation), 86, 87
Décade internationale de la musique expérimentale, 185, 193
December 52 (1952 open form piece) (Brown), 179
Dechelle, François, 343
Decoust, Michel, 225
Defurne, Philippe, 333
de Gaulle, Charles, 21
Deguy, Sylvie, 339
Dejoux, Jacques, 240
Delage, Bernard, 362n45
Delalande, François, ii, 59, 67, 68–69, 101, 115, 122, 127, 135–36, 139–40, 144, 146, 156, 218, 220, 221, 223, 229, 233
Deleuze, Gilles, 106, 154, 243
Delgado, Raoul, 239
Deliège, Celestin, 52
Delorme, Danièle, 366n8
Delume, Caroline, 334
Demy, Jacques, 196
Density 21.5 (1936 composition) (Varèse), 41
Déotte, Jean-Louis, 107, 107–8
Depalle, Philippe, 239
Derrien, Marcelle, 366n8
Desbazeille, Renaud, 304
Descartes, René, 62–63
Deschamps, Gérard, 195
Deschênes, Marcelle, 271
Déserts (1954) (Varèse), 42, 111, 161, 173
Deshays, Daniel, 279
Desnos, Robert, 193
détournement, 195–96

Deupree, Taylor, 335
Devilleneuve, Hélène, 302
Dhomont, Francis, 40, 51, 150
Diaghilev, Georges, 37
Diamorphoses (Xenakis), 175, 201, 218
Die Klangwelt der Elektronische Musik program, 180
Diennet, Jacques, 236
Dietrich, Luc, 11
Digidesign, 91, 92
Digital Equipment Corporation (DEC), 86, 87, 235
digitization, 100, 101, 141, 232, 312, 313, 314, 315, 318–19, 378n13
Dine, Jim, 195
Dionysus, 124
disco, 238
displacement phenomenon, 44, 298
La Divine comédie (Bayle and Parmegiani), 83, 242
DJ Olive, 124, 216, 330
DJ Scanner, 124, 330
DMDTS (Directorate of Music, Dance, Theater, and Performances), 227, 228
Le Domaine Musical, 198, 211, 226
Donato, François, ii, 96, 116, 128
Donaueschingen festival, 171–72, 177, 268
Donin, Nicolas, 146
Doniol-Valcroze, Jacques, 196
dreaming, 18, 31, 37, 38, 44, 119, 164, 165, 242, 244, 279, 327, 328, 332
Drever, John Levack, 329
Drew, Daniel, 340
Drogoz, Philippe, 239
Drouet, Jean-Pierre, 213
dualism, 124–25
Dubedout, Bertrand, 274, 324
Dubuffet, Jean, 151
Ducarme, Jean-Louis, 56
Duchamp, Marcel, 36, 41
Duchemin, Noémi, 227, 269
Duchenne, Jean-Marc, 298, 307, 336
Dufour, Denis, 6, 124, 143, 238, 239
Dufourt, Hugues, 226
Dufrêne, François, 42, 198
Dufrêne, Jacques, 42, 195
Duhamel, Georges, 25
Dumaurier, Elisabeth, 282
Dumay, Alain, 87
Dunkelmann, Stéphan, 156, 291
Durand, Daniel, 304

Durand, Gilbert, 75
Durey, Louis, 354n94
Dürr, Bernard, 232, 235, 239, 245–46, 265, 287, 359n71, 374n12, 376n38
Dury, Rémi, 302
Dutilleux, Henri, 322–23
Dutrieu, Pierre, 331
DVD, 101
Dyaxis system, 142, 299
dynamic species concept, 58
Dynamophone, 41

École de Paris, 5
École Normale de Musique de Paris, 40
L'Écran transparent (1973 piece), 244
Ecuatorial (1934 piece) (Varèse), 41
Edison, Thomas, 28
Eggling, Viking, 20
EIMCP (Ensemble Instrumental de Musique Contemporaine de Paris), 202, 206, 212, 216
Eimert, Herbert, 180, 181, 182, 185, 198
Einfeldt, Thomas, 335
electro, 3, 48, 49, 122, 271, 340
Electronic Music Foundation (EMF), 92, 295, 319, 343
Elementa Harmonica (Aristoxenus), 56
Eliot, T. S., 213
Eloy, Christian, 276, 293, 304, 323, 324, 378n17
Eloy, Jean-Claude, 181, 226, 273, 307, 322, 359n87
Eluard, Paul, 21
EMF (Electronic Music Foundation), 92, 295, 319, 343
Emmanuel, Pierre, 224
Endresen, Sidsel, 335
Ensemble Instrumental de Musique Contemporaine de Paris (EIMCP), 202, 206, 212, 216
Ensemble InterContemporain (EIC), 238
Enthusiasm (1930 film) (Vertov), 35
Entr'acte (1924 silent short) (Claire), 37
Eötvös, Peter, 181, 226, 238
Epitaph für Aikichi Kuboyama (Eimert), 198
Erda (1972 work) (Schwarz), 243
eRikm, 124, 340
Erlih, Devy, 213
Ernst, Max, 39
Erosphère (Bayle), 86–87, 88
Erro, 195
Espaces inhabitables (1966 piece) (Bayle), 203, 121, 214
Esthétique des arts relais (Schaeffer), 43–44
Estrada, Julio, 40

Étude 1 (Stockhausen), 171
Étude 1 & 2 (Boulez), 170
Étude aux accidents (Ferrari), 201, 202
Étude aux allures (Schaeffer), 195, 201
Étude aux chemins de fer (Schaeffer), 39
Étude aux objets (Schaeffer), 201
Étude aux sons animés (Shaeffer), 201
Étude aux sons tendus (Ferrari), 201, 202
Étude floue (Ferrari), 201
Études de bruits (1948) (Schaeffer), 48, 99, 101, 130, 161, 163, 164, 177, 186
Études I and II (Boulez), 39
Étude vocale (Rollin), 39
Evangelisti, Franco, 181
Éveil à la musique (radio program), 69
L'Expérience Acoustique (1970–73 work) (Bayle), 83, 113, 231, 241
experimental music, 47, 65–66, 238
Exposition des Musiques Expérimentales, 121, 213, 215

Faber et Sapiens (Schaeffer), 76
Fabriquedecouleurs, 335
Falco, Fabrice di, 331
Falla, Manuel de, 33, 40
Fantasy in Space (1952 piece) (Luening), 182
Farabet, René, 248, 373nn43,45
Fauré, Gabriel, 40
Favory, Jean, 140, 142
Favotti, Gino, 294, 303, 323, 324, 342
Favre(-Marinet), Marc, 237, 239, 250, 254, 307, 329
Favreau, Emmanuel, 88, 91–92, 145, 287, 292, 343
Faÿ, Eric, 331
Feedback Ensemble, 238
Feldman, Morton, 5, 51, 179
Felix, Catherine, 26
Fenn, Karen, 327
Fennesz, Christian, 124, 326, 332, 335, 344
Ferès, Maria, 166
Ferrari, Giordano, 18
Ferrari, Luc, 1, 6, 51, 52, 70, 97, 124, 125, 126, 161, 187, 188, 195, 196, 202, 204, 205, 206, 207, 208, 211, 213, 214, 215–16, 226
Ferreyra, Beatriz, 40, 51, 56, 97, 205, 209, 215, 221, 236, 251, 308, 337
Festival de la Recherche, 199
Fèvre, Véronique, 333, 334
Feyler, Pierre, 334
Filippi, Lionel, 239, 251
Fillioud, Georges, 101, 289

Les fils de l'eau (1953 film) (Rouch), 171
filter (FLT), 87
First Decade of International Experimental Music event, 185
Fischer-Karwin, Heinz, 171
Fischinger, Oskar, 20
Five Studies of Noises (1948 composition) (Schaeffer), 34
Flamand, Paul, 12
Fleuret, Maurice, 261, 247
Fleury, Patrick, 277
Flon, Suzanne, 350n18
FLT (filter), 87
Fluxus movement, 199
Fober, Dominique, 343
Fognini, Mireille, 239
Folio (1953 piece) (Brown), 179
Forest, Lee de, 31
Formosa, Marcel, 140
Fort, Bernard, 237
Fourès, Henry, 324
Fourneau, Alain, 236
fragmentation, 163
Fragments pour Bayle (Thomas), 138
Framework Plan, 222–23
Francès, Robert, 56, 223
Franco-Quebec integrated research program, 229
Franju, Georges, 196
Frapat, Monique, 69
Freinet, Celestin, 68
Frémiot, Marcel, 58, 140, 214, 236
French National Education, 53, 123, 144, 145
French National Radio, 3, 9
French New Wave, 196
French Overseas Radio, 70
French Radio and Television, 21, 50, 185, 249, 289, 340, 347
Fresnais, Gilles, 239
Frigon, Michel, 271
Frima, Roger, 239
Frisell, Bill, 233
Fritsch, Johannes, 181, 238
Frize, Nicolas, 237, 239
Fromanger, 195
Frydman, Armand, 239
Frydman, Sylvain, 303
Fuente, Benjamin de la, 377n22
Fulchignoni, Enrico, 171, 185, 193, 211
Fusco, Ottavia, 330
futurism, 9, 32–36, 44

Futuristie (1975 composition) (Henry), 34
Futurist Orchestra, 32–33

Gabriel, Georges, 298
Gabrieli, Andrea and Giovanni, 104
Gagneux, Renaud, 250
Galpérine, Alexis, 334
Garcia, Jean-Baptiste, 317
Garcia, Xavier, 237, 306, 307, 333
Garcia Lorca, Federico, 305
Garrett, Jean-Wilfrid, 109, 165, 185
Garside, Katie Jane, 340
Gaudibert, Pierre, 373n41
Gaudreault, André, 351n31
Gaussin, Alain, 239
Gauthier, Théophile, 109, 366
Gayou, Évelyne, 253–54, 267, 277, 283, 284, 285, 317, 332, 333, 342, 344, 368n43
GEC (Groupe d'Études Critiques), 73, 204
Gehlhaar, Rolf, 238
Geoffroy, Jean, 304
GER (Groupe d'Enseignement Recherche), 222
Germain, Jean, 370
German radio, 29, 180, 200, 216, 225–26, 249
Germinal (1985 collaboration) (Studio 123), 88, 277–78, 285
Gervasoni, Pierre, 376n15
Gerzon, Michaël, 115
Gesang der Jünglinge (1954) (Stockhausen), 82, 182
Geslin, Yann, 74, 87, 144, 238, 239, 256, 264, 272, 273, 277, 279, 285–86, 290, 292, 310, 323, 332, 341, 342, 345, 377n23
Ges of Vierzon, 237
Giomi, Francesco, 332
Giovannoni, David, 28
GIRATEV, 232–33, 235
Giroudon, James, 269, 372n17
Giugno, Giuseppe di, 225, 234
Glass, Philip, 40
globalization, ix, 49, 54, 314, 324
Globokar, Vinko, 225, 239
GMAP (Groupe de Musique Algorithmique de Paris), 210
GMEA (Groupe de Musique Expérimentale d'Albi), 73, 237
GMEB (Groupe de Musique Expérimentale de Bourges), 73, 104, 114, 236–37
Gmebaphone, 104, 114, 236–37
Gmebogosse, 236
GMEM (Groupe de Musique Expérimentale de Marseille), 115, 236

GMR Tools (software), 53
GMVL (Groupe de Musique Vivante de Lyon), 73, 237
Gobeil, Gilles, 271, 329, 342
Gobin, Pascal, 140
Godard, Jean-Luc, 196, 197, 347
Godebert, Georges, 26
Goethe, Johann Wolfgang von, 275, 329
Goeyvaerts, Karel, 181
Goff, Philippe le, 339
Gogol, Nicolas, 25
Goldsmith, 195
Goléa, Antoine, 185
Gontcharova, Natalia, 17, 34
Gonzales-Arroyo, Ramon, 301
Goodman, Lanie, 275, 299
Gouillard, Eric, 333
Grame, 269–70, 325, 339
Grand, Gilles, 278
Grand Auditorium of Radio France, 113, 271, 274–76, 279–80, 288, 289, 297, 331–40, 363n57, 376n1
Granulométrie (1967 work) (Henry), 42, 198
graphic symbols library, 141
Graton, Jean-Loup, 239
Grätzer, Carlos, 302
Gravesano, 183–84, 211
Green, Julien, 26
Greie, Antye, 337
Greimas, Algirdas-Julien, 107
Grémillon, Jean, 22, 166, 185
Grenier, Marcel, 373n43
GRI (Groupe de Recherche Image), 73, 204, 224
Grieg, Edvard, 40
Grillo, Alex, 341
Grippe, Ragnar, 239
Grisey, Gérard, 227
GRM Expérience, 124
GRM Tools (software), 51, 73–74, 75, 85, 88, 89, 90–93, 141, 145
Gropius, Walter, 16
La grotte (graphic transcription) (Mion), 148
Groult, Christine, 294
Groupe de Musique Algorithmique de Paris (GMAP), 210
Groupe de Musique Expérimentale d'Albi (GMEA), 73, 237
Groupe de Musique Expérimentale de Bourges (GMEB), 73, 104, 114, 236–37
Groupe de Musique Expérimentale de Marseille (GMEM), 115, 236

INDEX   401

Groupe de Musique Vivante de Lyon (GMVL), 73, 237
Groupe d'Enseignement Recherche (GER), 222
Groupe de Recherche de Musique Concrète (GRMC), 9, 12, 47, 50, 131, 167, 186–87, 193, 194, 217
Groupe de Recherche Image (GRI), 73, 204, 224
Groupe de Recherches Musicales (GRM), 9, 12, 47, 50, 203–4; audiences, 110, 111, 113–15, 117–18, 120–27; concepts, 57–64; concerts, 109–20, 127–28; diversification, 122–23; electroacoustic concerts, 117–20; experimental music, 120–21; installations and, 126–27; as name, 47–49; pedagogy, 64–73; presentations, 120; spaces, 104–17; as studio/as school, 49–52; as style, 52–54; techno music and, 123–26; tools, 73–102; traditional concerts, 121–22
Groupe de Recherche Technique (GRT), 73, 79, 204, 224, 232, 240
Groupe d'Études Critiques (GEC), 73, 204
Group for Visual Research (GRI), 73
Group of Six, 38
GRT (Groupe de Recherche Technique), 73, 79, 204, 224, 232, 240
Grünenwald, Jean-Jacques, 23, 26, 39, 163
Grutter-Rauch, Janine, 171
Guattari, Felix, 106, 243
Guertin, Marcelle, 229
Guide des objects sonores (Chion), 56–57, 62, 63–64
Guignebert, Jean, 351n41
Guilbaud, Théodore, 370n36
Guilbert, Denis, 239
Guillet, Michel, 341
Guiot, Dominique, 239
Guitry, Sacha, 162–63
Gurdjieff, Georges-Ivanovitch, 10–12, 15, 51, 70
Guttman, Newman, 200

Haas, Max de, 202
Habault, Daniel, 237
Hainz, Raymond, 195, 206
Hampson, Robert, 337
The Hands, 95–96
Hanno, Yoshihiro, 337
Harbonnier, Christoph, 337
Harrison, Jonty, 304
Hartung, Hans, 195
Harvey, Jonathan, 293
Hassel, Jon, 337
Hatwell, Yvette, 150

Haubenstock-Ramati, Roman, 175
Haut voltage (1956 ballet) (Henry), 82, 176
Haydn, Joseph, 114, 230
Heckel, Jürgen, 340
Heidsieck, Bernard, 42, 340, 354n112
Heleau, Anne-Marie, 304
Hendrix, Jimi, 142, 143, 233
Henry, Pierre, 1, 6, 26, 39, 40, 42, 44, 47, 50, 51, 52, 70, 77, 78, 79, 82, 99, 107, 109, 110, 111, 114, 120, 121, 123–24, 125, 127, 163–64, 165–66, 168–69, 171–72, 173, 175, 177–78, 185, 186, 187–89, 193, 194, 196, 197–98, 202, 204, 216, 217, 238
Hermon, Michel, 242
Hétérozygote (1963 piece) (Ferrari), 148, 195, 196, 212, 216
Hidalgo, Juan, 205
Higaki, Tomonari, 337
Hiller, Lejaren, 181, 200
Hindemith, Paul, 29
Histoire d'une bouchée de pain (1925 film) (Vertov), 35
Hodeir, André, 50, 53, 70, 161, 167, 170, 186
Hoérée, Arthur, 19, 20
Höller, York, 181, 226
Holophon (software), 236
Holt, Robert H., 149, 151
Honegger, Arthur, 19, 27, 33
Hoog, Emmanuel, 312
Hoogewerf, Wim, 334
Hooreman, Paul, 39
Hörspiel, 14, 19, 29, 43, 195, 216, 249
Hurel, Philippe, 342
Hussenot, Olivier, 12
Husserl, Edmund, 58, 150, 223
Husson, Raoul, 185
Huteau, Alain, 333, 334
Hyperprism (1923 piece) (Varèse), 41

ICMC (International Computer Music Conference), 91, 142, 292, 320
IDHEC (Institut des hautes études cinématographiques), 21
Ikeda, Ryoji, 83–84, 337
Illiac Suite for String Quartet (computer composition) (Hiller and Isaacson), 181, 200, 262
imaged-thought, 149–56
L'image musique (Cullin), 149
Imaginary Landscape No. 1 (Cage), 42
IMEB (Institut International de Musique Electroacoustique de Bourges), 236, 325, 339, 378n19

INA (Institut National de l'Audiovisuel). *See* Institut National de l'Audiovisuel (INA)
INA-GRM, 253, 273, 278, 301, 320, 327, 331
In Search of Musique Concrète (Schaeffer), 184–85
L'Instant mobile (Parmegiani), 82, 121, 213
Institut d'Electronique Fondamentale d'Orsay, 232
Institut de Recherche et Coordination Acoustique/Musique (IRCAM). *See* IRCAM (Institut de Recherche et Coordination Acoustique/Musique)
Institut des hautes études cinématographiques (IDHEC), 21
Institut International de Musique Electroacoustique de Bourges (IMEB), 236
Institut National de l'Audiovisuel (INA), 2, 3, 25, 30, 85, 92, 101, 101–2, 122, 219, 224–25, 249, 253, 260–61, 282, 283, 287, 288, 289, 295, 303, 307, 311, 312–13, 314–15, 318–19, 333, 344, 367n27, 378n13
instrumentarium, 53, 110, 196, 237
Intégrales (1925 piece) (Varèse), 41
Interludes (Cage), 179
internal/external space concept, 107–8, 116
Internationale Situationniste, 42, 134, 195–96
International Society for Contemporary Music (ISCM), 322–23
International Telecommunication Union (ITU), 24
Ionesco, Petrika, 275
Ionisation (1931 piece) (Varèse), 41
IRCAM (Institut de Recherche et Coordination Acoustique/Musique), 5, 85, 132, 146, 183, 219, 224–25, 231, 234, 260, 262, 263, 264, 265, 268, 272, 274, 279, 287, 292–93, 320, 322–23, 343, 351n27, 359n82, 374nn4,13, 376n9, 377nn24,12
Isaacson, Leonard, 181, 200
ISCM (International Society for Contemporary Music), 322–23
Isou, Isidore, 42, 353n73
Italian futurism, 32–34, 35, 36
Ivernel, Daniel, 350n18
Ivsic, Radovan, 175

Jacomucci, Claudio, 339
Jacquin, Marc, 25
Jaffe, David, 96
Jaffrenou, Pierre-Alain, 231, 232, 235, 239, 269–70, 280, 339, 372n17, 375n25
Jakobson, Roman, 59, 223
Jandl, Ernst, 301
Janin, Pierre, 171, 209

Jarre, Jean-Michel, 238–39, 258
Jarsky, Irène, 305, 326
Jassex (1966) (Parmegiani), 213
Jaubert, Maurice, 24
Jauss, Hans-Robert, 66, 357n33
Jazz et jazz (1951 mixed work) (Hodeir), 170
Jean, Monique, 271
Jeanneney, Jean-Noël, 101, 289
Jenny, Georges, 31
Jeune France, 12, 13
Jézéquel, Jean-Luc, 229, 281, 283
Jisse, David, 216, 307, 317, 341
Jodelet, Florent, 279, 300
Jodlowski, Pierre, 341
Johns, Jasper, 195
Johnson, David, 226, 238
Jolivet, André, 25–26, 350n9, 352n52
Jorn, Asger, 195
Joste, Martine, 327
Le joueur de bruits (Philippot), 170
Jours de ma vie (film) (de Haas), 202
Joyce, James, 203, 330
Justel, Elsa, 107, 236
Just So Stories (Kipling), 174

Kagel, Mauricio, 181, 226
Kaltenecker, Martin, 14
Kamler, Piotr, 206, 211
Kanaka, Atau, 340
Karatchentzeff, Pierre, 232
Karcher, Eric, 333
Karlsson, Erik-Mikael, 342
Kasparov, Yuri, 301
Kassovitz, René, 206
Kassowitz, Peter, 202
Kast, Pierre, 196
Kastler, Daniel, 361
Kaufmann, Dieter, 301
Kenan, Amos, 193
Kent, Lone, 340
Kergomard, Henri, 251, 256, 269, 310
Kessler, Thomas, 239, 300
keyboard phonogènes. *See* phonogènes
Khlebnikov, Boris, 17, 34
Kieffer, Anna-Maria, 305
Kientzy, Daniel, 278, 328
Kipling, Rudyard, 174
Kirsanoff, Dimitri, 19, 352n53
KK Null (Kazuzuki Kishino), 341
Klangfarbenmelodie, 63
Klangfiguren No. 2 (1954 work) (Stockhausen), 182

Klasen, Peter, 195
Klavierstück XI (1956 piano piece) (Stockhausen), 179
Klee, Paul, 308
Klein, Yves, 151, 195
Knobelsdorff, Leopold von, 181, 199
Kodály, Zoltán, 323
Koechlin, Olivier, 142, 143–44, 233, 290
Koenig, Gottfried Michaël, 51, 181, 182, 198–99
Kontakte (1960 work) (Stockhausen), 82, 182, 198, 199, 211, 370n42
Koré (1972 work) (Bokanowski), 243
Kotonsky, Vladimir, 239
Kott, Jean, 268
Koulechov, Lev, 34, 353n76
Kraftwerk, 125, 271
Krenek, Ernst, 181
Kropfl, Francisco, 368n49
Kubler, Françoise, 327
Kuffler, Eugénie, 239

Lachenmann, Helmut, 326
Lagniel, Christine, 303
Laing, Ronald D., 275
Lallemand, Jean-Claude, 230
Laloux, René, 206
Lancino, Thierry, 307
Landowski, Marcel, 226, 227, 239
Landy, Leigh, 340, 343
Lang, Jack, 261
Langlois, Philippe, 19–20, 342, 372n36
Lapeyre, Jean-Louis, 367
Lapoujade, Robert, 203, 212
Laracine, Albert, 66
Larionov, Mikhail, 34
Larivière, Régis Renouard, 214
Larsen, Ellen, 275
Larvor, Francis, 304
Laubeuf, Vincent, 335
Laubier, Serge de, 84–85, 92, 95, 291, 299, 302, 340
Lauras, Marc, 237
Leacock, Richard, 197
Le Duc, Jean-Claude, 114, 258
Leenhardt, Roger, 12, 14
Lefèvre, Adrien, 144, 148, 287, 317
Léger, Fernand, 276
Leipp, Émile, 206
Lejeune, Jacques, 221, 239
Lemaître, Maurice, 33, 42, 353nn73–74
Lencement, Olga, 6, 25
Léonard, Jean-Claude, 350n18

Léonard de Vinci (1953 film) (Fulchignoni), 171
Leonardo da Vinci, 356n10
Lepeuve, Monique, 174
Leroi-Gourhan, André, 76
Le Roux, Maurice, 166
Leroux, Philippe, 274, 279, 293, 297, 342
Leroy, Thierry, 289
Lescot, Andrée, 171
Letort, Bruno, 341, 362n45
Letterism, 42, 121, 198, 353n73
Letterist International, 42, 195
Letz, Stéphane, 343
Levaillant, Denis, 277, 279, 303
Levin, Thomas, 20
Levinas, Michaël, 227, 293, 307, 322–23
Lewis, Georges, 233
L'Herbier, Marcel, 21
Liberation of Paris, 21–22
Liberovici, Andrea, 330, 334, 342
Lichenstein, Roy, 195
Liet, Jean-Marc, 333
Ligeti, György, 181, 198, 203
Limouzin, Vincent, 333
Lindemann, Eric, 263
Lintz Maués, Igor, 338
listening conducts, 135–36
Littérature et sensation (Richard), 137–38
Livsic, Benedikt, 17
Lonchampt, Jacques, 336
Lonsdale, Michaël, 273
looping, 28, 163
Lorain, Denis, 375n20
Lortat-Jacob, Bernard, 118–19, 308, 362n36
Losa, Diego, 101, 315–16, 318, 331, 332, 334
Louet, Pierre, 275
Luciani, Annie, 96
Luening, Otto, 41, 181–82
Lumière noire (Parmegiani), 88
Lussier, René, 329
Lusson, Matthieu, 302
Lynch, Francés, 299
Lyon, Marianne, 228, 308

Machado, Milton, 300
Mâche, François-Bernard, 70, 187, 202, 206, 208, 211, 214, 217, 240, 281, 308, 322, 327, 359n87, 360n93
Mâche, Geneviève, 98, 99, 271, 282, 315, 360n93
Machover, Tod, 268
Maciunas, George, 199
MacSoutiLs, 84–85, 291, 299

Madaule, Jacques, 15
Maderna, Bruno, 170, 181, 182, 183
Maffesoli, Michel, 3, 119–20, 122, 125, 349n3
Mager, Jörg, 31, 47, 350n15, 352n65
Magison, 295
Magritte, René, 39
Maheu, Jean, 227, 228
Maïakovski, Vladimir, 17, 34, 353n75
Maiguashca, Mesías, 226, 238
Mailliard, Bénédict, 78, 86, 86–88, 87–88, 232, 234, 235, 247, 263–64, 272, 273, 277–78, 279, 285, 290, 292, 359n76, 374n11
Main, 335
Maison de la Radio, 94, 96, 196, 211, 213, 215, 224, 229, 232, 235, 246, 250, 260, 271, 274, 275, 280, 285, 289, 296, 314–15, 343, 352n57, 369n10, 370n41, 373nn38,50. *See also* Grand Auditorium of Radio France
Makino, Katori, 239
Malapa, Eweda, 272
Malbosc, Pierre, 140
Malec, Ivo, 39, 66, 175, 202, 204, 205, 206, 207, 208, 211–13, 215, 246, 247, 271, 274, 281, 282, 288, 291, 296, 297, 315, 326, 344, 345
Malevich, Kasimir, 34
Malina, Judith, 330
Mallarmé, Stéphane, 36, 245, 273
Malraux, André, 174
Mandelbrojt, Jacques, 140–41
Mandolini, Ricardo, 298
Manessier, Alfred, 195
Mangin, Bernadette, 330
Manifesto of Futurist Musicians (Pratella), 32
Manoury, Philippe, 279, 293, 307, 343
Manuel, Roland, 322
Manuel de résurrection (1998) (Mâche), 261
Man with a Movie Camera (1929 film) (Vertov), 34, 35
Marchetti, Lionel, 249
Margolis, Al, 341
Marie, Jean-Étienne, 40, 52, 164, 169, 170, 174, 177, 206, 207, 208, 217, 237, 354nn102–103, 357n12, 357n36
Marinetti, Filippo-Tomaso, 32, 34, 36, 353nn72,73,75,84
Markeas, Alexandros, 310
Marker, Chris, 196
Marquis, Sylvain, 335
Martelli, Henri, 322
Martenot, Maurice, 27, 31, 185, 206, 352n66
Martin du Gard, Roger, 25

Martinoir, Brian de, 370n35
Martinucci, Mauricio, 335
Mataix, Christiane, 300
Matalon, Martin, 293
Mathews, Max, 86, 95–96, 182, 200, 201, 225, 231, 232, 234, 241, 262–63, 344–45, 359n85, 374n4
Mathieu, Georges, 144, 336
Matmos, 34, 341
Matsumoto, Hinoharu, 239
Matthews, Kaffe, 341
Maufras, Robert, 26
Maurer, Michel, 303
Mavena (Malec), 175
Maximin, Bérangère, 338
Mayuzumi, Toshiro, 51
Mazeron, Liliane, 280
McDonald, Henk, 200
McLaren, Norman, 271
McLuhan, Marshall, 134, 244
Mégevand, Denise, 305, 331
Melisson, 237
Melochord, 31, 177, 180
Menez, Tanguy, 333
Mercier, Arnaud, 337
Mercure, Pierre, 271
Merleau-Ponty, Maurice, 223, 356n15
Merlet, Claire, 334
Mersenne, Marin, 63
Mesens, 39
Messiaen, Olivier, 39, 40, 44, 50, 53, 63, 77, 111, 131, 135, 161, 163, 167–68, 169–70, 185, 192, 202, 215, 217, 253, 254, 323, 350n9, 354n104, 376
Meston, Olivier, 268
Méta-instrument, 95
Meyer, Alexandre, 333
Meyer-Eppler, Werner, 177, 180, 181, 182, 367n32, 368n45
MFA (*Musique française d'aujourd'hui*), 228
Michaux, Henri, 11, 276
Le microphone bien tempéré (1951) (Henry), 165–66
MIDI (Musical Instrument Digital Interface), 84
MIDI Formers, 74, 84–85, 90, 92
MIDI systems, 58
Miereanu, Costin, 295, 297
Mignon, Paul-Louis, 25
Mikrophonie I (Stockhausen), 198, 211
Milan Studio, 183, 199
Milhaud, Darius, 40, 53, 70, 131, 168, 174, 186, 323, 354n94
Mille Plateaux (Deleuze and Guattari), 106–7

Miller, Joan, 225
MIM (Musique Informatique de Marseille), 32, 140–41, 156, 291
Minali Bella, Jean-Paul, 331
Minard, Robin, 342
Ministry of Culture, 122, 123, 225, 227, 228, 247, 269, 310, 325
Minjard, Jean-François, 300
Mion, Philippe, 84, 137, 148, 239, 249, 250, 251, 262, 269, 273, 276, 277, 280, 281, 282, 285, 293, 305, 307, 338, 376n11
Miroglio, Thierry, 335
Mitry, Jean, 175, 188, 193
Mitterand, François, 373n40
modality, 40
Moholy-Nagy, Laszlo, 29, 350n15
Moles, Abraham, 79, 106, 107, 116, 117, 185, 206, 217, 349n2, 355n3, 368n52
Molino, Jean, 136–37, 343
Mollard, Claude, 354
Molvaer, Nils Petter, 340
Mondrian, Piet, 33
Monnet, Marc, 239
Monory, Kasimir, 195
Montel, Geoffroy, 336
Montès-Baquer, José, 244
Montessori, Maria, 68
Montreal School, 270–71
Moog, Robert, ix, 352n65
Moog Synthesizer, 82, 229, 245, 359n71
Moore, Richard, 225
Moorer, Andy, 263
Moreau, Jeanne, 26
Moreux, Serge, 164
morpho-concept, 58
morphological description of sounds, 62–63
Morphophone, 366n6
morphophones, 79, 80
Mossy, 193
Moullet, Patrice, 256
Mouloudji, 215
Mounier, Emmanuel, 12
Mouse on Mars, 340
Moyal, Marie-Noëlle, 139, 239, 273
Müller, Volker, 181, 293
Mundinger, Carol, 303
Murail, Tristan, 227, 377n24
Musée de l'Homme, 99
Le Muse en Circuit, 216
musical objects, 43
Musica su due dimensioni (Maderna), 170

Music I, II, III, VI programs, 200–201, 262
Les musicographies (CD-ROM), 146, 148, 233
Music V program, 86, 87, 201, 225, 232, 234, 241
music writing, 184; about, 129–30; analysis and transcription, 130–34; graphic symbols library, 141; listening conducts, 135–36; revolutions in, 139–40; simulacrum concept, 134–35; sound images (i-sounds), 138–39; temporal semiotic units, 140–41; thematic analysis, 137–38; tripartition theory, 136–37
Musique 1; *Musique 2* (1925 tracts), 39
Musique acousmatique (Bayle), 108
musique concrète: avant-garde period, 161–89, 219; influences, 9, 32–42; invention of, 9, 42–45; term usage, 46–47, 49. *See also* GRM (Groupe de Recherches Musicales)
Musique et arts plastiques (Bosseur), 29
La musique concrète et le XXè siècle (Henry), 47–48
Musiques & Recherches, 270, 360
Musiques en fête (Lortat-Jacob), 118
Les musiques électroacoustiques (2000 CD-ROM), 69, 123, 144, 253, 258, 344
Musique vivante (Marie), 52, 164
Musseau, Michel, 303, 304
Mussorgsky, Modest, 40
Mutations (1969 work) (Risset), 201, 231, 241

Nadah, Nadia, 378n14
Nadja (Breton), 39, 215
Nakamura, Toshimaru, 340
Naon, Luis, 310, 323, 330, 342
Narita, Kazuko, 333
National Conservatory of Paris. *See* CNSMP
Nattes, Marc-Olivier de, 331
Nattiez, Jean-Jacques, 136–37, 152, 223, 229, 282, 295, 363n15
natural/cultural debate, 61, 74, 125, 253
Neill, Alexander, 68
New York School, 51, 178–80, 212, 215, 370n40
N'guyen Van Tuong, 205, 206, 207, 208
Niblock, Phil, 341
Nicolai, Carsten, 340
Nietzsche, Friedrich, 14, 124–25, 125, 219
Nikolais, Alvin, 238
Nodaïra, Ichiro, 293
Noetinger, Jérôme, 340
Nono, Luigi, 183
Normandeau, Robert, 271, 301
Nouaille, Xavier, 141–42, 233–34, 372n21
Nouvelle Vague, 212

406   INDEX

Nuit Blanche (Mâche), 214, 217

Objeu (1961 film) (Lepeuve), 174
Oca, Pierre, 215
OCORA (Office de Coopération Radiophonique), 187, 217
Octandre (1924 piece) (Varèse), 41
L'Œil écoute (1970 work) (Parmegiani), 107, 229, 241, 243, 244
Office de la Radio Télévision Française (ORTF), 10, 61, 70, 77, 100, 112, 120–21, 190, 194, 203–4, 219, 220, 224, 233, 243, 244, 245, 248, 358n68, 369n25
Offrandes (1921 piece) (Varèse), 41
Ohana, Maurice, 239
Ollivier, Albert, 12
Olson, Harry, 200
Omaggio à Joyce (Berio), 203
Onda, Aki
Ondes Martenot, 27, 31, 41
Opium (Cocteau), 38
optical film, 34
Orage, A. R., 11
L'oreille en colimaçon (radio program), 69
Orion, Thibault d'
Orlarey, Yann
Orphée (ballet) (Béjart and Henry), 172
Orphée 51 ou Toute la lyre (Schaeffer and Henry), 78, 79, 107, 109, 166, 171, 361n13
Orphée 53 (Schaeffer and Henry), 161, 166, 171, 177
Ortega, Sergio
ORTF (Office de la Radio Télévision Française). See Office de la Radio Télévision Française (ORTF)
Ossart, Roland, 237
Ouspensky, Peter-Demianovitch, 11
Ouzounoff, Alexandre
Oxygène (1977 piece) (Jarre), 238–39

Pacqueteau, Jean-Marie, 231
Paik, Nam June, 199
Pallandre, Jean, 309, 342
Pan Sonic, 326, 337, 344
Parade (1917 ballet) (Satie), 37
Paranthoën, Yann, 249, 317
Paris, François, 237
Paris, Roberto, 330
Paris Conservatory, 31, 49, 50, 90, 92, 216–17, 249–51, 268–69, 270, 285, 309–10, 323
Parmegiani, Bernard, 1, 51, 52, 66, 82, 83, 88, 97, 107, 121, 125, 137–38, 188, 202, 205, 206, 207, 208, 213, 214, 216, 229, 241–45, 244, 246–47, 275, 276, 281, 282, 285, 286, 297, 298, 307, 308, 315, 344–45, 352n51
Parmerud, Åke, 299, 303, 337
La Partition intérieure (Siron), 153
Pascal, Michel, 205, 307, 329
Passacaille (Haubenstock-Ramati), 175
Passerone, Félix, 163
Patou, Marie-Claude, 301
Paulhan, Jean, 11
Pauwels, Louis, 11
Péguy, Charles, 14
Peignot, Jérôme, 64, 186, 193, 357n31, 368n54
Peirce, Charles-Sanders, 138–39, 308
La pensée et le mouvant (Bergson), 105, 149
La perception de la musique (Francès), 56
Pérez, Anne, 175
Petit, Arnaud, 274, 277, 377n24
Petit, Roland, 273
Petitdidier, Patrice, 334
Petitot, Bernard, 281
Petitot, Léonce, 281, 297
Pfenninger, Rudolf, 20
phenomenology, 58, 60, 62, 223
Philippe, Gérard, 366n8
Philippot, Michel, 6, 50, 53, 70, 131, 161, 167, 170, 178, 186, 187, 202, 204–5, 206, 207, 210, 216–17, 233, 239, 315, 345
Philipps, Tom, 151
phonautographs, 28
phonogènes, 27, 53, 73, 79–82, 166, 168, 172, 175, 185, 210, 234
Phonophani, 341
Phonothéque, 99–100
Phonurgia Nova, 25
Piaget, Jean, 105, 223
Picabia, Francis, 36, 37, 41, 353n90
Picasso, Pablo, 37
Piéplu, Claude, 239
Pierce, John, 200, 234, 262
Pierret, Marc, 252
Pierrot Lunaire (Schoenberg), 21, 41
Piscator, Erwin, 16
playback, 28–30, 81, 237
Poème électronique (1958) (Varèse), 103, 112, 174, 201
Poisson, Agnès, 327, 330
Poliakoff, Serge, 195
Polychrome Portraits, 144, 145, 147, 148
Pomey, Bernard, 336
Pompidou, Georges, 225

INDEX   407

Ponge, Francis, 174, 329, 352n50
Pontefract, Jean-François, 30
pop art, 195, 212
Popp, Markus, 337
Porché, Wladimir, 24, 50, 186
Portal, Michel, 279
Portradium, 341
potentiometric control panel, 53
Pottier, Laurent, 115
Poulenc, Francis, 323, 354n94
Poullin, Jacques, 53, 76–77, 78–79, 110–11, 185–86, 188, 206, 210
Pound, Ezra, 213
Pour en finir avec le pouvoir d'Orphée (1972 work) (Parmegiani), 107, 243
Pousseur, Henri, 51, 181, 183, 226
Pradalié, Louis, 46–47
Prado, Cécile Le, 298
Prager, Jonathan, 330
Pratella, Francesco, 32, 361n28
Presley, Elvis, 367n34
Prévert, Jacques, 367n30
Project of Music for Magnetic Tape, 179
Pro Tools system, 75, 77, 91–92, 292, 303, 304, 305, 317, 320, 321
Prunières, Henry, 351n33
Pschenitschnikova, Natalia, 341
Psychology of Space (Moles and Rohmer), 106, 119
Puce Muse, 85
Puckette, Miller, 84, 263, 335, 359n85
Puech, Mathias, 317
Pythagoras, 56, 64

Les quatre saisons (Schwarz), 88
Queneau, Raymond, 303, 304

Rabelais, François, 245, 336, 373n44
Racot, Gilles, 250, 276, 277, 278, 279, 293, 302, 333, 341, 343, 344
Radelet, Patrick, 304
Radigue, Éliane, 51, 177–78, 344, 390
Radio Corporation of American (RCA), 22, 200
Radiodiffusion d'Outre-Mer, 187, 193
Radiodiffusion Française (RDF), 10, 12, 13, 14, 24, 44, 216
Radiodiffusion Nationale, 12, 14, 22
Radiodiffusion-télévision française (RTF). See Radio Télévision Française (RTF)
radio-ear concept, 35
Radio France, 121, 122, 224. *See also* Grand Auditorium of Radio France; Maison de la Radio
Radio Jeunesse, 12–13
Radio Paris IV, 165
Radio Television, 50
Radiotelevisione Italiana (RAI), 182, 183, 203
Radio Télévision Française (RTF), 24–25, 70, 73, 76, 76–78, 77, 94, 187, 190, 203, 216, 354n102, 369n2, 369n25
Radique, Eliane, ix, 51
Radkiewicz, Wiska, 69, 253, 342, 357nn42–43
RAI (Radiotelevisione Italiana), 182, 183, 203
Ralphs-Frisius, Korinna, 276
Rancillac, Bernard, 195
Rapt (1931 film) (Kirsanoff), 19, 352n53
Ratkje, Maja, 341
Rauschenberg, Robert, 195
Ravel, Maurice, 33, 323
Ray, Man, 36, 41, 288
Raysse, Martial, 195
RCA synthesizer, 41, 200
real/living debate, 60–61
Real Time System, 87
Rebotini, Arnaud, 126, 331, 335
Recio, Lucio, 333
recording, 28–30, 97
record player, 75, 75–76
Red Army Choir, 21
Redolfi, Michel, 236, 237, 272, 273, 280, 299, 307, 331, 337
reduced listening, 30–31, 43, 57–59, 139
Reflets (1961 piece) (Malec), 130, 213
Regnault-Bousquet, Cecile, 151
Régnier, Francis, 67, 85–86, 205, 215, 223, 231–32, 241, 359n76
Reibel, Guy, 6, 50, 51, 56, 58, 67–68, 69, 71, 82, 86, 97, 115, 121, 196, 205, 209, 214, 220, 229, 241, 242, 244–45, 246–47, 249, 250–51, 253–58, 264, 268–69, 272, 273, 274, 282, 285–86, 304, 308, 342, 345
Reich, Steve, 243
Reinaud, Mathieu, 144
Relâche (1924 ballet) (Picabia and Satie), 37
Rémus, Jacques, 253
Renard, Claire, 69, 239, 253, 357nn41,43
Renard-Barrault (theater company), 179
Renouard-Larivière, Régis, 214, 242, 293, 304, 308, 323, 339, 343, 375n31
Répons (Boulez), 225, 268
Requiem (1973 piece) (Chion), 83, 244, 258, 291
Research Department, 61, 72–73, 78, 190, 197, 203–6, 210, 219, 220, 224, 229, 240, 241, 244, 372n36. *See also* Service de la Recherche de l'ORTF

Research Festival, 121, 205–6
Resnais, Alain, 196
Respighi, Ottorino, 29, 253n61, 323
Restany, Pierre, 194–95
reverberation (REV), 87
reverse playback, 28, 163
Revue Musicale, 12
Reznikov, Hanon, 330
Richard, Jean-Pierre, 137–38, 193
Richter, Hans, 20
Ricœur, Paul, 5, 138
Rieflin, Bill, 340
Rigoni, Michel, 106
Riley, Terry, 243
Rilke, Rainer Maria, 120
Rimbaud, Arthur, 36, 272, 330
Rimbaud, Robin, 340
Risset, Jean-Claude, ix, 105, 108, 116, 201, 225, 231, 232, 241
Rissin, David, 209, 247, 258
Rist, Simone, 97, 209, 255
ritardando (RAL), 87
Rituel d'oubli (1969) (Mâche), 217, 240
Rivette, Jacques, 188, 196
La rivière endormie (Milhaud), 174
Robert, Élisabeth, 327
Robert, Jean-Pierre, 301
Robert, Marthe, 275
Rodet, Xavier, 287
Rogers, Carl, 68
Rohmer, Elisabeth, 119
Rohmer, Eric, 196, 197
Rollin, Monique, 39, 130, 165, 170
Romains, Jules, 36
rootedness/errancy concept, 106–7, 116
Roque-Alsina, Carlos, 239, 273
Rossellini, Roberto, 188
Rosset, Christian, 249, 307
Rossi, Tino, 121
Rotella, Mimmo, 195
Rouch, Jean, 171, 197
Roudier, Patrick, 250
Rouget, Gilbert, 124–25, 362n52
Roullier, Pierre, 331, 333
La Roue Ferris (1971 work) (Parmegiani), 242, 243
Roussel, Albert, 323
Rouveyrollis, Jacques, 273
Roux, Sébastien, 335
Rouxel, Jacques, 239, 372n29
Rouzoul, Nicole, 315
Rovere, Gilbert, 213

Roy, Christophe, 280, 301
Roy, Claude, 12
Roy, Stéphane, 132, 136–37, 152, 154, 271
RTF (Radio Télévision Française), 70, 73, 76, 77, 94, 187, 190, 203, 216
Rudnik, Eugeniusz, 181
Ruetsch, Christophe, 334, 341
Russian futurism, 34–36
Russolo, Luigi, 32–33, 35, 40–41, 196, 342, 353n71
Ruttmann, Walther, 20, 35, 42, 355n116

Saariaho, Kaija, 279, 293, 307
SACEM (Société des Auteurs Compositeurs et Éditeurs de Musique), 48, 131, 143, 228, 323, 324, 344, 371n1
Saint-Martin, Dominique, 318
Saint-Phalle, Niki de, 195
Sanders, Dirk, 138, 193
Sanguinetti, Eduardo, 213
S. A. R. E. G. company, 79
Satie, Erik, 36–37, 38, 40, 44, 270, 353n90
Saudrais, Charles, 213
Sauguet, Henri, 174
Saussure, Ferdinand de, 59, 223
Savouret, Alain, 58, 236, 239, 243, 247, 264, 273, 277, 279, 307, 309–10
Savy, Thomas, 331
Schaeffer, Elisabeth, 11, 15, 17, 45, 349n5
Schaeffer, Pierre: *10 ans d'essais radiophoniques 1942-1952* and, 25, 26; *À la recherche d'une musique concrète*, 184–85; art-science notion and, 55–56; avant-garde and, 176–80, 189; Bayle and, 174, 202, 211; Beaune Workshop and, 13–15, 50; *Bidule en ut*, 48, 163, 164, 165, 170, 272; *Bohor*, 211; Boulanger and, 40; on bridge arts, 13, 13–24; Cage and, 178–80; Cannes Film Festival, 186; CERPS and, 294–95; Chion and, 257, 258, 259; cinema and, 19–21; Club d'Essai and, 22–27, 42, 50, 190; CNSMP and, 221–22, 238–39, 249–51, 255, 268; Cologne Studio and, 199–200; on communication theory, 134–35; computers and, 210, 218, 231, 232; concepts and, 57–64; Concert Collectif, 206–7; *Concert de bruits*, 9, 42, 44, 120, 162–63; death of, 50, 288; détournement and, 195; Dufour and, 345; early life, 10, 131; *Étude aux objets*, 201, 290; *Études aux allures*, 195; *Études de bruits*, 39, 164; Ferrari and, 202, 212, 215; film music, 171; *First Decade of International Experimental Music*, 185; first musical works, 163–64;

first musique concrète concert, 164–65; first record releases, 186; *Five Studies of Noise*, 34; futurism and, 33–34; GER and, 222–23; GRM and, 9, 47, 48, 50, 190, 194, 203, 220, 222, 223; GRMC and, 9, 50, 131, 186–87, 193–94; GRM catalog and, 297; *Guide des objets sonores* and, 57, 282; Gurdjieff and, 10–12, 51; Henry and, 163, 187–88, 197, 246; "Hommage à Pierre Schaeffer" concert, 274, 290; INA and, 224; Jeune France network and, 12–13; *La Coquille à planètes*, 15–18; *La musique concrète*, 209; Liberation of Paris and, 21–22; listening transcriptions, 130, 131; modes of listening, 118, 135–36; musique concrète and, 1–2, 3, 37, 40, 42–54, 109, 133–34, 161, 216; *Orphée 51 ou Toute la lyre*, 109, 166, 171–72; Parmegiani and, 202; pedagogy and, 65–68, 70–73; Philippot and, 216; photos of, 167, 168; on popular art, 121; potentiometric control panel and, 109–10; publications about, 252, 343, 344; radio broadcast works, 165–66, 247, 248, 249; Reibel and, 250, 251, 254, 255; Research Department and, 197, 203–4, 206, 218, 220, 223, 224; Research Festival, 205–6; retirement of, 50, 219, 220, 268; RTF and, 13, 187, 190, 193; simulacrum concept, 134–35, 149; Sorbonne conference, 184; SOS and, 56, 208–9, 309; as sound engineer, 12; sound library and, 318; sound object concept, 35; sound spaces and, 103, 104, 107; sound techniques, 27, 30–31; Stockhausen Concert, 211; Studio d'Essai and, 15–18, 33, 50; successors of, 50; *Symphonie pour un homme seul*, 34, 39, 109, 164, 170, 176; teaching, 49–51, 204, 238–39; television and, 220, 239; TOM and, 6, 32, 54, 56–57, 97, 130, 136–37, 149, 152, 186, 208–9, 222, 223, 247, 253, 254, 282, 347; tools and, 75–79, 82, 86, 89, 97, 99, 101, 102; *Trièdre fertile*, 202, 245–46; typo-morphology of, 140, 141, 156, 306; wife of, 15, 45; Xenakis and, 192, 210, 218, 232

Scherchen, Hermann, 41, 70, 168, 173, 177, 183–84, 185, 186, 211
Schloezer, Boris, 185
Schloss, Andy, 96
Schmitt, Christian, 336
Schmitt, Florent, 323
Schmitt, M. C., 340
Schoenberg, Arnold, 21, 40, 41, 63, 161, 181, 216, 322, 323
Schöffer, Nicolas, 127, 175, 197, 238
Schumann, Robert, 40
Schütz, Heinz, 181
Schwarz, Jean, 67, 88, 99, 101, 221, 223, 239, 243, 249, 275, 277, 279, 285, 286, 294, 296, 308, 315, 341
Schwitters, Kurt, 42
Scott de Martinville, Edouard-Léon, 28, 352n56
Sée, Jean-Claude, 367n14
Seigner, Louis, 26
sequence-play, 58
Sequential Drum/Radio Baton, 95–96
Service de la Recherche de l'ORTF, 47, 65, 82, 97, 100, 120–21, 134, 190, 194, 197, 203–4, 210, 220, 224, 241, 258. *See also* Research Department
Settel, Zack, 340
Les Shadoks (1968 animation series) (Borg), 121, 239–40
Shapira, Claire, 239
Shryer, Claude, 271
Sighicelli, David, 336
Sighicelli, Samuel, 336, 377n22
signed listening (*écoutes signees*), 146
Sikora, Elzbieta, 239, 276, 328
Le silence est d'or (1946 film) (Clair), 19, 351n32
Silent Film Sound (Altman), 19
SIM (Société Internationale de Musique), 33
Simon, Emilie, 34
Simonovic, Konstantin, 202, 206, 216, 255
Simpson, David, 334
simulacrum concept, 44, 134–35
Siron, Jacques, 153
Situationist International, 42, 134, 195–96, 212
Six pièces en un acte (Sauguet and Tardieu), 174
Skyvington, William, 231
slide phonogènes, 79, 168
Smalley, Denis, 58, 88, 276, 305, 307, 308, 343, 344, 390
Société de Radiodiffusion de la France d'Outre-mer (Sorafom), 187
Société des Auteurs Compositeurs et Éditeurs de Musique. *See* SACEM
Sogar, 335, 340
Sola, Madeleine, 279
Solfège de l'objet sonore (SOS), 6, 56, 61, 82, 97, 99, 194, 208–9, 220, 309, 356n6, 370n32
Solomos, Makis, 343
Sonagraphe, 147, 365n40
Sonatas (Cage), 179
Sonic Contours (1952 piece) (Ussachevsky), 182
sonogram, 93, 132, 142, 147, 233
Sonothèque, 97–99

Sorafom (Société de Radiodiffusion de la France d'Outre-mer), 187
Sorbonne Conference, 184
SOS (Solfège de l'objet sonore). See *Solfège de l'objet sonore* (SOS)
Soudan, Jérôme, 330, 340
Soulages, Pierre, 279
"Sound and Space" (Poullin), 111
sound capture, 28–30
Sound Designer II, 142
sound images (i-sounds), 138–39
sound object, 43, 57–59
sound scenery, 43
sound synthesis, 27, 31–32, 200–201
Sound Tools II, 91
Souris, André, 39, 354n98
Souzay, Gérard, 350n18
space/interaction concept, 108, 116
Spatiodynamisme (Schöffer and Henry), 175
Spectral School, 226–27
spectro-morphology, term usage, 58
speed variation, 28
Sphärophon, 31
Spoeri, 195
Spooky Tooth (band), 121, 238
Sprenger-Ohana, Noémie, 146
Srawley, Stephen, 274, 294
Stefani, Gino, 295
STEIM Institute, 95, 332
Steinberg (film) (Kassowitz), 202
Steinberg, Saul, 202
Stella, Frank, 195
Sten, Helge, 335
Stibler, Jacques, 239, 333
Stiegler, Bernard, 150–51, 289, 322, 365n56, 376n3
Stieglitz, Alfred, 41
Stimmung (Stockhausen), 211, 221
Stockhausen, Karlheinz, ix, 49, 51, 82, 104, 106, 125, 161, 167–68, 171, 177, 179, 180, 180–81, 182, 183, 198–99, 200, 211, 221, 225, 226, 253, 271, 274, 303, 315, 370n42
Stohl, Frédéric, 279
Stone, Carl, 337, 340
Stravinsky, Igor, 33, 40, 55, 173, 185, 225, 254
Stroppa, Marco, 181, 376n10
Studie I (1954 electronic étude) (Stockhausen), 182
Studie II (1954 electronic étude) (Stockhausen), 182
Studio 11, 199
Studio 38, 28
Studio 39, 28
Studio 52, 210, 230, 369n10
Studio 54, 82–83, 196, 210, 229–30, 245, 247, 374n12
Studio 116, 51, 84, 85, 95, 96, 196, 229, 235, 245, 267–68, 289, 320–21, 342, 369n10, 373n50, 374n72
Studio 123, 85–90, 232, 235
Studio d'Essai, 9, 14, 15–22, 25, 27–29, 33, 44, 50, 61, 65, 75, 94, 177, 193, 197, 249, 351n41, 366n8, 377n6
St. Werner, Jan, 337, 340
Suite 14 (Schaeffer), 48, 52
Suite for 14 instruments (Schaeffer), 163, 164
Suite pour Edgar Poe (1972 work) (Reibel), 242–43
Surrealism, 6, 9, 31, 37, 38–39, 44, 164, 165, 176, 215, 272, 353n90, 354n97
Sylvestre, Gaston, 273
Symphonie des bruits (Schaeffer), 52
Symphonie mécanique (Boulez), 175
Symphonie pour un homme seul (1950 composition) (Schaeffer and Henry), 34, 39, 48, 52, 109, 163, 164, 166, 170, 176, 177, 186
synthesizers, 58, 82, 83, 84–85, 210, 229, 236, 238, 241
SYNTOM project, 232
Syter (SYstème TEmps Réel), 73–74, 86, 88, 88–92, 141, 234–35, 246, 260, 264, 265–67, 276–77, 278–79, 280, 285, 286, 289, 290–92, 297–98, 299, 300, 302, 304, 305, 320, 341, 359n84, 377n23
Szendy, Peter, 107–8, 116

Tachism, 121
Taillefer, Germaine, 354n94
Tainter, Charles Sumner, 28
Tamayo, Claudia, 271
Tanada, Fuminori, 327
Tanaka, Yumiko, 340
Tanguy, Agnès, 56, 247, 248, 283, 375n35
Tanguy, Yves, 336
tape recorders, 76–77, 180, 197
Tarasti, Eero, 295
Tardieu, Jean, 21, 22, 24, 109, 174, 190, 193, 215, 352n50
Tardy, Pierre, 250
Techno, 121, 123–26, 238
Technological Research Group (GRT). See GRT (Groupe de Recherche Technique)
Tegzes, Maria, 326
Télémaque, 195
Telharmonium, 31, 352n64
Temporal Semiotic Units (UST), 58, 140–41, 156, 214

Teruggi, Daniel, 50, 88, 128, 187, 251, 275, 276, 277, 278, 279, 280, 285, 289, 290, 291, 296, 297, 300, 305, 306, 307, 311, 320, 322–23, 332, 341, 373n51, 376n5
Terzieff, Laurent, 243
Tessier, Roger, 227
Texier, Marc, 250
Texte 2 (Boucourechliev), 203
Tezenas, Suzanne, 178, 179
Théâtre des Champs-Élysées, 33, 42, 120, 172, 173
thematic analysis, 137–38
Theremin, 31, 352nn65–66, 375n37
Theremin, Leon, 31
"The Sound World of Electronic Music" lectures, 180
Thévenot, Jean, 177
Thiébault, Jean-Baptiste, 365n43
Thigpen, Benjamin, 326
Third Sonata (1955–57) (Boulez), 179
Thom, René, 118, 308
Thomas, Jean-Christophe, 137–38, 146, 147, 148, 229, 281, 282, 306, 315, 343, 365n39
Thoresen, Lasse, 141, 156
"Three Key Ideas for Early Learning Music Pedagogy" (Delalande), 69–70
Tibaudeau, Jean, 272
Tides (Smalley), 88
timbre, 63
Timbres-Durées (1952 piece) (Messiaen), 39, 77, 111, 168–69
Times Five (Brown), 212
Tinguely, Jean, 195
Toch, Ernest, 29
Toch, Nicole, 300
Toch, Noël, 300
Toeplitz, Kasper, 330, 337
TOM (*Traité des objets musicaux*). See *Traité des objets musicaux* (*TOM*)
Tomasi, Henri, 27
tonal rupture, 40
tools: Acousmaline database, 101–2; Acousmographe, 141–49; Acousmonium, 96–97; analog sound synthesis, 82–84; analog synthesis, 75–77; computer music, 85–92; keyboard phonogènes, 78–79; limited versus open-ended instruments, 92–94; MIDI tools, 84–85; phonogènes, 77–78; slide phonogènes, 79; studio ergonomics, 94–96; universal phonogènes, 79, 81–82
Topart, Jean, 26, 202
Toscane, Jean, 120
Toth, Catherine, 350
Toulier, Jean-Pierre, 87, 235
Toupie dans le ciel (Bayle), 148, 272
Touraine, Geneviève, 166
Touzet, Vincent, 302
Traité des objets musicaux (*TOM*), 6, 32, 47, 54, 56, 57, 58–60, 61, 62, 64, 65, 67, 72, 76, 82, 89, 97, 130, 133, 136–37, 149, 152, 186, 190, 194, 208–9, 220, 223
trames orientées (guided grids), 175
transcriptions, 130–34
transposition, 28, 78–79, 81, 163
Trautonium, 31, 177, 352n64, 368n46
Trautwein, Friedrich, 31, 350n15, 368n46
Tremblement de terre très doux (1978) (Bayle), 39, 272
Trémollières, Benoît, 330
Trièdre fertile (Schaeffer), 202
Trio GRM+; TRIO-GRM-PLUS; Trio TM+, 238, 256, 272, 285, 286, 345
tripartition theory, 136–37
Triptyque électroacoustique (1974 piece) (Reibel), 83
Trois portraits d'un Oiseau-Qui-N'Existe-Pas (1962 film score) (Bayle), 203, 212
Truffaut, François, 196
Trutat, Alain, 197, 248, 373n43
Tudor, David, 179, 199
Tulio-Marin, Sergio, 239
Turcotte, Roxane, 271
Tutschku, Hans, 326
Tzara, Tristan, 33, 36

Ucla, Bernard, 255
Underdown, Walter, 245
Une larme du diable (Garrett, Bernhart, and Clair), 165
Une larme du diable (Gautier), 109
Une tour de Babel (1999 piece) (Henry), 123
Unités Sémiotiques Spatio-Temporelles (USST), 141
universal phonogènes, 79, 81, 81–82
Ussachevsky, Vladimir, 41, 181–82
USST (Unités Sémiotiques Spatio-Temporelles), 141
UST (Temporal Semiotic Units), 140–41, 214
UST-USST (Unités Sémiotiques Spatio-Temporelles), 141

Vaggione, Horacio, 298, 307, 308
Vaïnio, Mika, 124, 326, 332, 344

Valéry, Paul, 18, 65, 105
Valette, Denis, 87, 235
Valse molle (1973 work) (Savouret), 243
Vande Gorne, Annette, 51, 270, 282, 327
Vandelle, Romuald, 205, 206, 207
Vandenbogaerde, Fernand, 239, 294, 323, 324
Van den Broek, Viviane, 373
Van Wymeersch, Brigitte, 63
Varèse, Edgar, 33, 40–42, 44, 48, 56, 104, 111–12, 131, 173–74, 180, 185, 200, 201, 215, 303, 352n52, 354n109, 363n15, 367n24
Variations en étoile (Reibel), 214
Variations pour une porte et un soupir (Henry), 144, 197
Variations sur une flûte mexicaine (Schaeffer), 163
Vasto, Lanza del, 11
Vayne, Catherine, 308
VCA (Voltage Control Amplifier), 82, 236, 359n71
Veitl, Anne, 227, 269, 342
Vejvoda, Goran, 341
Vercken, François, 187
Vérin, Nicolas, 294, 305, 342
Verne, Jules, 273, 328
Vertov, Dziga, 34–36, 42, 353n80
Vian, Boris, 193
Vidal, Jack, 221, 247, 253, 283–84
Vidolin, Alvise, 307
Viel, Tanguy, 336
Vigo, Jean, 19
Villeglé, Jacques de la, 195
Villeneuve, Christian, 239
Viñao, Alejandro, 293, 299
Vinet, Hugues, 88, 91, 142, 268, 287, 290, 292, 343, 359n82
vinyl discs, 28
Violostries (Parmegiani), 213
Vis, Lucas, 246
Visage (Berio), 198
Visage III; *Visage IV*; *Visage V* (Ferrari), 202
Vitet, Bernard, 213
Vivo, Raphaël de, 236
Voile d'Orphée (Henry), 79, 172, 186
Voyage au centre de la tête (Bayle), 273, 274
Voyage au cœur d'un enfant (Béjart), 176
Vrod, Jean-François, 340

Wagner, Richard, 38, 40, 107
Waisvisz, Michel, 95, 332, 337
Wallumrod, Christian, 335
Wanderley, Marcello, 343

Wang, Miao-Wen, 301
Wangermée, Robert, 354n98
Wargnier, Francis, 254
Warhol, Andy, 195, 238
Watson, Chris, 341
WDR (Westdeutscher Rundfunk) (Cologne), 180, 182–83, 198–200, 203, 225–26, 244, 262, 293. *See also* Cologne Studio
Webern, Anton, 40, 63, 161, 185, 198, 323
Weekend (1930 sound film) (Ruttmann), 35, 355n116
Wessel, David, 263, 374n8, 377n24
Weyergans, 198
Whitehead, Alfred North, 308
Whitman, James, 226
Wiener, Jean, 165
Wiener, Norbert, 182
Wilcox, Fred L., 368n47
Williams, Paul, 179
Williams Mix (1952 compostition) (Cage, Brown, and Wolff), 179
Winckel, Fritz, 206
Wishart, Trevor, 293, 329
Wittman, Christian, 337
Wolff, Christian, ix, 5, 51, 179, 370
Wolman, Gil, 42
Wyatt, Robert, 121

Xenakis, Iannis, ix, 1, 6, 51, 52, 62, 77, 93, 112, 151, 162, 174, 175, 177, 178, 181, 184, 188, 192–93, 198, 201, 202, 204, 206–7, 210, 211, 216, 217–18, 226, 232–33, 274, 308, 323, 359n87, 361n21
Xing, Rufeng, 305
Xu, Shuya, 305

Yamaha, 84, 262, 263, 321, 322
Yanowski, Jean, 206
Yuasa, Yuasa, 359n87

Zanési, Christian, ii, 124, 128, 239, 250, 264, 273, 276, 277, 280, 282, 283, 284, 285, 296, 297, 303, 308, 317, 326, 332, 335, 341, 344, 361n33
Zavoloka, 340
Zazou, Hector, 340
Zéro de conduite (1937 film) (Vigo), 19
Z'ev, 341
Zorn, John, 233
Zuccheri, Marino, 183
Zurbrugg, Nicholas, 351n25

Founded in 1893,
UNIVERSITY OF CALIFORNIA PRESS
publishes bold, progressive books and journals
on topics in the arts, humanities, social sciences,
and natural sciences—with a focus on social
justice issues—that inspire thought and action
among readers worldwide.

The UC PRESS FOUNDATION
raises funds to uphold the press's vital role
as an independent, nonprofit publisher, and
receives philanthropic support from a wide
range of individuals and institutions—and from
committed readers like you. To learn more, visit
ucpress.edu/supportus.

www.ingramcontent.com/pod-product-compliance
Lightning Source LLC
Chambersburg PA
CBHW051241300426
44114CB00011B/847